Forestry: Management and Conservation of Forests

Forestry: Management and Conservation of Forests

Edited by Malcolm Fisher

SYRAWOOD
PUBLISHING HOUSE

New York

Published by Syrawood Publishing House,
750 Third Avenue, 9th Floor,
New York, NY 10017, USA
www.syrawoodpublishinghouse.com

Forestry: Management and Conservation of Forests
Edited by Malcolm Fisher

© 2018 Syrawood Publishing House

International Standard Book Number: 978-1-68286-576-7 (Hardback)

Cataloging-in-Publication Data

Forestry : management and conservation of forests / edited by Malcolm Fisher.
    p. cm.
Includes bibliographical references and index.
ISBN 978-1-68286-576-7
1. Forests and forestry. 2. Forest management. 3. Forest conservation. I. Fisher, Malcolm.
SD373 .F67 2018
634.9--dc23

TABLE OF CONTENTS

PREFACE

Forestry refers to the study and practice of preserving, maintaining, repairing, creating, and using forest resources in an optimum way. It includes natural water quality management, landscape protection, watershed management, etc. This book provides comprehensive insights into the field of forestry. It discusses in detail about the various topics related to this field. This text includes some of the vital pieces of work being conducted across the world, on various topics related to forestry. Selected concepts that redefine the field have been presented in it. The book will serve as a valuable source of reference for graduate and post graduate students and researchers.

This book has been the outcome of endless efforts put in by authors and researchers on various issues and topics within the field. The book is a comprehensive collection of significant researches that are addressed in a variety of chapters. It will surely enhance the knowledge of the field among readers across the globe.

It gives us an immense pleasure to thank our researchers and authors for their efforts to submit their piece of writing before the deadlines. Finally in the end, I would like to thank my family and colleagues who have been a great source of inspiration and support.

Editor

Structural Characterization of *Prosopis africana* Populations (Guill., Perrott., and Rich.) Taub in Benin

Towanou Houètchégnon,[1] **Dossou Seblodo Judes Charlemagne Gbèmavo,**[1,2]
Christine Ajokè Ifètayo Nougbodé Ouinsavi,[1] **and Nestor Sokpon**[1]

[1]*Faculty of Agronomy, Laboratory of Forestry Studies and Research, University of Parakou, BP 123 Parakou, Benin*
[2]*Laboratoire de Biomathématiques et d'Estimations Forestières, Faculté des Sciences Agronomiques, Université d'Abomey-Calavi, 04 BP 1525 Cotonou, Benin*

Correspondence should be addressed to Towanou Houètchégnon; houetchegnon@gmail.com

Academic Editor: Piermaria Corona

The structural characterization of *Prosopis africana* of Benin was studied on the basis of forest inventory conducted in three different vegetation types (savannah, fallow, and field) and three climate zones. The data collected in 139 plots of 1000 m^2 each related to the diameter at breast (1.3 m above ground), total height, identification, and measurement of DBH related *P. africana* species height. Tree-ring parameters such as Blackman and Green indices, basal area, average diameter, height of Lorey, and density were calculated and interpreted. Dendrometric settings of vegetation type and climate zone (Guinea, Sudan-Guinea, and Sudan) were compared through analysis of variance (ANOVA). There is a significant difference in dendrometric settings according to the type of vegetation and climate zone. Basal area, density, and average diameter are, respectively, 4.47 m^2/ha, 34.95 stems/ha, and 37.02 cm in the fields; 3.01 m^2/ha, 34.74 stems/ha, and 33.66 cm in fallows; 3.31 m^2/ha, 52.39 stems/ha, and 29.61 cm in the savannahs. The diameter distribution and height observed at the theoretical Weibull distribution show that the diameter and height of the populations of the species are present in all positively skewed distributions or asymmetric left, a characteristic of single-species stands with predominance of young individuals or small diameters or heights.

1. Introduction

A wide range of perennial woody species distributed in the humid tropics meet many needs of indigenous populations [1]. In Benin, we count 172 species consumed by the local population as food plants [2] and 814 as medicinal plants [3]. With the continual increase in demand for products derived from these species, traditional collection methods have gradually given way to irrational methods of collection [4]. Those woody species of great usefulness to local communities are threatened in their distribution areas because of pressure exerted on them and/or their habitats: *Adansonia digitata* [5], *Afzelia africana* and *Khaya senegalensis* [6], *Garcinia lucida* [7], *Anogeissus leiocarpa* [8], *Pentadesma butyracea* [9], and *Prosopis africana* [10] are edifying cases. The frequency of *P. africana*, for example, is becoming weaker in its range because of excessive overexploitation by cutting the stems and branches of it, which limits its natural regeneration capacity

[11]. *P. africana* enriches the soil by fixing nitrogen; its leaves are rich in protein, and sugar pods are used as foodstuffs for feeding ruminants in Nigeria [12]. The pulp of the pods contains 9.6% protein, 3% fat, and 53% carbohydrate and provides energy value 1168J [13]. In some areas, its fermented seeds are used as condiment in preparations in Nigeria [14, 15]. As *Parkia biglobosa* the *Prosopis africana* seeds are fermented and used as condiments [16]. The *P. africana* seeds are used in Nigeria and Benin in the preparation as a local condiment [17, 18]. Similarly, *P. africana* is used in the preparation of foods such as soup and baked products and in the manufacture of sausages or sausages and cakes. The pods of some mesquite species are used as a staple food by many native populations in the desert of Mexico and the Southwest United States (Simpson [19] quoted by Geesing et al. [20, 21]). In Kaka and Seydou [22] cited by Geesing et al. [20, 21], tasting panels have found that a partial substitution of corn flour, sorghum, or millet flour mesquite at a rate of

10% does not affect the taste of traditional dishes but helps to elevate the flavor. The pods are very palatable to cattle in Burkina Faso [23, 24]. Despite the recognized importance of the species for the rural population, the report of Benin on food tree species has clearly mentioned near absence of information and scientific data on its ecology, its production, and its management in traditional agroforestry systems [4]. *P. africana* is often found in fallow, on sandy clay soil above the laterite. The strong anthropic pressure due to slash-and-burn agriculture practiced by 70% of the agricultural population of Benin and fallow periods increasingly reduced locally affects the population structure of *P. africana*. This is compounded by the fact that until today the species exists in natural stands and has not a planning study or regeneration study in Benin while structure, regeneration, and the likely risk of the disappearance of the species are still less studied. However, the acquisition of reliable data on the ecology, distribution, and the structure of a forest species are necessary for the development of an optimal development plan and conservation that are effective [7, 25, 26]. The purpose of this study is to describe the characteristics of dendrometric populations of *P. africana* in different plant communities for future development. It is a specific way

(1) to determine the dendrometric characteristics of the different plant formations (savannahs and fallow fields) to *Prosopis africana* and different climatic zones (Guinean, Sudano-Guinean, and Sudan) of Benin,

(2) to determine the structure of *P. africana* trees in each of different plant formations and climatic zones and compare between them. We made the assumptions that (i) the dendrometric characteristics of *P. africana* vary from plant formation to another and from one climatic zone to another and (ii) structural *P. africana* trees vary among different plant formations (savannahs and fallow fields) and different climate zones.

2. Material and Methods

2.1. Study Area.
Benin is situated between $9°30'$N and $2°15'$E with an annual mean rainfall of 1039 mm and a mean temperature of $35°$C. It covers a surface area of 114763 km^2 with a population size of 6769914 inhabitants dominated by women (3485795) [27]. Three climatic zones associated with their vegetation types can broadly be distinguished (Figure 1).

(1) *The southern zone gathering the coastal and Guineo-Congolese zones*: from the coast up to the latitude 7°N, the climate is subequatorial with two rain seasons alternating with a long dry season from December to February. The coastal one is dominated by mangrove swamps with predominant species such as *Ipomea pescaprae, Remirea maritime, Rhizophora racemosa, Avicennia germinans*, and *Dalbergia ecastaphyllum*. The Guineo-Congolese zones are dominated by semideciduous forests with predominant species such as *Dialium guineense, Triplochiton scleroxylon, Strombosia glaucescens, Cleistopholis patens, Ficus mucuso,*

Cola cordifolia, Ceiba pentandra, Trilepisium madagascariense, Celtis spp.*, Albizia* spp.*, Antiaris toxicaria, Diospyros mespiliformis, Drypetes floribunda, Memecylon afzelii, Celtis brownii, Mimusops andogensis, Daniellia oliveri, Parkia* spp., and *Vitellaria paradoxa* [28–31].

(2) *The transition zone*: this zone is situated between the latitudes 7°N and 9°N. The climate becomes tropical one and subhumid with a tendency to a pattern of one rainy season and one dry season. The two rainfall peaks' pattern indicates a unimodal rainfall regime. Dominant vegetation types are galleries and savannahs with predominant species such as *Isoberlinia doka, I. tomentosa, Monotes kerstingii, Uapaca togoensis, Anogeissus leiocarpa, Antiaris toxicaria, Ceiba pentandra, Blighia sapida, Dialium guineense, Combretum fragrans, Entada africana, Maranthes polyandra, Pterocarpus erinaceus, Terminalia laxiflora*, and *Detarium microcarpum* [31].

(3) *The northern zone or Sudanian zones*: this zone is characterized by a tropical climate with a unimodal rainfall regime. The rain season lasts on average for seven months from April to October with its maximum on August or September. Dominant vegetation types are dry woodland and savannahs. Predominant species are *Haematostaphis barteri, Lannea* spp., *Khaya senegalensis, Anogeissus leiocarpa, Tamarindus indica, Capparis spinosa, Ziziphus mucronata, Combretum* spp., and *Cissus quadrangularis*. The high pressure of human activities on forests in this zone led to the extinction of species such as *Milicia excelsa, Khaya senegalensis, Afzelia africana*, and *Pterocarpus erinaceus* [31, 32]. This is the case of *Prosopis africana* which became rare in fallows according to von Maydell [16].

2.2. Data Collection.
The ecological and structural characterization of *P. africana* was done using inventory in three habitats of *P. africana* (farm, fallow, and savannah) according to climatic zones. Adults (DBH ≥ 10 cm) were measured within circular plots of 1000 m^2 size and regenerations were measured within 5 subplots of about 28 m^2 size. A standard distance of 100 m was observed between two plots in each of the vegetation types. Table 1 shows plots distribution according to ecological zones of the country. Variables measured on each tree included the diameter at breast height (DBH ≥ 10 cm) and the total and bole height.

2.3. Data Analysis.
To determine the dendrometric characteristic of *P. africana*, dendrometric parameters were calculated. These parameters are presented in Table 2.

The structural characterization of *P. africana* according to vegetation types and climatic zones was done using the diameter and height class-size distribution. Different histograms of frequency from the diameters and heights were adjusted to Weibull 3-parameter distribution using the software Minitab 16. This distribution was used as it is simple in usage [33].

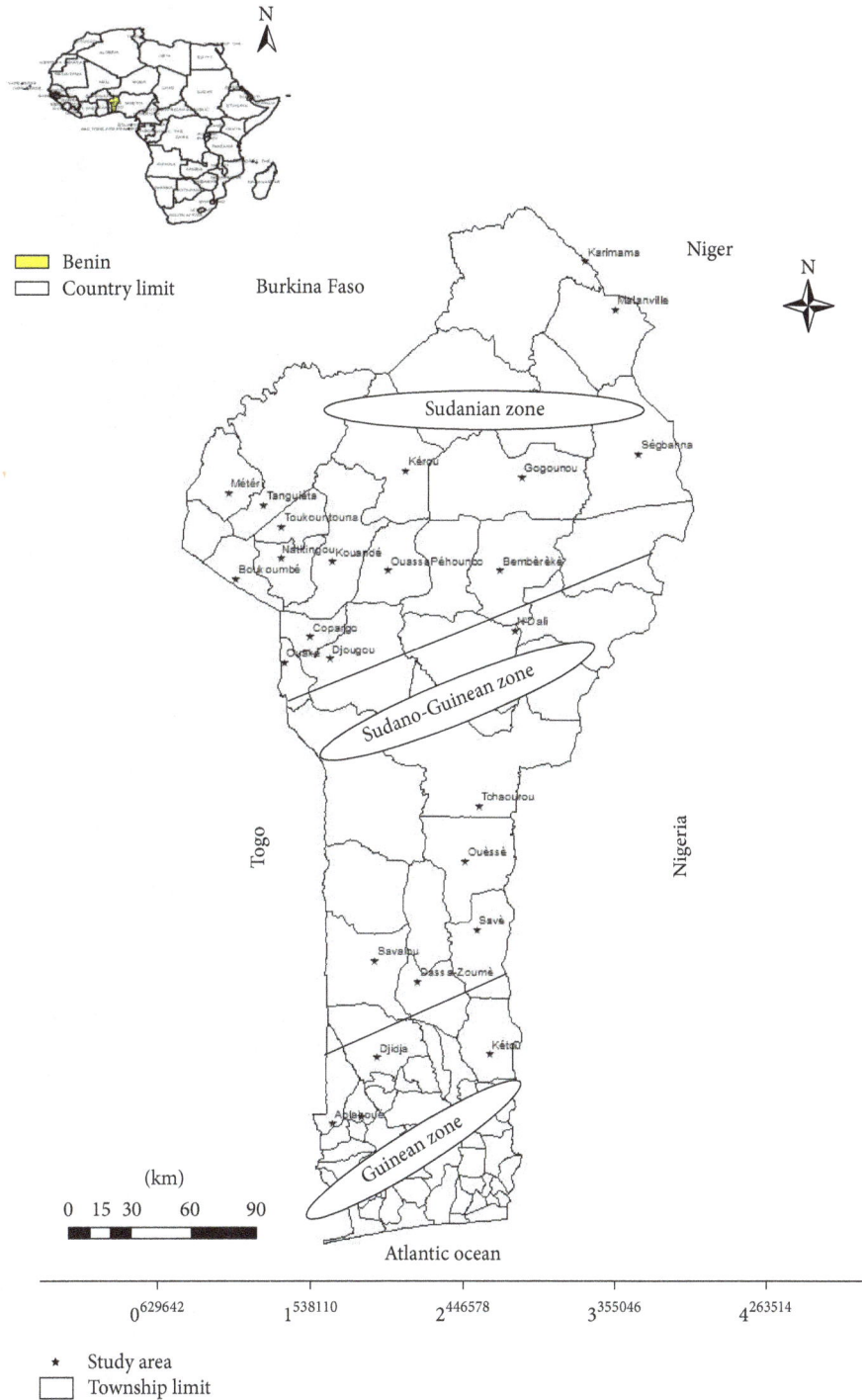

FIGURE 1: Map showing zones of study.

According to Rondeux [34], its probability density function is given by the following equation:

$$f(x) = \frac{c}{b} \left(\frac{x-a}{b} \right)^{c-1} \exp \left[-\left(\frac{x-a}{b} \right)^c \right]. \tag{1}$$

In this equation x denotes the diameter or height of trees; a, b, and c are, respectively, position parameter, scale parameter, and form parameter. Considering the form parameter, different forms of distributions can be distinguished. Table 3 shows the different forms of distribution using Weibull 3-parameter model.

TABLE 1: Plots distributions according to ecological zones.

Climatic zones	Farm	Fallows	Savannah	Total
Guinean	16	12	11	39
Sudano-Guinean	14	14	18	46
Sudanian	10	21	23	54
Total	40	47	52	139

TABLE 2: Dendrometric parameters and their formula.

Parameters	Formula
Density	$D_g = \sqrt{\dfrac{1}{n}\sum_{i=1}^{n} di^2}$
Medium basal surface area	$G = \dfrac{\pi}{40000s}\sum_{i=1}^{n} di^2$
Lorey height of individuals	$H_L = \dfrac{\sum_{i=1}^{n} gihi}{\sum_{i=1}^{n} gi}$
Diameter of tree with medium basal surface area	$D_g = \sqrt{\dfrac{1}{n}\sum_{i=1}^{n} di^2}$
Blackman index	$I_B = \dfrac{S_N^2}{N}$
Green index	$I_G = \dfrac{(I_B - 1)}{n - 1}$

Notes: n, total number of trees within one plot; di, diameter of the ith tree; S_N^2, variance of population trees; N, mean of population trees.

Dendrometric parameters according to vegetation types and climatic zones were compared using two-way ANOVA with the software Minitab 16.

3. Results

3.1. Dendrometric Parameters according to Vegetation Types. Table 4 shows dendrometric characteristics at *P. africana* populations and at all populations' levels according to vegetation types. Parameters' means compared with Student *t*-test revealed significant differences of mean ($P < 0.01$). In fact, the diameter, basal surface area, and Lorey height of *P. africana* populations range, respectively, from 30 to 37 cm, 3 to 4 m²/ha, and 9 to 11 m. High values of diameters and basal surface areas were observed from the farms whereas high values of heights were observed from fallows. As shown in Table 4, probability values indicate a significant difference of parameters (density, diameters, and basal surface area) according to vegetation types. Besides, the regeneration density was found to be high in habitats under low pressure.

3.2. Dendrometric Parameters according to Climatic Zones. Table 5 shows dendrometric characteristics at all vegetation types levels and at *P. africana* populations ones according to different climatic zones. Considering the whole populations, comparison of parameters means using Student *t*-test revealed a significant difference ($P < 0.01$) of parameters (density, diameters, and basal surface area). As for *P. africana* populations, the diameter and Lorey height were in average, respectively, 33 cm and 10 m. The table analysis showed that

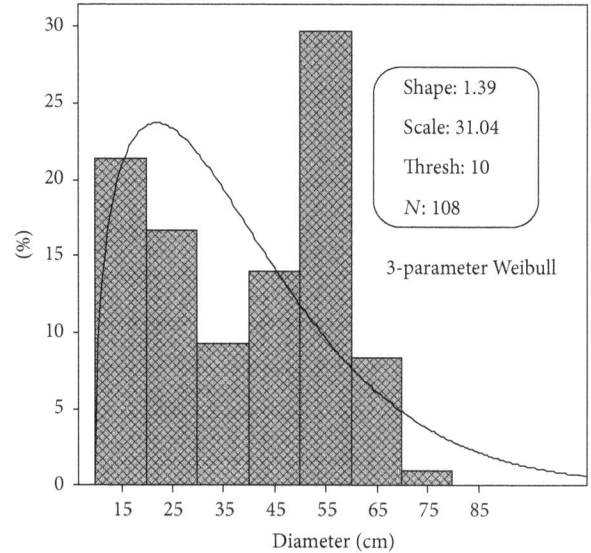

FIGURE 2: Diameter class-size distribution in the Guinean zone.

the diameter increases according to rainfall gradient. In fact, the diameter increases as vegetation becomes more and more watered. But climatic gradient did not affect trees height. Means of heights were, respectively, 11 m in the Guinean zone, 9 m in Sudano-Guinean zone, and 9 m in Sudanian zone. Parameters means compared using Student *t*-test revealed a significant difference ($P < 0.05$) of parameters (density, diameters, and basal surface area) according to climatic zones. The basal surface areas were, respectively, 3 m²/ha in the Guinean zone, 6 m²/ha in Sudano-Guinean zone, and 1 m²/ha in Sudanian zone.

Besides, Blackman index (IB) obtained was 32.20 and Green index was 0.054 which is near 0 and shows a random distribution of *P. africana* populations according to climatic zones.

3.3. Diameter Class-Size Distribution of P. africana Populations. Figures 2, 3, and 4 show the diameter class-size distribution of *P. africana* populations according to the three climatic zones of Benin. Figure 3 indicates a J inversed distribution of *P. africana* populations describing multispecific groups of species. The two others figures (Figures 2 and 4) indicate left skew distribution describing monospecific groups of trees dominated by subjects with small diameters. In fact, subjects with diameter ranging from 10 to 70 cm are the predominant in the Guinean zone. Besides, subjects with diameter over 80 cm are quasi-absent. Unlike the Guinean zone, small subjects with diameter class-size distribution of 10–20 cm were found to be the predominant in the Sudano-Guinean zone. As for the Sudanian zone, subjects with diameter ranging between 10 and 30 cm are the most abundant. Subjects with diameter over 90 cm are quasi-absent in this zone.

3.4. Height Class-Size Distribution of P. africana Populations. Figures 5, 6, and 7 show the height class-size distribution of *P. africana* populations to the three climatic zones of Benin.

Structural Characterization of Prosopis africana Populations (Guill., Perrott., and Rich.) Taub in Benin

5

TABLE 3: Distribution forms from 3-parameter Weibull model according to the parameter c.

Value of c	Types of distribution	References
$c < 1$	J inversed distribution describing multispecific groups of species	
$c = 1$	Exponential distribution describing populations in extinction	
$1 < c < 3.6$	Left skew distribution describing monospecific groups of trees with small diameters	[35]
$c = 3.6$	Bell shaped distribution describing monospecific groups or plantation species	
$c > 3.6$	Positive distribution describing monospecific groups of trees with big diameters	

TABLE 4: Dendrometric characterization of *P. africana* according to vegetations types.

Parameters	Farms (pl = 40)		Fallows (pl = 47)		Savannah (pl = 52)		P values
	M	SE	M	SE	M	SE	
P. africana							
Density (N, stems/ha)	34.95ab	5.54	34.74b	5.16	52.39a	4.97	0.022
Diameter (D_g, cm)	37.02a	2.36	33.66a	2.20	29.61a	2.12	0.067
Basal surface area (G, m²/ha)	4.47a	0.71	3.01a	0.67	3.31a	0.65	0.303
Lorey height (H_L, m)	9.25a	0.42	10.72b	0.39	8.66b	0.38	0.001
Contribution of Basal surface area (C_s, %)	86.99a	4.35	73.96ab	4.06	65.69b	3.90	0.002
Density of regeneration (N_r, stems/ha)	7.28a	12.3	23.21a	11.48	27.96a	11.05	0.438
Global							
Density (N, stems/ha)	58.23b	16.81	108.62ab	15.68	126.28a	15.09	0.010
Diameter (D_g, cm)	10.73a	0.66	8.91ab	0.62	8.57b	0.60	0.041
Basal surface area (G, m²/ha)	4.44a	0.89	5.47a	0.83	4.97a	0.80	0.696

Note: M = mean, SE = standard deviation.
The averages followed the same line of the same letters (a or b or ab) are not significantly different at the 5% level (test Tuskey).

TABLE 5: Dendrometric characterization of *P. africana* according to climate zone.

Parameters	Guinean zone (pl = 39)		Soudanian zone (pl = 54)		Soudano-guinean zone (pl = 46)		P values
	M	SE	M	SE	M	SE	
P. africana							
Density (N, stems/ha)	28.45b	5.59	35.41b	4.98	58.21a	5.09	0.000
Diameter (D_g, cm)	40.38a	2.38	22.63b	2.12	37.28a	2.17	0.000
Basal surface area (G, m²/ha)	3.33b	0.73	1.38b	0.65	6.08a	0.66	0.000
Lorey height (H_L, m)	11.25a	0.43	8.88b	0.38	8.51b	0.39	0.000
Contribution of Basal surface area (C_s, %)	69.38b	4.39	83.37a	3.91	73.89ab	4.00	0.051
Density of regeneration (N_r, stems/ha)	4.85a	12.44	11.10a	11.08	42.50a	11.31	0.051
Global							
Density (N, stems/ha)	108.15ab	16.98	61.71b	15.13	123.27a	15.45	0.014
Diameter (D_g, cm)	11.57a	0.67	6.70b	0.60	9.94a	0.61	0.000
Basal surface area (G, m²/ha)	5.07a	0.90	2.10b	0.80	7.72a	0.82	0.000

Note: M = mean, SE = standard deviation.
The averages followed the same line of the same letters (a or b or ab) are not significantly different at the 5% level (test Tuskey).

On the whole, the parameter of form (c) ranges between 1 and 3.6 indicating left skew distribution describing monospecific groups dominated by subjects with small heights. In fact, subjects with height ranging between 8 and 12 m are the predominant one in the Guinean zone. Unlike the Guinean zone, subjects with height ranging between 6 and 10 m were found to be the predominant in the Sudano-Guinean zone. As for the Sudanian zone, subjects with height ranging between 6 and 12 m are the most abundant. Subjects with diameter over 90 cm are quasi-absent in this zone. Besides, subjects with height over 21 m are quasi-absent in the Guinean zone. Those whose height is over 23 m are quasi-absent in the Sudano-Guinean zone and subjects with height over 22 m are quasi-absent in the Sudanian zone.

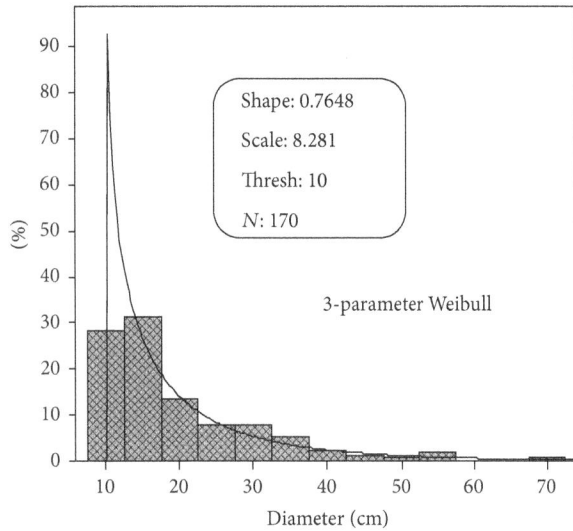

FIGURE 3: Diameter class-size distribution in the Sudano-Guinean zone.

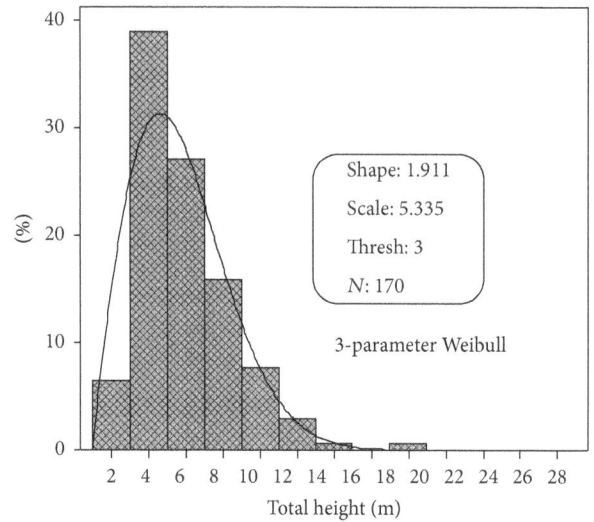

FIGURE 5: Height class-size distribution in the Guinean zone.

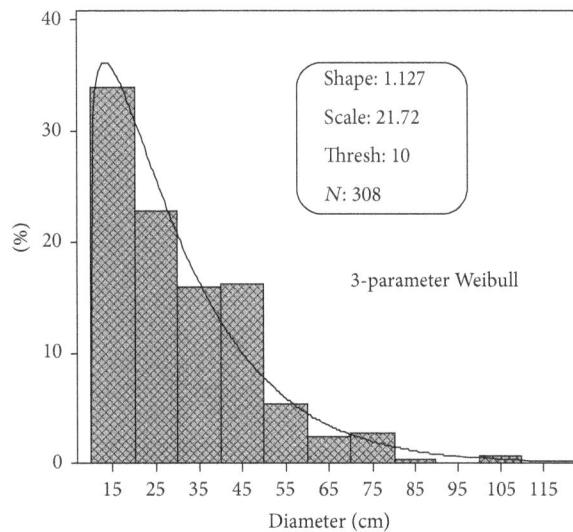

FIGURE 4: Diameter class-size distribution in the Sudanian zone.

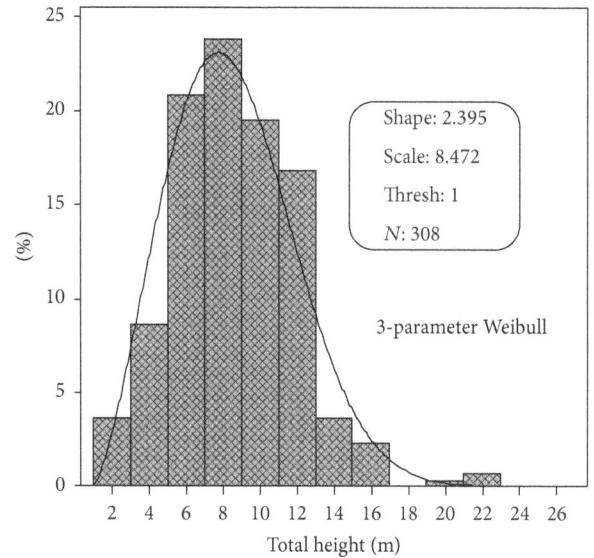

FIGURE 6: Height class-size distribution in the Sudano-Guinean zone.

4. Discussion

4.1. Dendrometric Characterization of P. africana Populations

4.1.1. Dendrometric Features. Dendrometric parameters are important tools used in forestry. Sokpon [36] reported that the average diameter of the tree is a useful parameter of interest and is often recommended in forestry. The average density values noted in savannah stands (52.39 trees/ha) are significantly lower than those obtained by Glèlè Kakaï et al. [37] in stands of *Pterocarpus erinaceus* (169.4 trees/ha), by Sagbo [38] in the stands dominated by *Isoberlinia* spp. (205 trees/ha), and by Ouédraogo et al. [39] in Burkina Faso (4000 individuals/ha). The values in the Sudano-Guinean zone (58.21 stems/ha) are also lower than Gbesso et al. [40] in Benin *Borassus aethiopum* (78 and 133 stems/ha) in this

same area. These differences may be partly due to inventory methods used and also because the inventoried stands are not exactly the same. They can also reduce, in part, the strong anthropic pressure from local populations on forest trees of value. The diameter of the populations of *P. africana* is higher in fields and in Guinean and Sudano-Guinean areas. This can be explained by the abundance of rainfall that could have a positive effect on the size of diameters. Note that conservation in the fields by local people [18] to human food purposes (because seeds are condiments that sell in markets of Effèoutè in Kétou, Dassa-Zoumé, Glazoué, Aplahoué, and Klouékanmè) in these climatic zones could have a positive effect. Trees have benefited interviews from crop in the fields. Regarding the basal area of the plants groups studied, it varies between 3.31 and 4.47 m^2/ha in

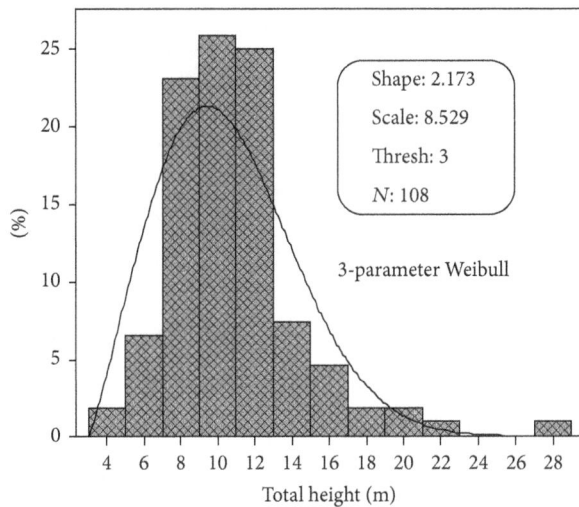

FIGURE 7: Height class-size distribution in the Sudanian zone.

the vegetation and then 1.38 and 6.08 m^2/ha in climate zones. This reveals the importance in the exploitation of *P. africana* in arboriculture. The number of plants per hectare is between 34.74 and 52.39 for adults trees in vegetation and 28.45 and 58.21 in climate zones. As for regeneration, the variation is between 7.28 and 27.96 in the vegetation and then between 4.85 and 42.50 in climate zones. These results indicate that the population is very dense in savannahs than in the anthropic formations, which can mean that the species is under pressure in anthropogenic environments. These results are similar to those of Assogbadjo et al. [41] in the forest reserve Wari-Maro which showed that dendrometric features have more values for *Anogeissus leiocarpa* in stands under low pressure. Such results were also obtained by Kiki [42] on *Vitex doniana* and Fandohan et al. [43] on the *Tamarindus indica* and have shown that human pressures have a negative effect on dendrometric parameters such as density and regeneration adult density but a positive effect on the mean diameter. The heights are higher in the wettest area (Guinea) than other areas. Thus we find individuals of 11.25 m. These results are lower than those of Ouinsavi et al. [44] who obtained, respectively, the palmyra over 15 m high in the Sudan region and those of Bonou et al. [45] (16.9 m) and Sinsin et al. [46] (17 m) of *Afzelia africana*.

4.2. Structures in Diameter and Height. The development of forest stands requires the mastery of the structure diameter and height of trees. These structures are indicative of events related to the life of stands [34]. Forest stands, according to whether they are single species or multispecies, even-aged, young, and old, have structures types. It is known that the structures in diameter and height of these forest types are adjusted to known theoretical distributions [35]. According to Rondeux [34], Philip [47], and McElhinny et al. [48], in even-aged structure, sizes by diameter classes have typical distribution often resembling a Gaussian curve that can become asymmetrical bimodal seen in certain circumstances. According to the same authors, in an even-aged stand, all

trees have the same age or close with a low variable height which is mainly explained by their social position (dominant, codominant). The horizontal structures of populations have mostly left asymmetrical characteristic of single-species stands with predominance of young individuals or small diameters or low heights. According to Arbonnier [49], Sudanian climate was suitable for the optimal development of the African mesquite trees. There is thus generally a relationship between species temperament and their stem diameter class distribution. However, the diameter structure of the Sudano-Guinean zone presented a distribution whose appearance is in "inverted J" which, according to Rondeux [34] and Husch et al. [35], is a characteristic of multispecies stands. According to the same authors, in an even-aged stand, all trees have the same age or close with a low variable height that is mainly explained by their social position (dominant, codominant). In the case of this study, only the diameter structure at the Sudano-Guinean zone has a distribution whose appearance is in "inverted J" feature of multispecies stands. This reflects a relative predominance of individuals with small diameters. Similarly, the structure of *P. africana* diameters, at this climatic zone, has a bell shaped appearance, a characteristic of single-species stands. This distribution is left-skewed (1 < c < 3.6), a characteristic of a relative predominance of young individuals or small diameters. But we can say that individuals of *P. africana* of this climatic zone are not all the same age or young, and the observed left asymmetries cannot be explained by the youth of the population of the species but by their disruption or vulnerability at certain stages of their development. Regarding the distribution of tree height, it generally has a Gaussian shape which may be asymmetrical in the conditions of life of the stand. As for the overhead structure, the assembly has a bell shape of a distribution to the left asymmetry characteristic of stands with predominance of individuals with low heights. According to Bonou et al. [45], the use of Weibull distribution probability density function is becoming increasingly popular for modeling the diameter distributions of both even- and uneven-aged forest stands. The popularity of Weibull is derived from its flexibility to take on a number of different shapes corresponding to many different observed unimodal tree-diameter distributions. In addition, the cumulative distribution function of Weibull exists in closed form and thus allows for quick and easy estimation of the number of trees by diameter class without integration of the probability density function once the parameters have been fitted. The bell shaped function obtained with the diameter or height classes distribution of the African mesquite with a left dissymmetry, a notable exception of the diameter structure of the Sudano-Guinean zone, corroborates the results of Cassou and Depomier [50] with the African fan palm population of Wolokonto in Burkina Faso and the results of Ouinsavi et al. [44], with the palm trees in Benin. Similar results were also obtained by Kperkouma et al. [51], with the Shea butter trees of Donfelgou in Togo. Also Bonou et al. [45] obtained the same distribution as far as *Afzelia africana* trees populations are concerned in Benin. However this structure might not be derived only from the species temperament but also from human pressure.

5. Conclusion

The structural characterization of populations of *P. africana* has helped dendrometric and horizontal structuring of *P. africana* stands groups, distinct in their specific traits induced by climatic conditions and vegetative strata that are the fields, fallow, and savannahs. The structural characteristics of populations varied greatly from one climate zone to another and from one formation to another plant. It can be concluded that the species is present in all climatic zones of Benin but with varied densities and that it is in the Sudan and Sudano-Guinean areas quite abundant. The density is an average of 126.28 stems/ha, 109 stems/ha, and 58 trees/ha, respectively, in savannahs, fallows, and fields. The diameter of the shaft means for these formations varies between 9 and 11 cm with the high value in the field. In terms of the basal area, it is an average of $4.44 \, m^2$/ha at the field level, $5 \, m^2$/ha at the savannah, and then 5 m/ha at the fallow. For the regeneration of density, the variation is between 5 and 43 stems/ha. It is higher in the Sudano-Guinean areas (43 individuals/ha) and Sudanese areas (11 individuals/ha). The lower regeneration density is in the Guinean zone (5 individuals/ha). The average diameter of the basal area of tree is most interesting in the Guinean area compared to other areas. The stands of this area offer a potential timber into lumber and service development which would draw added value from the sale of wood. Ecological structure of the African mesquite populations of Benin, adjusted to Weibull distribution, showed a bell shaped curve with a left dissymmetry proving the predominance of young trees within these populations.

Conflict of Interests

The authors declare that there is no conflict of interests regarding the publication of this paper.

Acknowledgment

The authors are grateful to the International Tropical Timber Organization for the financial support.

References

[1] FAO, FLD, and IPGRI, *Forest Genetic Resources Conservation and Management. Volume 1: Overview, Concepts and Some Systematic Approaches*, International Plant Genetic Resources Institute, Rome, Italy, 2004.

[2] J. T. C. Codjia, A. E. Assogbadjo, and M. R. Mensah, "Diversité et valorisation au niveau local des ressources forestières alimentaires du Bénin," *Cahiers Agricultures*, vol. 12, pp. 321–331, 2003.

[3] B. Sinsin and L. Owolabi, *Rapport sur la monographie de la diversité biologique du Bénin*, Ministère de l'Environnement et de l'Habitat et de l'urbanisme (MEHU), Cotonou, Bénin, 2001.

[4] O. Eyog-Matig, O. G. Gaoué, and B. Dossou, "Réseau 'Espèces Ligneuses Alimentaires,'" in *La première réunion du Réseau*, 241, Institut International des Ressources Phytogénétiques, Ouagadougou, Burkina-Faso, December 2002.

[5] M. Sidibe and J. T. Williams, *Baobab Adansonia digitata*, International Centre for Underutilised Crops, Southampton, UK, 2002.

[6] B. Sinsin, T. Sinandouwirou, and A. Assogbadjo, *Caractérisation écologique de deux essences fourragères du Bénin: Khaya senegalensis et Afzelia africana*, Rapport d'étude IPGRI/SAFORGEN, IPGRI, Cotonou, Bénin, 2000.

[7] N. M. Guedje, J. Lejoly, B.-A. Nkongmeneck, and W. B. J. Jonkers, "Population dynamics of *Garcinia lucida* (Clusiaceae) in Caméronien Atlantic forests," *Forest Ecology and Management*, vol. 177, no. 1–3, pp. 231–241, 2003.

[8] A. E. Assogbadjo, R. G. Kakaï, F. H. Adjallala et al., "Ethnic differences in use value and use patterns of the threatened multipurpose scrambling shrub (*Caesalpinia bonduc* L.) in Benin," *Journal of Medicinal Plants Research*, vol. 5, no. 9, pp. 1549–1557, 2011.

[9] C. Avocèvou-Ayisso, B. Sinsin, A. Adégbidi, G. Dossou, and P. van Damme, "Sustainable use of non-timber forest products: impact of fruit harvesting on *Pentadesma butyracea* regeneration and financial analysis of its products trade in Benin," *Forest Ecology and Management*, vol. 257, no. 9, pp. 1930–1938, 2009.

[10] J. C. Weber, M. Larwanou, T. A. Abasse, and A. Kalinganire, "Growth and survival of *Prosopis africana* provenances tested in Niger and related to rainfall gradients in the West African Sahel," *Forest Ecology and Management*, vol. 256, no. 4, pp. 585–592, 2008.

[11] N. M. Pasiecznik, P. J. C. Harris, and S. J. Smith, *Identifying Tropical Prosopis Species: A Field Guide*, HDRA, Coventry, UK, 2004.

[12] A. A. Annongu, J. K. Joseph, and F. Liebert, "Effect of anaerobic fermentation and Lyle treated *Prosopis africana* seed meal on the nutritional and hematological responses of Harco chicks," *Journal of Raw Materials Research*, vol. 1, pp. 33–41, 2004.

[13] FAO, *Non-Wood Forest Products and Nutrition*, Food Nutrition Division Publication, Food and Agricultural Organization of the United Nations, Quebec City, Canada, 2003.

[14] M. O. Aremu, A. Olonisakin, B. O. Atolaye, and C. F. Ogbu, "Some nutritional and functional studies of *Prosopis africana*," *Electronic Journal of Environmental, Agricultural and Food Chemistry*, vol. 5, pp. 1640–1648, 2006.

[15] A. Kalinganire, J. C. Weber, A. Uwamariya, and B. Kone, "Improving rural livelihoods through domestication of indigenous fruit trees in Parklands of Sahel," in *Indigenous Fruit Trees in the Tropics: Domestication, Utilization and Commercialization*, F. K. Akinnifesi, R. R. B. Leakey, O. C. Ajiyi et al., Eds., pp. 186–203, CAB International Publishing, Wallingford, UK, 2007.

[16] H. J. V. Maydell, "Arbres et arbustes du Sahel. Leurs caractéristiques et leurs utilisations," Gtz, 385p + annexes, 1983.

[17] J. T. Barminas, H. M. Maina, and J. Ali, "Nutrient content of *Prosopis africana* seeds," *Plant Foods for Human Nutrition*, vol. 52, no. 4, pp. 325–328, 1998.

[18] T. Houètchégnon, D. S. J. C. Gbèmavo, C. Ouinsavi, and N. Sokpon, "Ethnobotanical knowledge and traditional management of mesquite (*Prosopis africana* Guill., Perrot. et Rich.) populations in Benin, West Africa," *The Journal of Ethnobiology and Traditional Medicine*. In press.

[19] B. B. Simpson, *Mesquite: Its Biology in Two Desert Ecosystems*, Dowden, Hutchinson and Ross, Stroudsburg, Pa, USA, 1977.

[20] D. Geesing, M. Al-Khawlani, and M. L. Abba, *La Gestion des Espèces de Prosopis Introduites: L'Exploitation Économique Peut-Elle Juguler les Espèces Envahissantes?*, Food and Agriculture Organization, Quebec City, Canada, 2004.

[21] FAO, "Unasylva no. 217, Vol. 55; Menaces pour les forêts," *Revue Internationale des Forêts et des Industries Forestières*, vol. 55, no. 2, pp. 36–44, 2004.

[22] S. Kaka and R. Seydou, *Tests d'Utilisation du Prosopis en Alimentation Humaine à N'Guigmi et Bosso*, INRAN, FAO, Niamey, Niger, 2001.

[23] B. Toutain, "Le rôle des ligneux pour l'élevage dans les zones soudaniennes de l'Afrique de l'Ouest," in *Les Fourrages Ligneux en Afrique: État Actuel des Connaissances. Colloque International sur les Fourrages Ligneux en Afrique. Addis-Abeba, Ethiopie 8–12 Avril 1980*, H. N. Le Houérou, Ed., pp. 105–110, CIPEA, 1980.

[24] S. Saré, "Potentialités fourragères et effets de l'élevage extensif sur la biodiversité végétale dans la réserve biosphère de la mare aux hippopotames (Ouest Burkinabé)," Mémoire d'Ingénieur IDR/UPB 92p+annexes, 2004.

[25] The Nature Conservancy, *The Five-S Framework for Site Conservation. A Practitioner's Handbook for Site Conservation Planning and Measuring Success*, vol. 1, The Nature Conservancy, 2nd edition, 2000.

[26] C. Delvaux and B. Sinsin, "Les plantes médicinales dans la Forêt Classée des Monts Kouffé au centre du Bénin: Stratégie de conservation, de restauration et de production compatible avec le développement local," *SOMA (Italie)*, no. 1, pp. 73–81, 2002.

[27] INSAE-RGPH3, Synthèses des analyses, 47 pages, 2002.

[28] N. Sokpon, "Tenure foncière et propriété des ligneux dans les systèmes agroforestiers traditionnels au Bénin," *Annales de la Faculté des Sciences de Kisangani*, pp. 115–122, 1994.

[29] N. Sokpon and J. Lejoly, "Les Plantes à Fruits Comestibles D'une Forêt Semi-Caducifoliée de Pobè au Sud-est du Bénin," in *L'alimentation en Forêt Tropicale: Interactions Bioculturelles et Perspectives de Développement*, C. M. Hladik, A. Hladik, H. Pagezy, O. F. Linares, and A. Froment, Eds., vol. 1, pp. 115–124, UNESCO, Paris, France, 1996.

[30] C. A. Adomou, *Vegetation patterns and environmental gradients in Benin. Implications for biogeography and conservation [Ph.D. thesis]*, Wageningen University, Wageningen, The Netherlands, 2005.

[31] C. Ouinsavi and N. Sokpon, "Traditional agroforestry systems as tools for conservation of genetic resources of *Milicia excelsa* Welw. C.C. Berg in Benin," *Agroforestry Systems*, vol. 74, no. 1, pp. 17–26, 2008.

[32] A. C. Adomou, B. Sinsin, and L. J. G. van der Maesen, "Phytosociological and chorological approaches to phytogeography: a meso-scale study in Benin," *Systematics and Geography of Plants*, vol. 76, no. 2, pp. 155–178, 2006.

[33] N. L. Johnson and S. Kotz, *Distributions in Statistics: Continuous Univariate Distributions*, John Wiley & Sons, New York, NY, USA, 1970.

[34] J. Rondeux, *La Mesure des Arbres et des Peuplements Forestiers*, Les Presses Agronomiques de Gembloux, Gembloux, Belgium, 2nd edition, 1999.

[35] B. Husch, T. Beers, and J. J. R. Kershaw, *Forest Mensuration*, John Wiley & Sons, Hoboken, NJ, USA, 4th edition, 2003.

[36] N. Sokpon, *Recherche ecologique sur la forêt dense semi-décidue de pobè au Sud-Est du Bénin: groupements végétaux, structure, régénération naturelle et chute de litière [Ph.D. thesis]*, Universite Libre de Bruxelles, Brussels, Belgium, 1995.

[37] R. Glèlè Kakaï, B. Sinsin, and R. Palmr, "Etude dendrométrique de *Pterocarpus erinaceus* poir. des formations naturelles de la zone soudanienne au Bénin," *Agronomie Africaine*, vol. 20, no. 3, pp. 245–255, 2008.

[38] P. Sagbo, *Etude des caractéristiques dendrométriques des peuplements naturels à dominance Isoberlinia spp.: Cas de la forêt classée de l'Ouémé Supérieur au nord du Bénin [Mém. Ingénieur Agronome]*, Université d'Abomey-Calavi, Cotonou, Benin, 2000.

[39] A. Ouédraogo, T. Adjima, K. Hahn-Hadjali, and S. Guinko, "Diagnostic de l'état de dégradation des peuplements de quatre espèces ligneuses en zone soudanienne du Burkina Faso," *Science et Changements Planétaires/Sécheresse*, vol. 17, no. 4, pp. 485–491, 2006.

[40] F. Gbesso, H. Yedomonhan, B. Tente, and A. Akoegninou, "Distribution géographique des populations de rôniers (*Borassus aethiopum* Mart, Arecaceae) et caractérisation phytoécologique de leurs habitats dans la zone soudano-guinéenne du Benin," *Journal of Applied Biosciences*, vol. 74, no. 1, pp. 6099–6111, 2014.

[41] A. E. Assogbadjo, R. L. G. Kakaï, B. Sinsin, and D. Pelz, "Structure of *Anogeissus leiocarpa* Guill., Perr. Natural stands in relation to anthropogenic pressure within Wari-Maro Forest Reserve in Benin," *African Journal of Ecology*, vol. 48, no. 3, pp. 644–653, 2010.

[42] M. Kiki, *Structure et régénération naturelle despopulations de Tamarindus indica L. et de Vitexdoniana Sw. dans la Réserve de Biosphère Transfrontalière du W/Bénin: Cas de laCommune de Banikoara*, Mémoire d'Ingénieur des travaux, EPAC/UAC, 2008.

[43] A. B. Fandohan, A. E. Assogbadjo, R. L. G. Kakaï, B. Sinsin, and P. van Damme, "Impact of habitat type on the conservation status of tamarind (*Tamarindus indica* L.) populations in the W National Park of Benin," *Fruits*, vol. 65, no. 1, pp. 11–19, 2010.

[44] C. Ouinsavi, C. Gbémavo, and N. Sokpon, "Ecological structure and fruit production of African fan palm (*Borassus aethiopum*) populations," *The American Journal of Plant Sciences*, vol. 2, no. 6, pp. 733–743, 2011.

[45] W. Bonou, R. Glèlè Kakaï, A. E. Assogbadjo, H. N. Fonton, and B. Sinsin, "Characterisation of *Afzelia africana* Sm. habitat in the lama forest reserve of Benin," *Forest Ecology and Management*, vol. 258, no. 7, pp. 1084–1092, 2009.

[46] B. Sinsin, O. Eyog-Matig, A. E. Assogbadjo, O. G. Gaoué, and T. Sinadouwirou, "Dendrometric characteristics as indicators of pressure of *Afzelia africana* Sm. dynamic changes in trees found in different climatic zones of Benin," *Biodiversity and Conservation*, vol. 13, no. 8, pp. 1555–1570, 2004.

[47] S. M. Philip, *Measuring Trees and Forests*, Cabi Publishing, 2nd edition, 2002.

[48] C. McElhinny, P. Gibbons, C. Brack, and J. Bauhus, "Forest and woodland stand structural complexity: its definition and measurement," *Forest Ecology and Management*, vol. 218, no. 1–3, pp. 1–24, 2005.

[49] M. Arbonnier, *Arbres arbustes et lianes des zones sèches d'Afrique de l'Ouest*, CIRAD-MNHN, 2002.

[50] J. Cassou and D. Depomier, *In Annonce: Réunion Tri-Partie Sur l'Agroforesterie*, Sikasso, Ouagadougou, Burkina Faso, 1997.

[51] W. Kperkouma, B. Sinsin, K. Guelly, K. Kokou, and K. Akpagana, "Typologie et structure des parcs agroforestiers dans la Prefecture de Donfelgou," *Sécheresse*, vol. 1, no. 3, p. 8, 2005.

2

Livelihoods and Welfare Impacts of Forest Comanagement

Linda Chinangwa,[1,2] **Andrew S. Pullin,**[3] **and Neal Hockley**[1]

[1]*School of Environment, Natural Resources and Geography, Bangor University, Bangor, Gwynedd LL57 2DG, UK*
[2]*Institute of Advanced Study of Sustainability, United Nations University, No. 53-70, Jingumae 5-Chome, Shibuya-ku, Tokyo 150-8925, Japan*
[3]*Centre for Evidence-Based Conservation, School of Environment, Natural Resources and Geography, Bangor University, Bangor, Gwynedd LL57 2DG, UK*

Correspondence should be addressed to Linda Chinangwa; chinangwa@unu.edu

Academic Editor: Ilan Vertinsky

Comanagement programmes are gaining popularity among governments as one way of improving rural livelihoods. However, evidence of their effects on the livelihoods and welfare remains unclear. We used the sustainable livelihoods framework and stated preference techniques to assess the livelihoods and welfare impacts of forest comanagement on 213 households in Zomba and Ntchisi districts. The results show that approximately 63% of respondents perceive that, overall, comanagement has had no impact on their livelihoods. However, the programme is enhancing financial capital by introducing externally subsidised income generating activities and human and social capital among some community members through training programmes. A majority of households (80%) are willing to pay annual membership fees to participate in the programme (mean = 812 Malawi Kwacha), because of perceived potential future benefits. Education, gender of the household head, a positive perception of current livelihoods benefits, and a position on the committee increase household willingness to pay membership fees. However, the positive willingness to pay despite the negative perception of overall livelihoods impacts may also demonstrate the weaknesses of relying on stated preference surveys alone in estimating welfare effects.

1. Introduction

Forest comanagement approaches are promoted as one way of improving the livelihoods and welfare of rural communities [1]. However, the evidence for their livelihoods and welfare impacts has been found to be weak, due to limited rigorous impact evaluation studies [2]. Nevertheless, the approaches are gaining popularity and wider acceptance by governments and donors in the developing world as a prerequisite for conservation and development policies [3, 4]. The initiatives are also part of the larger economic and institutional reforms being pursued by many governments under IMF and World Bank lending conditionality since the 1990s [5]. Given this continued popularity, it is important to understand how comanagement affects the livelihoods and welfare of participating communities, to ensure effective and efficient implementation and resource allocation [1]. Therefore, using the case of a forest comanagement programme in government forest reserves in Malawi, we assess the programme's current livelihoods impacts and estimate its perceived welfare benefits among participating communities.

To assess the impact of comanagement programmes on livelihoods and welfare, we adopted the sustainable livelihoods framework, for example, [6–8], combined with a contingent valuation question (stated preference technique) to estimate household's "willingness to pay" (WTP) to participate in the programme, for example, [9]. The combination of approaches was essential for obtaining a comprehensive view of the livelihoods and welfare impacts, which would otherwise be difficult to achieve if each method was used on its own. For example, although stated preference techniques are widely used in valuing natural resource and environmental welfare benefits, due to their hypothetical nature they are vulnerable to hypothetical and strategic biases and do not provide explicit evidence for what a household has actually gained [10–12]. Furthermore, household WTP may reflect either present welfare benefits accrued or expected future benefits [13]. Therefore, by using both the livelihoods framework and stated preference techniques, it is possible to externally validate the benefit estimated by the contingent valuation [14]. This combined approach is novel in the forest comanagement literature. Therefore, this paper makes a useful empirical and methodological contribution to the existing literature on the livelihoods and welfare benefits assessment and forest comanagement approaches.

2. Measuring Livelihoods and Welfare Impact

2.1. Sustainable Livelihoods Framework. Livelihoods have been defined as "means, activities, capabilities, assets and entitlements by which people build a living" [15–17]. Thus they comprise both material and social resources [15]. Therefore, when assessing livelihood impacts of development policies and projects, both economic and social aspects of human wellbeing should be considered [16]. The sustainable livelihoods framework developed by Department for International Development (DFID) [17] emphasises the role that development policies and programmes play in human social and economic wellbeing; hence it offers a logical point of reference for assessing forest comanagement programmes, because they aim at improving both the social and economic wellbeing of local communities [7].

The sustainable livelihoods framework highlights five capitals upon which livelihoods impacts can be assessed. These are natural, financial, physical, social/political, and human capital (Table 1). The framework emphasises that, for households to achieve positive livelihood outcomes, a range of capital categories are required, because no single category can sufficiently meet households' multiple and varied livelihoods needs [17]. Therefore, at a given time households may draw on the different capital base to pursue a range of livelihood strategies, so as to yield positive livelihoods outcomes [16, 17, 28].

A household's choice of livelihoods strategy is determined by the household's preferences and priorities, as well as trends (e.g., population and resources trends), shocks (e.g., droughts), and seasonality (e.g., shift in prices and employment opportunities), which are beyond their control [18]. Additionally, availability of and access to the different forms of capital are regulated by the existing transforming structures and processes (i.e., institutions and policies) [17]. Therefore, household livelihood strategies and outcomes are influenced by opportunities and capabilities to access and acquire capital within the context and dynamics of vulnerabilities, transforming structures, and processes [15]. Thus, although a comanagement programme is not a capital in itself, it has the potential to provide opportunities and capabilities for accessing the different forms of capital that forest and forest systems provide and support [19]. By using the opportunities, capabilities, and activities provided by the programme, households can develop livelihoods strategies that respond to their needs and constraints and eventually translate into positive outcomes [20]. For example, compared

to state forest management, comanagement gives communities legal rights to access and use forests sustainably, hence potentially providing them with new livelihoods opportunities and sources [7]. Additionally, access to and sustainable use of forests can reduce the risks and vulnerabilities that local communities face, since forests resources are an important safety net in stress periods such as crop failure and drought [21]. Studies on livelihoods, for example, [8, 19, 22, 23], have identified and described various indicators for assessing opportunities and capabilities that those comanagement programmes can provide to communities in order to improve their livelihood (Table 2).

We assessed impacts on natural capital based on local people's perceived changes in the availability, quantity, and quality of forest resource stocks (i.e., timber trees, NTFPs, and improved forest conditions) and changes in access to forest resources. Impacts on financial capital were evaluated in terms of perceived changes in income sources, income levels, ability to access loans, employment opportunities, and ability to accumulate savings. We assessed physical capital at both community and household level (Table 2). At community level, we evaluated the differences in infrastructure developments (e.g., roads) before and after the programme was initiated, whilst at household level we identified the various assets that households have acquired because of their participation in the programme.

Currently there is no consensus on the indicators for measuring social capital, due to its multidimensional nature and ability to change with time and contexts [25]. Furthermore, DFID [17] suggest that measuring social capital may be difficult for an outsider and may require a lengthy analysis over time. Thus in an attempt to assess the impact of forest comanagement on social capital, we evaluate the degree of participation in communal activities (i.e., collective action and cooperation) before and after the programme started.

Although human capital comprises education, knowledge and skills, health, and food security, in this paper we present impacts on human capital based on perceived changes in training and knowledge development before and after the implementation of the programme [19]. The health aspect of human capital was excluded because direct impacts of forest comanagement activities on health are likely to be limited and difficult to quantify [8]. Furthermore, important elements of human health such as vaccinations and provision of health care are not part of the programme under study in Malawi [26]. Additionally, because different forms of capital are linked and can be converted into each other [17, 20], human capital in terms of food and nutrition can be reflected in the assessment of natural capital through changes in access to and availability of forest products including fruits and vegetables and financial capital through income effects. Therefore to avoid duplication and double assessment, food and nutrition were considered to be directly linked and reflected in the natural capital benefits in terms of access and availability of forest products.

Livelihoods are sustainable if they can cope with and recover from stresses and shocks and maintain or enhance the current and future capital base, without undermining the natural resource base [27]. As such households may use

TABLE 1: Livelihoods capitals and their definitions.

Capitals	Definition	Reference
Natural	(i) The natural resources and environmental services that form the basis for human survival and economic activities (e.g., forests, water, and pollution sinks)	[17, 24]
Financial	(i) Capital bases that enable a household to pursue particular livelihoods strategy (e.g., cash, credit, income, and savings)	[17, 25]
Physical	(i) Basic infrastructure (e.g., transport, communications), housing, and equipment of production	[17, 24]
Social/political	(i) Aspects of the society or community upon which households depend, when pursuing livelihoods strategies that require coordinated actions (e.g., networks, social relations, associations, norms, and trust)	[17, 25]
Human	(i) Skills, knowledge, labour, good health, and physical capability that enable one to pursue livelihoods strategies	[17]

TABLE 2: Indicators of forest comanagement opportunities and capabilities for the different livelihood assets.

Livelihood capitals	Indicators of comanagement opportunities and capabilities
Natural	(i) Improved availability of and access to forest resources: (e.g., timber, firewood trees, and poles)
Financial and income	(i) Increased livelihoods and income sources (ii) Increased income levels (iii) Access to loans (iv) Employment (v) Ability to accumulate savings
Physical capital	(i) Development projects (e.g., road building) (ii) Accumulation and acquisition of assets (e.g., land, house, household, and farm assets)
Social capital	(i) Friendly relationships and social organization (ii) Degree of participation in local communal activities (i.e., collective action and cooperation)
Human capital	(i) Training and knowledge development

different combinations of available capitals and activities in order to reduce their vulnerability to stresses and shocks [20, 24]. Thus, it is difficult to draw a conclusion on the overall impact of the comanagement based on the changes in the individual capitals alone [28]. Therefore, in addition to perceived changes in the different livelihoods capital indicators, local peoples' perceptions of the overall impact of the comanagement programme on their livelihoods were also sought.

2.2. Stated Preference Techniques: The Contingent Valuation Method. The contingent valuation method (CVM) is a survey-based stated preference technique used to value goods and services that are not traded on the market [29]. The approach uses hypothetical scenarios with a defined payment vehicle to elicit respondents' willingness to pay (WTP), which estimates the utility gained from the described service [9]. The underlying assumption is that although respondents are presented with a hypothetical scenario, their behaviour and responses reflect their behaviour in real situations [10]. We used the contingent valuation method, rather than choice experiments, because the study was not interested in exploring and valuing different attributes and levels of comanagement but rather valuing the existing programme as a whole [30].

Due to their hypothetical nature, stated preference surveys are prone to biases including hypothetical bias, strategic bias, and social desirability bias. Hypothetical biases arise

when the hypothetical situation presented to respondents fails to reflect the real situation; as such the respondents do not consider budget constraint, and hence the resultant values usually overstate the real value [10]. We minimised the occurrence of hypothetical biases by using a payment vehicle common and familiar to our study communities and also by prompting the respondents to consider their budget constraint as they respond to the question [31]. Strategic bias occurs when respondents respond to the question with intent to influence the study outcome in their favour, if they believe that the hypothetical scenario may become a reality [11]. Social desirability bias, usually associated with face-to-face interviews, occurs when respondents give responses that they perceive as culturally acceptable or to be liked by the interviewer, with a desire to appear to relate to the socially desirable attributes of the programme [32]. To avoid strategic and social desirability biases, participants in our survey were made aware that the situation being presented to them is hypothetical and developed for the purpose of the study and was not directly connected to the programme implementers. However, in order to ensure that consent was informed, it was explained that the outcomes of the study would be made available to the programme coordinators for their reference. Other limitations of CVM surveys pertinent to this study include the difficulty in validating the estimated values externally and uncertainties associated with using the method in developing countries because of the low income and illiteracy of respondents [9]. However, despite the biases and limitations, CVM surveys remain a useful method for estimating welfare impacts of environmental management policies in both developed and developing countries if it is properly designed [10–12].

3. Methods

3.1. Study Area.

Zomba-Malosa and Ntchisi forest reserves are two of the 12 forest reserves where the Malawi Government is implementing the Improved Forest Management and Sustainable Livelihoods Programme (IFMSLP) through the Department of Forestry [33]. The programme aims to address forest degradation and poverty through promoting community involvement in the management of government owned forest reserves. The programme, which was in its 7th year at the time of the study, was implemented for 14 years from 2005 to 2014.

Zomba district is located in the southern region of Malawi and covers 2,580 square kilometres. Zomba-Malosa forest reserve is the only gazetted forest in the district and covers 15,756 hectares, consisting of Miombo woodlands and pine plantations. It is a catchment for major lakes and rivers in the country (e.g., Lake Malawi) and a significant source of water both for domestic and for agricultural use. The reserve is also a source of wood energy (charcoal and firewood) to households in the district as well as neighbouring districts. This has accelerated deforestation and degradation of the reserve. Additionally, the reserve is being encroached upon in the peripheral areas for settlement and agriculture, resulting in further deforestation.

Ntchisi district is located in the central region of Malawi and covers 1,655 square kilometres. Ntchisi district has 3 gazetted forest reserves, namely, Ntchisi, Kaombe, and Mndilasadzu forest reserves, with Ntchisi forest reserve being the largest covering 9,720 hectares. The reserve is located in a remote and rural part of the district, approximately 32 km from the district centre. The reserve is a source of nontimber forest products including mushrooms and edible caterpillars and water for communities living around the reserve. Tree cutting in search of edible caterpillars is said to be a significant cause of deforestation and degradation in the reserve.

3.2. Data Collection

3.2.1. Questionnaire Survey Design and Procedure.

Prior to the household survey, pretesting was done in order to assess the acceptability of the payment vehicle and the response rate to the open-ended CVM question. A total of 20 households participated in the pretesting survey, conducted with communities living around the Dzalanyama forest reserve, in Lilongwe. These communities were participating in a community forest management programme, but not under the IFMSLP. Before the start of each survey session, focus group discussions with community members and key informant interviews with members of the committee, traders and representatives of the community-based organizations, were conducted in each study community to gather general

information about the programme and its impacts on livelihoods. A systematic random approach was used in selecting the households to participate in the survey interviews, from a village register provided by the communities' village heads. The village list formed the sampling frame from which every fourth household on the list was selected to form part of the study. A total of 213 household heads in participating study communities were interviewed (114 in Zomba-Malosa, 99 in Ntchisi), representing approximately 32% of the total household population in the selected study communities.

3.2.2. Socioeconomic and Livelihoods Questionnaire.

The questionnaire first gathered the socioeconomic characteristics of the respondents, including age, education, location, income source, and wealth indicator. Key informants and focus group participant revealed that households' house or dwelling characteristics, that is, type of walls, roof, floor, and window, can reflect the wealth status of an individual or household. Following this information, different parts of a house or dwelling were assigned a score depending on the type of material they are made from, with 1 being the lowest score and 4 being the highest score. Hence household wealth indicator was created based on aggregate scores assigned to different household characteristics. Wealth indicator ranged from 4 to 11, with a score of 4 representing the poorest and 11 being the richest household.

Household socioeconomic characteristics were tested as predictors for (1) perceived overall programme impact and access to programme benefits and (2) households' responses to the contingent valuation question, in order to determine benefit distribution across community members and factors affecting access to benefits. Respondents were also asked to indicate their perceived changes in the different livelihood capitals before and after the programme was initiated (Table 1) and the programme's current overall impact on their livelihoods (i.e., whether they were benefiting or not). The response to perceived current overall impact on their livelihoods was also tested as a determinant for the households' WTP.

3.2.3. Contingent Valuation Survey.

We used annual membership fee as a payment vehicle to elicit household willingness to pay values for forest comanagement programme. Many individuals in the study area belong to small village groups (e.g., village banking group, irrigation farming groups), to which they are required to pay an annual membership fee, to show commitment. Therefore, respondents are familiar with the payment vehicle adopted, hence minimising the occurrence of hypothetical bias. The hypothetical scenario was presented as follows:

> *Imagine that the Government and its partners will no longer be in a position to fund some of the activities of the programme, thus they would like to ask each community member to contribute in the form of a membership fee, so as to ensure that the activities of the programme continue in the community.*

After presenting the payment vehicle, respondents were allowed to ask questions to ensure that they had understood the scenario, before presenting them with the WTP questions. The questions asked by the majority of the respondents with regard to the payment vehicle were (1) what will the money be used for, and who will use it? and (2) will payment translate to increased access to forest resources? This reflects that respondents were able to understand the questions and process the issues and constraints as they would in a real situation before making a payment decision.

Following presentation of the hypothetical scenario, respondents were asked whether or not they would be willing to pay a membership fee. If the response was "no," they were asked to give reasons for their response, and the interview was terminated. All such responses were considered as zero WTP. If the respondents answered "yes," they were then asked how much they would be willing to pay per year. An open-ended question was used because (1) it would have been impossible to have sufficient sample size within forest comanagement implementing communities for a dichotomous question; (2) during the preliminary survey we observed that respondents did not have problems in stating the WTP amount since the payment vehicle is common and familiar to most communities; and (3) being a heterogeneous community open-ended questions provide more information on WTP that would enable us to assess the credibility of the responses [34].

3.3. Data Analysis

3.3.1. Probit Model. A probit regression model was used to explore factors that predict whether households (1) perceive positive overall livelihoods impact of comanagement, (2) accessed new income sources initiated by the programme, and (3) were willing to pay membership fees to participate in forest comanagement. According to Wooldridge [35] the probit model equation is specified as

$$\Pr(Y = 1 \mid X) = \Phi(X\beta), \quad Y = \{1, 0\}, \tag{1}$$

where Y (dummy variable) is equal to 1 for households giving a positive response and zero if otherwise. Φ is a cumulative density function, X are household and individual characteristics, and β are parameters to be estimated.

3.3.2. Ordinary Least Square (OLS) Regression and Tobit Regression Model. Factors affecting open-ended WTP estimates can be explored using Ordinary Least Squares (OLS) regression. However, the use of OLS regression might lead to biases in parameter estimates and misleading inferences depending on the number of zero WTP responses in the data set [35, 36]. If zero responses are excluded, the use of OLS on the censored data set may also result in sample selection bias, as the remaining data with positive WTP only is unlikely to be a random sample, even if the initial sample (all included) was random, and as such it may provide inconsistent parameter estimates [35]. Therefore, in case of relatively large numbers of zero WTP, the censored regression model, known as Tobit, is the theoretically preferred model [36]. A Tobit model with selectivity allows decomposition of the data set to examine

more closely the effects of the independent variables on positive WTP observations [36]. However, so far there is no clear guide in the literature as to what number of zero WTP observations require the use of Tobit regression in place of OLS. Therefore both OLS regression (including the zero WTP) and Tobit regression (censored at zero WTP) results are presented in this paper.

The general description of the OLS model is

$$Y^* = X_i\beta + \varepsilon_i, \quad \varepsilon_i \approx N\left(0, \sigma^2\right), \tag{2}$$

where Y^* is the amount that the household indicated that they are willing to pay. OLS regression assumes that the dependent variable Y^* is linear and continuous. X are characteristics of the household and the head of household and β are parameters to be estimated. The error term ε is assumed to be normally distributed with mean zero and variance σ^2. The Tobit model follows the OLS regression equation; however, the observed willingness to pay (Y^*) represents the latent variables censored at WTP greater than zero. Therefore the Tobit equation follows:

$$Y_i^* = X_i\beta + \varepsilon_i, \quad Y_i = \begin{cases} Y_i^* & \text{if } Y^* > 0, \\ 0 & \text{if } Y^* \leq 0. \end{cases} \tag{3}$$

For both the OLS and Tobit regression models, the dependent variable is the annual amount households are willing to pay as a membership fee, measured on a continuous scale. The Variance Inflation factor (VIF) scores were less than 10 and Tolerance scores ranged from 0.64 to 0.91, indicating weak correlation between the explanatory variables [37]. For all the regression models, bootstrapping (1000 resamples) was used in estimating the coefficients, to correct for any distributional and asymptotic errors and to ensure that the results are valid, accurate, and closer to the population parameters [35]. Data were analysed using STATA version 11.2.

4. Results and Discussion

4.1. Perceived Forest Comanagement Livelihoods Impacts. Approximately 43% (Zomba-Malosa) and 28% (Ntchisi) of the respondents perceive that the comanagement programme has had or is having a positive impact on their livelihoods and approximately 57% (Zomba-Malosa) and 71% (Ntchisi) perceive that the programme has had no impact on their livelihoods. This difference in number of households perceiving a positive impact and those perceiving negative impact could suggest inequalities or elite capture in benefit sharing. This support the findings in [38] that found that a majority of participants in comanagement programme, in Malawi, perceive benefit distribution as unfair and that only a few influential members of the community, for example, committee members and chiefs, share the benefits.

A majority of respondents, 76% (Zomba-Malosa) and 73% (Ntchisi), perceive that the availability and accessibility of firewood and timber trees have reduced since the programme started (Table 3). This could be attributed to small levels of harvestable stock in the forest reserves due

TABLE 3: Perceived livelihoods status before and after comanagement programme in Zomba-Malosa and Ntchisi.

Livelihoods capitals	Indicators	Percentage response by district				Notes
		Zomba-Malosa ($n = 106$)		Ntchisi ($n = 99$)		
		Before comanagement	After comanagement	Before comanagement	After comanagement	
Natural capital	Better availability of and access to firewood and NTFP	55	24	51	27	
	Better availability of and access to timber and pole trees	56	33	71	19	
Financial capital	Accessed loans	5	35	0	30	Village banks initiated by the programme
	Saving	7	39	3	24	
	Access to new income sources	N/A	31	N/A	32	
	(i) Wage labour	N/A	43	N/A	17	During firebreak and forest road constructions
	(ii) Forest based income generating activities	N/A	19	N/A	70	Transport, initial inputs, and materials provided by project
	(iii) Irrigation agriculture	N/A	39	N/A	14	Perceived improvements in water flow due to improved forest condition
Physical capital	Have acquired assets	N/A	36		29	Through participation in wage labour or forest-based business initiated by the programme
Social capital	Participation in communal activity	10	39	20	49	
Human capital	Training and skill development	12	76	15	63	

to the general declining trend in forest resources over the years. Furthermore, tree populations take time to respond to new management approaches [7]; hence the period of comanagement implementation at the time of the study (7 years) would not have been long enough to allow forest rehabilitation to a level yielding adequate harvestable stock. Nevertheless, approximately 32% (Zomba-Malosa) and 24% (Ntchisi) of respondents attributed the reduction in access to forest resources to the strict laws and regulations being enforced under the comanagement programme (under the comanagement programme, the forest reserve is divided into coupes to facilitate selective tree forest resources harvesting; harvesting is done following the strict harvesting and management plans, rules, and regulation; the strict laws and regulations include selective harvesting of trees for timber, only collecting dead wood for firewood, and harvesting of both timber and nontimber forest products should only be done upon acquisition of permit from the management committee at a cost; furthermore, noncompliance with the laws and regulation attracts sanctions and penalties, e.g., community work or fines [38]). Since noncompliance with the laws and regulation attracts sanctions and penalties, for

example, community work or fine [38], thus to avoid penalty communities comply. Furthermore, harvesting permits are obtained upon paying a loyalty fee, which might be costly for the rural poor with limited disposable income, thus reducing their access to forest resource. Therefore, the perceived reduction in access to and availability of forest resources is a result of both comanagement (i.e., strict accessing rules and regulations) and preexisting poor forest condition. The strict enforcement of rules and regulations is necessary to allow for the regeneration of the forest [26]; however there is a need for balance so that the achievement of community livelihoods goals is not constrained.

Approximately, 31% (Zomba-Malosa) and 32% (Ntchisi) of respondents indicated that the programme has helped them to attain new income sources, such as (a) wage labour during firebreak construction and maintenance; (b) income generating activities, for example, timber sales, firewood sales, pottery (clay pots) sales, bee-keeping, and mushroom farming; and (c) indirect benefits in the form of dry season irrigated agriculture (Table 3). Similarly, there was an increase in the number of households accessing loans and saving in the local village banks since the programme started.

The estimated amount in annual savings ranged from MK 500 (US$ 1.7) (the exchange rate at the time of study was Malawi Kwacha (MK) 288.7347 = US$ 1) to MK 10000 (US$ 34.4) in Zomba-Malosa and MK 500 (US$ 1.7) to MK 6000 (US$ 20.8) in Ntchisi district. The loans are linked to the programme's enhancement of social capital within the participating communities, as the loans are accessed from a local community bank initiated by the programme. Additionally, 31% (Zomba-Malosa) and 32% (Ntchisi) of the respondents have accessed new income sources (e.g., wage labour and forest-based business groups initiated by the programme). Thus approximately 36% (Zomba-Malosa) and 29% (Ntchisi) of the respondents indicated that they have managed to acquire assets (e.g., household utensils, furniture, bicycles, and farm equipment) as a result of their participation in income generating activities initiated by the programme. However, the income generating activities that are being promoted in the area (e.g., firewood sale and pottery) are of low value [21], hence the minimal impact on household income levels. Also, it is important to note that the income generating activities are externally subsidised by the programme donors (e.g., initial transport to market), hence the current far-from-universal positive impact on livelihoods creates uncertainties for the programme's long term livelihoods impacts when the donor or external funding is withdrawn.

We observed that the programme's investment in physical capital at community level is limited. Both communities highlighted accessible forest roads as a major infrastructure development that they require. The communities were of the view that even if the forest reserves were to have significant high value timber trees, with potential to generate high revenue, the current poor roads would limit access to economically viable markets. This would limit the programme's potential to positively improve livelihoods, both during and even beyond its implementation period.

A majority of community members perceive that training and skill development activities have increased since the programme started (Table 3). Although the programme does not provide formal education, it contributes to the development of human capital, by facilitating training in forest and tree management techniques. The programme also facilitates and enhances the development of social capital through establishment of village committees and initiating regular community meetings, where issues relating to forest management and other developmental issues are discussed. The committee meetings allow for regular interaction with government forest staff and other stakeholders, hence increasing their social network base and ability to contribute to forest policies that affect their livelihoods [27]. Furthermore, social capital enhances human and financial capital among households since communities are able to form village banking groups and further access loans from the banks [38].

4.2. Who Has Benefited? The probability of perceiving a positive overall impact of comanagement on livelihoods is 89% higher for households in Zomba-Malosa compared to those in Ntchisi (Table 4, A) and 49% higher for households

that perceive better access to and availability of firewood. Firewood and NTFPs are essential for day-to-day livelihood strategies for rural households in Malawi; therefore improved access to and availability of forest resources directly and positively affect households' livelihoods.

The results show that households that perceive increased participation in communal activities because of the programme are 34% more likely to perceive the overall programme impact as beneficial than those who did not (Table 4, A), therefore suggesting that communities not only measure perceived benefits in terms of economic benefits but also measure perceived benefits in social and noncash benefits, contrary to Phiri et al. [23] who suggested that communities perceive the benefits of forest comanagement programmes as minimal because they only measure benefit in terms of monetary or tangible economic benefit. Access to new income sources increases a household's probability of describing the overall programme impact as positive by approximately 92% (Table 4, A). This is expected because access to new income sources can potentially translate into increased income levels and improved livelihoods [3, 8]. Furthermore, new income sources may diversify household livelihood sources hence reducing household's vulnerability to shocks and stresses (e.g., failure in crop production) [24].

The income generating activities initiated by the programme are forest-based and group-based; hence it is plausible that the probability of accessing new income sources is higher for households that perceived better access to and availability of forest resources (30%) and better participation in communal activities (40%) (Table 4, B). Households whose head is a committee member is approximately 60% more likely to access new income sources and to perceive the overall impact as beneficial (Table 4, B). Although this raises questions about the equity of benefits sharing among community members, there is no further evidence from the probit model to suggest that access to new income sources is influenced by household characteristics or social status (e.g., wealth status, gender, and age). Lastly, access to new income sources is positively and significantly related to access to loans. This is expected as usually households opt for loans for investment purposes, for example, small businesses, rather than consumption [39].

4.3. WTP to Participate in the Forest Comanagement Programme

4.3.1. Are Households Willing to Pay a Membership Fee? Although a majority perceive that currently they are not benefiting from the programme, approximately 83% (Zomba-Malosa) and 81% (Ntchisi) of respondents are willing to pay membership fees to participate in the forest comanagement programme. The mean annual willingness to pay amount is approximately MK 1,000 (US$ 3.5) in Zomba-Malosa and MK 400 (US$ 1.4) in Ntchisi, respectively. These values are approximately five times (Zomba-Malosa) and two times (Ntchisi) the minimum daily wage rate (daily wage rate in urban communities is estimated at MK 200; however the Malawi Government Employment Act stipulates MK 98

TABLE 4: Probit regression result on factors affecting perception of programmes overall impact and accessing new income sources.

Covariates	Perceived overall impact A		Accessing new income sources B	
	Coefficients	Bootstrapped Std. errors	Coefficients	Bootstrapped Std. errors
District (Ntchisi = 1; Zomba = 0)	−0.89****	(0.25)	−0.24	(0.24)
Better access to and availability of timber (1 = yes; 0 = no)	0.06	(0.13)	0.32**	(0.13)
Better access to and availability of firewood (1 = yes; 0 = no)	0.49****	(0.14)	−0.18	(0.13)
Better training and skill development (1 = yes; 0 = no)	0.07	(0.15)	−0.21	(0.14)
Better participation in communal activity (1 = yes; 0 = no)	0.34**	(0.16)	0.40**	(0.18)
Committee member (1 = yes; 0 = no)	0.40*	(0.23)	0.687***	(0.21)
Acquired assets (1 = yes; 0 = no)	0.50*	(0.26)		
Accessed new income sources (1 = yes; 0 = no)	0.92****	(0.23)		
Accessed loans (1 = yes; 0 = no)	0.69	(0.81)	0.78*	(0.56)
Saving (1 = yes; 0 = no)	−0.14	(0.22)	0.19	(0.21)
Married (1 = yes, 0 = no)	0.09	(0.12)	−0.23	(0.13)
Gender of household head (1 = female, 0 = male)	−0.07	(0.23)	−0.33	(0.21)
Age of household head (in years)	−0.01	(0.01)	0.01	(0.01)
Household size (number of adults and children)	0.03	(0.05)	−0.10	(0.05)
Land size (in hectares)	0.09	(0.03)	0.02	(0.03)
Wealth indicator (ordinal scale, 4–11)	0.03	(0.06)	−0.03	(0.06)
_cons	0.04	(0.78)	−0.37	(0.71)
Prob > chi2	0.00		0.01	
Number	213		213	
Pseudo R2	0.24		0.14	
Log pseudo likelihood	−101.11		−110.57	

Significance levels (*: 10%; **: 5%; ***: 1%; and ****: 0.01%).

as the daily wage rate in rural areas [40]) and represent approximately 6% (Zomba-Malosa) and 4% (Ntchisi) of the average estimated annual earning of the respondents (the estimated annual earning for respondents is MK 15000 in Zomba-Malosa and MK 9000 in Ntchisi). Considering that rural Malawi is characterised by high poverty levels, high unemployment rates, heavy reliance on smallholder agriculture, susceptibility to shocks, and limited disposable income such that 20% of the rural population struggle to even afford the daily recommended food requirements [41] (approximately 75% of Malawians live under the poverty threshold of under US$ 1.25 a day; 28% of the rural households (which is 85% of total population) are characterised as ultrapoor, with limited access to employment as 75% earn their living only from smallholder farming, and for those on wage employment the income is so minimal; hence the disposable income is very low; additionally people have limited or even no access to financial services such as credits, which further limits their economic growth and spending pattern [41]), the willingness to pay values represent a substantial

proportion of households' annual income. This suggests that communities are optimistic about substantial future livelihoods and welfare benefits from the programme and thus willing to invest in the programme's activities. However, it is doubtful that communities will accrue positive livelihoods and welfare benefits from the programme in future (especially after withdrawal of external support), when it is failing to provide the majority with positive benefits at present, despite the externally subsidised income generating activities it is currently implementing. Therefore, it can be argued that the estimated willingness to pay is due to respondents' optimism (optimism bias) of future benefits that forest recovery could potentially provide. Interestingly, a majority of those not willing to pay in both Zomba-Malosa (87%) and Ntchisi (72%) attributed their decision to lack of benefits from the programme and lack of trust in the leadership with regard to financial accountability and inability to pay. Chinangwa [38] also observed a general lack of trust in leadership with regard to financial accountability among communities participating in forest comanagement programme in Zomba-Malosa and

TABLE 5: Factors affecting households' willingness to pay a membership fee to participate in comanagement programme in Zomba and Ntchisi districts.

	Probit model (WTP = 1)	
	Coefficients	Bootstrapped Std. errors
Perceived overall impact dummy (1 = benefiting, 0 = not benefiting)	0.43*	(0.23)
Wealth indicator (ordinal scale, 4–11)	0.11*	(0.08)
Number of years in school	0.07**	(0.03)
Committee member (1 = yes, 0 = no)	0.36*	(0.25)
Land size (in hectares)	0.03	(0.03)
District (1 = Ntchisi, 0 = Zomba)	0.17	(0.22)
Gender of household head (1 = female, 0 = male)	−0.20	(0.22)
Married (1 = yes, 0 = no)	0.07	(0.12)
Household size (number of adults and children)	0.03	(0.05)
Age of household head (in years)	−0.00	(0.01)
_cons	−0.82	(0.79)
Prob > chi2	0.02	
Number of observations	213	
Pseudo R2	0.10	
Log pseudo likelihood	−95.13	

Significance levels (*: 10% and **: 5%).

Ntchisi, thus negatively affecting how they perceive and access benefit of the programme and consequently how they value the programme's welfare benefits relative to state management regime.

4.3.2. Factors Affecting Household Decision to Pay Membership Fee. The probit regression shows that households that perceive comanagement to have a positive impact on their livelihoods are 43% more likely to be willing to pay than those that perceive otherwise (Table 5), so as to secure continued livelihoods and benefit flows [9, 13]. An increase in households' wealth indicator by 1 point increases household probability for willingness to pay by 10% (Table 5). This suggests that wealthier households are more appreciative of forest comanagement and conservation, as they are less dependent on the forest for their livelihoods, than poor households. Furthermore, this shows that they are more able to invest in comanagement activities for the expected potential future benefits.

An extra year of schooling increases household probability of being willing to pay by 7% (Table 5). Mekonnen [9] suggests that more years in formal education enhance peoples' ability to understand and respond to the willingness to pay hypothetical questions. Therefore, the positive effect of schooling on willingness to pay may not necessarily indicate that educated people value the programme more than uneducated households but indicate that they understood the

hypothetical question better. Committee members are more likely to be willing to pay to participate in comanagement (Table 5) because they are trained in forest management and in constant contact with forest staff and hence have a broader knowledge of the overall benefits of the programme and as such they are more appreciative of the programme. However, the committee member's likelihood of being willing to pay could be attributed to the greater benefits they access through programmes (e.g., Table 4, B). Furthermore, committee members' willingness to pay may be affected by social desirability bias, as they may view being willing to pay as an acceptable answer, since they are programme coordinators at community level.

4.3.3. Factors Affecting How Much Households Are Willing to Pay in Membership Fees. Although Halstead et al. [36] argue that the sign of coefficients estimated using Tobit analysis may differ from those estimated using OLS, we found no such differences, probably due to the small proportion of zero WTP (approximately 20%, [42]). Our results show no significant difference between the OLS and Tobit models except for the size of coefficients (Table 6).

Both the OLS and Tobit regressions suggest that respondents' district, wealth indicator score, gender of household head, and land size significantly affect households' decision on how much they are willing to pay (Table 6). Households in Zomba-Malosa (mean WTP = MK 989 per year) are willing to pay more than households in Ntchisi (mean WTP = MK 400 per year). This may be attributed to socioeconomic variation across communities and how dependent the communities are on the forest for their livelihoods. For example, whilst an estimated 80% of the Ntchisi district economy and livelihoods are said to be agriculture based, it is estimated that 90% of Zomba-Malosa population are dependent on forests resources [35, 43]. Therefore it is plausible that communities in Zomba-Malosa are likely to be willing to pay more so as to secure their livelihoods and welfare flow.

An increase in households' wealth indicator by 1 point increases the WTP value by approximately MK 298 (OLS) or MK 412 (Tobit model). Households' WTP is associated with ability to pay [44]. As households' wealth status improves with increased income levels and asset base, they are likely to have disposable income and as such capable and likely to be willing to pay more for forest activities compared to poorer households [45]. Similar trends have been observed by Chikwuone and Okorji [44], for forest management in Nigeria.

Female-headed households are willing to pay approximately MK 298 (OLS) or MK 610 (Tobit model) less than male-headed households. Due to cultural norms, female-headed households in rural Malawi have limited access to forest management programme's financial benefits and resources [46]; thus the limited welfare benefits from the programme could be reducing their willingness to pay and value of the programme. This is in contrast to findings by Chikwuone and Okorji [44], who show that female-headed households are likely to be willing to pay more for community forestry compared to male-headed ones, because women

TABLE 6: Factors affecting how much households' willingness to pay a membership fee to participate in comanagement programme in Zomba and Ntchisi districts (in Malawi Kwacha).

	OLS model		Tobit model	
	Coefficients	Bootstrapped Std. errors	Coefficients	Bootstrapped Std. errors
District (1 = Ntchisi, 0 = Zomba)	−697.60**	(304.77)	−646.07**	(347.16)
Household size (number of adults and children)	−11.56	(44.93)	−26.00	(56.93)
Wealth indicator (ordinal scale, 4–11)	298.69**	(132.17)	412.54**	(191.84)
Gender of household head (1 = female, 0 = male)	−520.47**	(322.91)	−610.48**	(−587.61)
Number of years in school	−15.41	(73.70)	−32.15	(79.03)
Married (1 = yes, 0 = no)	146.18	(128.76)	210.23	(152.19)
Age of household head (in years)	−4.66	(17.73)	−11.26	(21.29)
Committee member (1 = yes, 0 = no)	−366.79	(269.37)	−280.87	(318.15)
Land size (in hectares)	60.59*	(35.10)	89.54*	(48.23)
Perceived overall impact dummy (1 = benefiting, 0 = not benefiting)	172.68	(338.46)	333.89	(421.54)
_cons	−15.13	(690.11)	−1479.25	(1293.83)
sigma_cons			2736.98****	(831.17)
Prob > chi2	0.01		0.02	
Number	213		213	
R-squared	0.07			
Root MSE	2435.68			
Pseudo R2			0.01	
Log likelihood			−1599.19	

Significance levels (*: 10%; **: 5%; and ****: 0.01%).

depend on forest resources for their livelihoods more than men. However it is plausible to argue that female-headed households have limited access to benefits and lower income levels and are more prone to risks and uncertainties in terms of income sources compared to male-headed households [47], hence likely to be willing to pay less than male-headed household.

The amount households are willing to pay is significantly and positively related to land size (Table 6). This could be because households with small land holdings may be encroaching into the forest to increase their land holdings and hence may not be engaged or interested in the conservation activities under comanagement, as they may be viewed as conflicting with their individual goals. Similar findings were found by Chinangwa [38] who observed that households with bigger land sizes are more likely to perceive forest conservation as criteria for measuring success of forest comanagement programme, compared to those with small land holdings in Malawi.

5. Conclusion

The findings of this study suggest that forest comanagement programmes can potentially improve household livelihoods by introducing profitable income generating activities; facilitating local lending and savings; enhancing social capital; and development of human capital through training. The positive effect on a household's likelihood of accessing new income sources from the programme when the household head is a committee member, coupled with the positive effects on household's WTP by households land holding sizes and wealth status, suggests that access to and distribution of programme benefits may be affected by households' socioeconomic status. Livelihoods diversification away from traditional agriculture through access to new forest-based and non-forest-based income sources could reduce household's vulnerability to stresses and hence eventually result in protection of the forest resources through reduced pressure and increased management and conservation activities by the participating communities. However, these efforts should be complemented with investment in physical capital and financial incentives, at community level, to enable community members to access economically viable markets and ensure that the programme's impacts are sustainable beyond the programme. The impacts of forest comanagement programmes often take a long time to materialise because there is a need to reestablish the condition of the forests to yield harvestable stock, as well as the need to develop effective management practices that are appropriate to the needs of the community. Therefore, although the current livelihoods impacts of the programme are minimal, this does not imply that the comanagement programme is a failure.

There is a potential for better or higher livelihood benefits from the programme in future, if proper management and utilization strategies are followed. The management and utilization strategies could include adherence to the selective tree harvesting and implementing afforestation programme and development of explicit user rights and benefit sharing procedure. Furthermore, to enhance accountability in management and utilization, public hearing and audit session should be introduced.

Although the livelihood impacts of comanagement are currently minimal and restricted to a subset of the community, community members may be willing to pay a membership fee to participate in a forest comanagement programme because of their perceived future benefits of the programme. This also demonstrates the danger of relying on stated preference surveys alone to estimate welfare effects, because the WTP values given by respondents could represent a number of things and may not always reflect respondents' present gains from the policy change or programme. Furthermore, this shows that community's investments of time and labour in the forest comanagement programme could be based on an overly optimistic view that in future the net welfare benefits from the programme will increase, which puts them at a risk of being taken advantage of by programme initiators in setting up CFM projects. Therefore, although contingent valuation methods remain important in estimating the economic value of environmental management policies like comanagement, the sustainable livelihoods framework seems more reliable at representing the real current impacts or benefits of comanagement on community livelihoods. Given the paucity of empirical evidence on comanagement, this paper makes useful empirical and methodological contributions towards the evidence base of forest comanagement livelihoods impacts that are likely to be applicable to other comanagement projects and studies as well as to other forms of CBFM initiatives at regional as well as global level.

Competing Interests

The authors declare that there is no conflict of any form regarding the publication of this paper.

Acknowledgments

This work was supported by the International Centre for Research in Agroforestry (ICRAF), Nairobi. The authors also acknowledge the staff of the Department of Forestry, Malawi, Mr. T. Kamoto, Mrs. S. Gama, Mr. A. Munyenyembe, Mr. Magagula, Mr. Nangwale, Mr. Makupete, Mr. Goneta, and all ground staff, for their assistance during the field work. They also acknowledge the support of their research assistants, Mr. W. Chinangwa, Mr. M. Munyenyembe, and Mr. G. Chipofya. Finally, the authors acknowledge the households and communities in their study sites, Zomba-Malosa and Ntchisi.

References

[1] E. A. Tsi, N. Ajaga, G. Wiegleb, and M. Muhlenberg, "The willingness to pay (WTP) for the conservation of wild animals: Case of the Derby Eland (*Taurotragus derbianus gigas*) and the African wild dog (*Lycaon pictus*) in North Cameroon," *African Journal of Environmental Science and Technology*, vol. 2, no. 3, pp. 51–58, 2008.

[2] D. E. Bowler, L. M. Buyung-Ali, J. R. Healey, J. P. G. Jones, T. M. Knight, and A. S. Pullin, "Does community forest management provide global environmental benefits and improve local welfare?" *Frontiers in Ecology and the Environment*, vol. 10, no. 1, pp. 29–36, 2012.

[3] T. Gobeze, M. Bekele, M. Lemenih, and H. Kassa, "Participatory forest management and its impacts on livelihoods and forest status: the case of bonga forest in Ethiopia," *International Forestry Review*, vol. 11, no. 3, pp. 346–358, 2009.

[4] L. A. Wily, "Reconstructing the African commons," *Africa Today*, vol. 48, no. 1, pp. 77–99, 2001.

[5] S. Tole, "Reforms from the ground up: a review of community-based forest management in tropical developing countries," *Environmental Management*, vol. 45, no. 6, pp. 1312–1331, 2010.

[6] T. Ali, M. Ahmad, B. Shahbaz, and A. Suleri, "Impact of participatory forest management on vulnerability and livelihood assets of forest-dependent communities in northern Pakistan," *International Journal of Sustainable Development and World Ecology*, vol. 14, no. 2, pp. 211–223, 2007.

[7] O. P. Dev, N. P. Yadav, O. Springate-Baginski, and J. Soussan, "Impacts of community forestry on livelihoods in the Middle Hills of Nepal," *Journal of Forest and Livelihood*, vol. 3, no. 1, pp. 64–77, 2003.

[8] V. G. Vyamana, "Participatory forest management in the eastern arc mountains of tanzania: who benefits?" *International Forestry Review*, vol. 11, no. 2, pp. 239–253, 2009.

[9] A. Mekonnen, "Valuation of community forestry in Ethiopia: a contingent valuation study of rural households," *Environment and Development Economics*, vol. 5, no. 3, pp. 289–308, 2000.

[10] I. J. Bateman, R. T. Carson, B. Day et al., *Economic Valuation with Stated Preference Techniques: A Manual*, Edward Elgar Publishing, Cheltenham, UK, 2002.

[11] R. C. Mitchell and R. T. Carson, *Using Surveys to Value Public Goods: The Contingent Valuation Method*, Resources for the Future, Washington, DC, USA, 1989.

[12] D. Whittington, "Administering contingent valuation surveys in developing countries," *World Development*, vol. 26, no. 1, pp. 21–30, 1998.

[13] N. Hanley, E. R. Wright, and B. Alvarez-Farizo, "Estimating the economic value of improvements in river ecology using choice experiments: an application to the water framework directive," in *Environmental Value Transfer: Issues and Methods*, S. Navrud and R. Ready, Eds., pp. 111–130, Springer, Amsterdam, The Netherlands, 2007.

[14] W. M. Hanemann, "Valuing the environment through contingent valuation," *Journal of Economic Perspectives*, vol. 8, no. 4, pp. 19–43, 1994.

[15] R. Chambers and G. Conway, "Sustainable rural livelihoods: practical concepts for the 21st century," IDS Discussion Paper 296, IDS, Brighton, UK, 1991.

[16] N. Das, "Impact of participatory forestry program on sustainable rural livelihoods: lessons from an Indian province," *Applied Economic Perspectives and Policy*, vol. 34, no. 3, pp. 428–453, 2012.

[17] Department for International Development (DFID), *Sustainable Livelihoods Framework: Guidance Sheets*, Department for International Development (DFID), London, UK, 1999.

[18] P. Baumann and S. Sinha, "Linking development with democratic processes in India: political capital and sustainable livelihoods analysis," Natural Resource Perspectives 68, Overseas Development Institute (ODI), London, UK, 2001.

[19] T. K. Nath and M. Inoue, "Impacts of participatory forestry on livelihoods of ethnic people: experience from Bangladesh," *Society and Natural Resources*, vol. 23, no. 11, pp. 1093–1107, 2010.

[20] T. Shimizu, "Assessing the access to forest resources for improving livelihoods in West and Central Asia countries," Food and Agriculture Organization, Livelihoods Support Programme, Working Paper 33, FAO, Rome, Italy, 2006.

[21] M. Fisher, "Household welfare and forest dependence in southern Malawi," *Environment and Development Economics*, vol. 9, no. 1, pp. 135–154, 2004.

[22] R. Goswami and M. Paul, "Using sustainable livelihood framework for assessing the impact of extension programmes: an empirical study in the context of joint forest management," in *Proceedings of the National Extension Education Congress*, Goa, India, 2011.

[23] M. Phiri, P. W. Chirwa, S. Watts, and S. Syampungani, "Local community perception of joint forest management and its implications for forest condition: the case of Dambwa forest reserve in southern Zambia," *Southern Forests*, vol. 74, no. 1, pp. 51–59, 2012.

[24] F. Ellis, *Rural Livelihoods and Diversity in Developing Countries*, Oxford University Press, London, UK, 2000.

[25] M. Woolcock, "The place of social capital in understanding social and economic outcomes," *Canadian Journal of Policy Research*, vol. 2, no. 1, pp. 11–17, 2001.

[26] Malawi Government, *Improved Forestry and Sustainable Livelihoods Programme: Mid-Term Review Report*, Department of Forestry, Malawi Government, Lilongwe, Malawi, 2008.

[27] T. K. Nath, M. Inoue, and J. Pretty, "Formation and function of social capital for forest resource management and the improved livelihoods of indigenous people in Bangladesh," *Journal of Rural and Community Development*, vol. 5, no. 3, pp. 104–122, 2010.

[28] M. R. Maharjan, T. R. Dhakal, S. K. Thapa, K. Schreckenberg, and C. Luttrell, "Improving the benefits to the poor from community forestry in the churia region of Nepal," *International Forestry Review*, vol. 11, no. 2, pp. 254–267, 2009.

[29] G. Garrod and K. G. Willis, *Economic Valuation of the Environment: Methods and Case Studies*, Edward Elgar, Cheltenham, UK, 1999.

[30] J. T. Bishop, Ed., *Valuing Forests: A Review of Methods and Applications in Developing Countries*, IIED, London, UK, 1999.

[31] J. Loomis, "What's to know about hypothetical bias in stated preference valuation studies?" *Journal of Economic Surveys*, vol. 25, no. 2, pp. 363–370, 2011.

[32] M. L. Loureiro and J. Lotade, "Interviewer effects on the valuation of goods with ethical and environmental attributes," *Environmental and Resource Economics*, vol. 30, no. 1, pp. 49–72, 2005.

[33] Malawi Government, *Baseline Report: Improved Forest Management for Sustainable Livelihoods (IFMSLP)*, Department of Forestry, Lilongwe, Malawi, 2007.

[34] N. Jones, K. Evangelinos, C. P. Halvadakis, T. Iosifides, and C. M. Sophoulis, "Social factors influencing perceptions and willingness to pay for a market-based policy aiming on solid waste management," *Resources, Conservation and Recycling*, vol. 54, no. 9, pp. 533–540, 2010.

[35] J. M. Wooldridge, *Econometric Analysis of Cross Section and Panel Data*, MIT Press, London, UK, 2nd edition, 2002.

[36] J. M. Halstead, B. E. Lindsay, and C. M. Brown, "Use of the tobit model in contingent valuation: experimental evidence from the Pemigewasset Wilderness Area," *Journal of Environmental Management*, vol. 33, no. 1, pp. 79–89, 1991.

[37] P. D. Allison, *Logistic Regression Using the SAS System Theory and Application*, SAS Institute, Cary, NC, USA, 1999.

[38] L. L. R. Chinangwa, *Does co-management programme reconcile community interests and forest conservation: a case study of Malawi [Ph.D. thesis]*, Prifysgol Bangor University, Wales, UK, 2014.

[39] S. Vermeulen and L. Cotula, *Making the Most of Agricultural Investment: A Survey of Business Models That Provide Opportunities for Smallholders*, IIED, London, UK; FAO, Rome, Italy; IFAD, Rome, Italy; SDC, Bern, Switzerland, 2010.

[40] S. Snyman, "Household spending patterns and flow of ecotourism income into communities around Liwonde National Park, Malawi," *Development Southern Africa*, vol. 30, no. 4-5, pp. 640–658, 2013.

[41] Malawi Government, *Malawi Employment Act, 2000*, Malawi Government, Lilongwe, Malawi, 2000.

[42] C. Wilson and C. Tisdel, "OLS and Tobit analysis: When is substitution defensible operationally?" Working Papers on Economic Theory, Applications and Issues, Paper 15, The University of Queensland, Brisbane, Australia, 2002.

[43] Malawi Government, *Ntchisi District Socio-Economic Profile 2005–2010*, Ntchisi District Assembly, Department of Planning and Development, Lilongwe, Malawi, 2005.

[44] N. A. Chikwuone and C. E. Okorji, "Willingness to pay for systematic management of community forests for conservation of non-timber forest products in Nigeria's rainforest region: implications for poverty alleviation," in *Economics of Poverty, Environment and Natural Resource Use*, R. B. Dellinkand and A. Ruijs, Eds., pp. 117–137, Springer, Amsterdam, The Netherlands, 2008.

[45] M. Hatlebakk, "Regional variation in livelihood strategies in Malawi," *South African Journal of Economics*, vol. 80, no. 1, pp. 62–76, 2012.

[46] C. Mawaya and M. P. Kalindekafe, "Access, control and use of natural resources, in southern Malawi: a gender analysis," in *Natural Resource Management: The Impact of Gender and Social Issues*, F. Flintan and S. Tedla, Eds., pp. 88–125, Fountain Publishers, Ottawa, Canada, 2007.

[47] A. R. Quisumbing, L. R. Brown, H. S. Feldstein, L. Haddad, and C. Pena, "Women: the key to food security," Food Policy Report for the International Food Policy Research Institute-(IFPRI), IFPRI, Washington, DC, USA, 1995.

Role of Forest Resources to Local Livelihoods: The Case of East Mau Forest Ecosystem, Kenya

D. K. Langat,[1] E. K. Maranga,[2] A. A. Aboud,[2] and J. K. Cheboiwo[3]

[1]*Kenya Forestry Research Institute, P.O. Box 5199, Kisumu 40108, Kenya*
[2]*Department of Natural Resources, Egerton University, P.O. Box 536, Njoro 20115, Kenya*
[3]*Kenya Forestry Research Institute, P.O. Box 20412, Nairobi 00200, Kenya*

Correspondence should be addressed to D. K. Langat; dkipkirui@yahoo.com

Academic Editor: Piermaria Corona

Forests in Kenya are threatened by unsustainable uses and conversion to alternative land uses. In spite of the consequences of forest degradation and biodiversity loss and reliance of communities on forests livelihoods, there is little empirical data on the role of forest resources in livelihoods of the local communities. Socioeconomic, demographic, and forest use data were obtained by interviewing 367 households. Forest product market survey was undertaken to determine prices of various forest products for valuation of forest use. Forest income was significant to households contributing 33% of total household income. Fuel wood contributed 50%, food (27%), construction material (18%), and fodder, and thatching material 5% to household forest income. Absolute forest income and relative forest income (%) were not significantly different across study locations and between ethnic groups. However, absolute forest income and relative forest income (%) were significantly different among wealth classes. Poor households were more dependent on forests resources. However, in absolute terms, the rich households derived higher forest income. These results provide valuable information on the role of forest resources to livelihoods and could be applied in developing forest conservation policies for enhanced ecosystem services and livelihoods.

1. Introduction

Forests are important in the livelihoods of local people in most developing countries. Local people depend on forests resources for various products such as fuel wood, construction materials, medicine, and food. Globally, it is estimated that between 1.095 billion and 1.745 billion people depend to varying degrees on forests for their livelihoods and about 200 million indigenous communities are almost fully dependent on forests [1]. Moreover, 350 million people who live adjacent to dense forests depend on them for subsistence and income [1, 2]. It is estimated that 20–25% of rural peoples' income is obtained from environmental resources in developing countries [3] and act as safety nets in periods of crisis or during seasonal food shortages [4, 5]. Deforestation and degradation of forest ecosystems, in Kenya, is widely acknowledged and, despite the widespread degradation, there is dearth of quantitative information on the role of forest resources to livelihoods and dependence to guide sustainable use. This paper analyzed the role of forest resources in local livelihoods and determined the forest dependence in East Mau forest ecosystem, Kenya.

2. Materials and Methods

2.1. Study Site. This study was undertaken in East Mau Forest situated about 50 km south of Nakuru Town at 35°58′00″E and 00°32′00″S, altitude range of 1200 and 2600 m (Figure 1). It has an area of approximately 280 km^2 and has the highest number of indigenous forest dwellers—the Ogiek community. East Mau forest forms an important watershed within the Mau Forest Complex, feeding major rivers and streams that make up the hydrological systems of Lake Victoria and inland Lakes of Nakuru, Baringo, and Natron. The forest is home to endangered mammals like the yellow-backed duiker (*Cephalophus sylvicultor*) and the African golden cat (*Felis aurata*) and other important fauna such as Giant Forest Hog, Gazelle, Buffalo, Leopard, Hyena, Antelope, Monkey, and

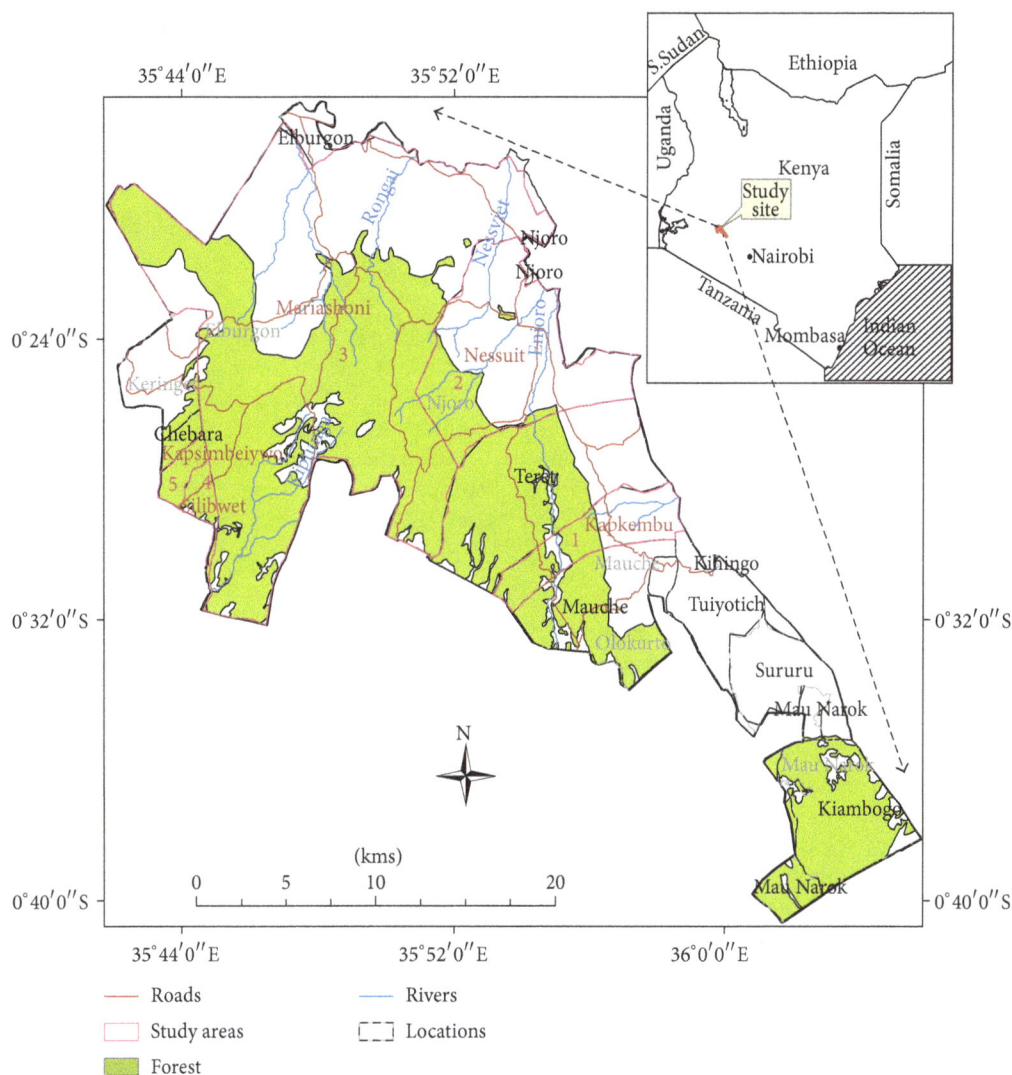

FIGURE 1: Map of the study area in East Mau forest in Kenya.

small animals like the Giant African Genet, Tree Hyrax, and Honey badger [6]. This makes the forest ecosystem an important resource base for the local communities and national and international community. The total forest area was originally about 66,000 ha but more than half of it was excised for human settlement in 2001 [7].

The area is comprised of the escarpments, hills, rolling land, and plains with slopes ranging from 2% in the plains to more than 30% in the foothills and geological studies have shown that the area is mainly composed of quaternary and tertiary volcanic deposits [8]. In the lowlands, the top soils are of mainly clay loam (CL) to loam (L) in texture and the subsoil texture ranges from silty clay loam (SCL) to clay loam (CL) and clay (C), with pH values ranging from 5.6 to 6.4, making them slightly to moderately acidic in nature [9]. In the lowland, Luvisol, Vertisol, Planosol, Cambisol, and Solonetz soils from the Holocene sedimentary deposits are primarily prevalent and occur in saline and sodic phases. In the upland areas, however, the soils have a high content of silt and clay

predominantly Ferrasols, Nitisols, Cambisols, and Acricsols [9]. The adjoining settlements have gentle slopes with deep-fertile-volcanic soils which are suitable for maize, wheat, potatoes, horticultural crops, and livestock keeping [10].

The climate is characterized by a trimodal precipitation pattern with the long and intense rains from April to June; short rains in August; and shorter, less intense rains from November to December with mean monthly rainfall between 30 mm and 120 mm and total annual precipitation of 1200 mm. The mean annual temperatures are in the range of 12 to 16°C, with greatest diurnal variation during the dry season [11].

2.2. Data Collection. Household data was collected from respondents from the month of January to May 2013 and September to December 2013. All households within and adjacent to East Mau forest totaling 43,257 households from 17 administrative units (locations) [12] constituted the research population. Five administrative units were selected in consultation with local administrative officials using two main criteria: age of settlement and ethnic composition of residents. The following administrative locations were selected: Mariashoni representing an old settlement predominantly occupied by Ogiek indigenous community, Kapkembu—representing a recent settlement with a homogenous community of the Kipsigis, Nessuit—representing a recent settlement with a heterogeneous population of indigenous and immigrant ethnic groups and Kapsimbeiywo, and Silibwet—representing a relatively old settlement with a homogenous community of the Kipsigis community.

Study villages in all the five locations were randomly selected from the list of villages provided by local administrative officials and village elders. Respondent households from each village were randomly selected from detailed households' lists (with names of household head and assigned numbers for use in random sampling). In polygamous unions, households were listed according to the wife's name and each considered a separate household. The simple size for each study village and location was determined using the most recent national census data [12] and applying the method by O. Mugenda and A. Mugenda [13]. In total 367 households were selected for the study. Sociodemographic data were collected using structured and semistructured questionnaires. To improve the confidence of the respondents and quality of data, local trained research assistants conversant with local languages interviewed the respondents in the presence of village elders. In most cases, the head of the household was interviewed and, in the absence, the wife or the eldest son was interviewed. The following socioeconomic data were collected from each household: sources of cash income, resources endowment (land size, livestock size, and physical assets), literacy levels (education level), household size, resident years, ethnicity, and distance from the forest. Forest utilization data included consumption patterns of forest products (including their sources, average quantity per month, and household monthly consumption), collection and type of forest products, and other associated information. The information obtained from respondents was triangulated using key informants and focus group discussions.

The market survey captured the prices of various forest products traded in local markets and prices used to value the household forest-product consumption and determined monetary contribution of the forest products to the total household income.

2.3. Data Analysis. The collected field data were compiled and analyzed using the statistical package IBM SPSS version 21 (2013) and Microsoft Office Excel 2010. The household incomes were calculated without accounting for local labour costs because of substantial variation in costs for each activity and the possibility of multiple tasks by households [14]. The household incomes were computed using the formulae (1) to (4) as shown below.

Household annual income = (forest Income + agriculture income + return to wealth + wage income):

$$Y_{\text{tinc}} = \sum_{i=1}^{n} [s_i], \qquad (1)$$

where Y_{tinc} is total household income and s_i is income source I.

Forest income = (fuel wood annual income + wild fruits income + poles income + thatching grass income and forest grazing, etc.):

$$Y_f = \sum_{i=1}^{n} [F_i P_i - (K_i)], \qquad (2)$$

where Y_f is total forest income, F_i is quantity of product collected I, P_i is market price of forest product I, and K_i is production costs of forest product i.

The value of forest grazing was estimated by substitute approach (the Appendix).

Crop income: this was summation of value of yield from various crops grown by a household less all costs of production. Total crop income was calculated as

$$Y_c = \sum_{i=1}^{n} [C_i P_i - (K_i)], \qquad (3)$$

where Y_c is total crop income, C_i is yield of crop I, P_i is market price of crop I, and K_i is production costs of crop i.

Livestock income = (cattle sale income + goats income + sheep income + donkeys income + chicken income) + income from livestock products that is

$$Y_l = \sum_{i=1}^{n} [N_i P_i - (K_i)] + \sum_{i=1}^{n} [Q_i P_i - (K_i)], \qquad (4)$$

where Y_l is total livestock income, N_i is number of livestock in category I, Q_i is quantity of product from livestock I, P_i is market price of livestock I, and K_i is cash costs of keeping livestock i, like pay for herder, costs of medicines, feeds.

Income from off-farm income/employment: this was the total value of earnings through hiring out of labour on other households' lands for agricultural or any other economic activity.

2.3.1. Statistical Tests. Socioeconomic data presents a challenge in a heterogeneous community where extreme income values from individual households are expected. Data was subjected to normality tests (box-plot, histogram). All the identified outliers in the data set were removed to conform to normal distribution. It was then that parametric tests (analysis of variance (ANOVA)) were applied [15]. In all statistical tests, $p \leq 0.05$ level of significance was used. Tests were conducted on socioeconomic characteristics, χ^2 test being for association of locations and sources of forest products, wealth, education level, and ethnicity. Comparison of means and one-way ANOVA were used to test the difference on forest incomes, relative forest incomes on locations, ethnicity, and wealth class and separation of means undertaken using Tukey B.

2.3.2. Measuring Forest Dependence. The forest dependence was measured using the relative forest income. Relative forest income (RFI) was computed as a share of net forest income to total household income accounts derived from consumption and sale of forest environmental resources. This was derived as

$$\text{RFI} = \frac{\text{TFI}}{\text{TI}}, \tag{5}$$

where TI is the total household income and TFI is total forest environmental income.

To test the level of forest dependence of income groups, sampled households were categorized into 3 income groups based on their level of total households income in Kenya Shillings: Poor, 0–156,000, Moderately Poor, 156001–270,000, and Rich, >271,000. The categories were based on local conditions and do not reflect the general poverty levels in the study area and Kenya.

3. Results

3.1. Socioeconomic and Demographic Characteristics of Households. The gender distribution of household heads showed that 62.6% (n = 243) were males while 37.4% (n = 145) were females. The mean age of household head was significantly different (p < 0.001) for female (53.35 ± 1.9) and male-headed households (47.56 ± 1.2). The majority of the respondents in the Kapsimbeiywo and Silibwet location were immigrants (100%) while in Nessuit there was an equal presence of indigenous (Ogiek—50%) and nonindigenous people (50%). In Mariashoni and Nessuit, the majority of households were of Ogiek tribe (65%) and Nessuit (50%). In Kapkembu, the area was inhabited mostly by nonindigenous group of Kipsigis (92.5%) and a small proportion of Ogiek at 7.5% (Table 1).

The majority of households were not born in the current place of residence (64.8%) and only about one-third (35.2%) were born in current place of residence. Results on the highest educational level attained by heads of households revealed that 73.4% have at least primary level of education, while 20% have attained secondary level of education and only 6.9% have completed postsecondary education with the lowest 2.4% and 4.9% in Nessuit and Mariashoni, respectively (Table 1).

3.2. Livelihood Activities of Households. Most of the households (90.5%) interviewed were farmers (n = 344) relying mostly on rain-fed agriculture and livestock keeping. Other livelihood activities were small scale retail business, wage employment, and sale of forest products. The total household income ($F_{(4,372)}$ = 5.10; p ≤ 0.001) was significantly different across study location and between indigenous and nonindigenous groups ($F_{(1,372)}$ = 7.82; p = 0.05). The total household income in 3 locations of Kapsimbeiywo, Nessuit, and Kapkembu was significantly different. However, in Kapkembu total household income differed significantly from Silibwet and Mariashoni (Table 1). Agricultural income was significantly different across locations ($F_{(4,382)}$ = 2.55, p = 0.05). Tukey B test separation of means showed that households in Kapsimbeiywo differed significantly from the households in other locations. However, agricultural household income in Silibwet, Kapkembu, Nessuit, and Mariashoni was not significantly different. In addition, income from sale of

forest products was not significantly different across location ($F_{(4,72)}$ = 1.23; p = 0.05) and between indigenous and nonindigenous groups ($F_{(1,75)}$ = 1.62; p = 0.05).

3.3. Assets

3.3.1. Livestock. Livestock keeping is an important economic activity undertaken by households. The average number of cattle, sheep, goats, donkeys, and hens was 5.0, 4.0, 2.0, 1.0, and 7.0, respectively, and the mean Tropical Livestock Unit (TLU) per household was 4.65 units. Total livestock units per household across locations were significantly different ($F_{(4,367)}$ = 11.86; p < 0.05). Separation of means by Tukey B test showed that TLU for households in Nessuit (Mean = 3.49, standard deviation (SD) = 2.81 and Kapsimbeiywo (Mean = 6.33, SD = 2.60)) were significantly different. However, households in 3 locations of Sililbwet (Mean = 4.99, SD = 1.84), Kapkembu (Mean = 5.02, SD = 1.71) and Marioshion (Mean = 5.10, SD = 2.46) were not significantly different in livestock units. Wealth group differed significantly in total livestock units ($F_{(2,367)}$ = 8.06; p < 0.05). Separation of means by Tukey B test showed that the poor households (Mean = 3.85, SD = 2.78) differed significantly from moderately poor (Mean = 5.23, SD = 2.41) and rich households (Mean = 4.76, SD = 2.54) in livestock holdings. Additionally, livestock holding (TLU) for indigenous and non-indigenous groups were not significantly different ($F_{(1,367)}$ = 0.410, p > 0.05).

3.3.2. Land. Most households in the study area allocate their land use to crops (both cash and food). Between 52% and 74% of the land holding is allocated for agricultural crops and less than 21% (14.2%–21%) was allocated to forest resources (planted or natural regeneration) (Table 1). Total land size, land under cash crops, and pasture were significantly different; however land under forests (planted and natural), food crops, and wastelands were not significantly different (Table 1). The ownership of land differs across locations with highest number of households indicating alternative ownership of land was highest in Kapsimbeiywo (73.3%) and least in Nessuit (4.0%). There was a strong association between alternative land ownership and location (χ^2 = 118.65, df = 4, p < 0.001).

3.4. Forest Use and Dependence

3.4.1. Sources of Forest Products. Diverse forest products were collected by households for home consumption and for sale (Table 2). Generally most of the products were obtained from public forest of East Mau forest. For example, most households reportedly obtained their firewood and charcoal from public forest compared to the other sources (72.9% and 67.3%, resp.) and this was similarly observed for all products (Table 2).

Households obtained foods products such as indigenous fruits (34.0%), mushrooms (49.3%), game meat (47.1%), and honey (51.6%) from public forest compared to other sources (own farms, neighbours, and markets). Overall, 45.5% households obtained various foods from the East Mau forest ecosystem. About fifty percent of the households obtained

TABLE 1: Socioeconomic and demographic characteristics of sampled households ($N = 367$).

Variable	Location					Sig (LSD)
	Kapsimbeywo	Silibwet	Kapkembu	Nessuit	Mariashoni	
Gender (HH) %						
Male	73.3	85.4	67.2	72.0	60.5	NS
Female	27.7	14.6	32.8	28.0	37.4	NS
Ethnicity (%)						
Indigenous	0.0	0.0	7.5	50	65	0.05*
Nonindigenous	100	100	92.5	50	35	0.05*
Education level (%)						
Primary	66.7	60.4	62.7	87.9	89.3	0.05*
Secondary	33.3	27.1	22.4	9.7	5.8	NS
Postsecondary	0.0	12.5	14.9	2.4	4.9	NS
Primary	66.7	60.4	62.7	87.9	89.3	0.05*
HH size						
Number	9.0 ± 2.4	10.1 ± 3.1	9.8 ± 2.6	8.9 ± 3.0	7.3 ± 3.6	NS
Adult equivalent	4.9 ± 1.4	6.0 ± 1.8	5.7 ± 1.5	5.1 ± 1.9	3.3 ± 1.6	NS
Land size and use						
Land size (Ha)	2.5 ± 1.2	2.1 ± 0.8	2.1 ± 0.9	1.7 ± 1.4	1.9 ± 1.5	NS
Natural forest	0.4 ± 0.3	0.4 ± 0.2	0.4 ± 0.2	0.3 ± 0.2	0.4 ± 0.3	0.00***
Planted forest	0.4 ± 0.2	0.3 ± 0.1	0.3 ± 0.1	0.3 ± 0.3	0.4 ± 0.2	NS
Food crops	0.8 ± 0.4	0.7 ± 0.3	0.7 ± 0.3	0.8 ± 0.7	0.8 ± 0.5	NS
Cash crop	0.5 ± 0.2	0.4 ± 0.3	0.4 ± 0.2	0.3 ± 0.2	0.6 ± 0.7	0.017*
Pasture land	0.6 ± 0.6	0.3 ± 0.1	0.3 ± 0.2	0.6 ± 0.6	0.8 ± 0.7	0.00***
Wastelands	0.4 ± 0.3	0.3 ± 0.1	0.3 ± 0.2	0.3 ± 0.2	0.4 ± 0.2	NS
Resident years	24.8 ± 2.0	23.0 ± 1.8	13.6 ± 5.4	14.8 ± 2.7	16.2 ± 4.5	NS
Food months	3.6 ± 0.4	4.7 ± 0.3	4.4 ± 0.3	4.3 ± 0.2	4.0 ± 0.2	NS
Age of HH (years)	44.8 ± 2.0	48.5 ± 1.9	40.3 ± 1.6	42.3 ± 1.4	42.4 ± 0.7	0.05*
Household cash incomes (KES)						
Total	170,075.85 ± 19,237.75[a]	259,363.80 ± 21,404.55[bc]	203,385.34 ± 9,506.64[ab]	212,286.69 ± 10,677.74[ab]	247,952.86 ± 9,448.39[bc]	0.01*
Agriculture	48,965.52 ± 7,841.79[a]	56,545.45 ± 7,899.30[ab]	65,530.30 ± 5,140.01[ab]	73,305.08 ± 4,626.89[ab]	58,817.39 ± 4,161.96[ab]	0.05*
Livestock	60,644.82 ± 7,599.54[ab]	86,521.67 ± 8,955.22[c]	62,231.34 ± 4,571.41[ab]	37,007.90 ± 3,642.59[a]	51,899.66 ± 4,710.23[ab]	0.01*
Forest product	18,666.67 ± 15,666.67[a]	7,937.50 ± 2,161.15[a]	5,100.00 ± 1,805.55[a]	25,982.14 ± 8,182.06[a]	19,720.00 ± 3,335.93[a]	NS
Off farm	127,789.65 ± 15,021.36[a]	141,563.11 ± 12,708.57[a]	130,873.13 ± 6,702.83[a]	119,698.18 ± 7,509.90[a]	114,714.56 ± 6,988.97[a]	NS

LSD is least significant difference; NS denotes no significant difference at $p \leq 5\%$ level.

Household incomes means (row) with a common superscript imply the mean difference is not significant at $p \leq 5\%$ level.

"∗" refers to significance level at 5%; "∗∗∗" denotes significance at 1%.

TABLE 2: Reported sources of forest products by of households ($N = 367$).

Product	Sources (% households)			
	Public forest	Own farm	Neighbours	Market
Firewood	72.9	21.6	3.4	2.1
Timber	58.0	16.6	6.2	19.2
Charcoal	67.3	8.2	7.6	16.9
Honey	51.6	13.8	9.7	24.9
Medicine	49.9	18.7	5.0	26.4
Poles	35.7	21.7	14.0	28.6
Thatch grass	30.6	35.0	6.2	28.2
Fruits	34.0	22.3	9.7	34.0
Animal fodder	66.7	31.2	1.8	0.3
Agricultural tools	42.8	18.9	1.3	37.0
Forest soils	45.2	21.8	7.3	25.7
Building stones	41.2	20.3	9.3	29.2
Mushrooms	49.3	14.4	8.1	28.2
Fibres	54.8	19.3	10.6	15.3
Meat	47.1	3.6	2.3	47.0

medicinal herbs from public forest. In the study area, 57.0%, 35.7%, and 54.8% of households reportedly obtained construction materials (timber, poles, and fibers, resp.) from the public forest (Table 2).

3.4.2. Quantities and Value of Forest Products. The extent of use and monetary value of various products is shown in Table 3.

Most households in the study area collected firewood (90.3%), herbal medicine (83.3%), poles (34.8%), and honey (27.4%) and the least collected product was building stones (5.7%) (Figure 2).

Firewood is the most collected product by households and each household collect an average of 122.00 backloads (4,100.00 kg) of firewood per year worth about KES 25,000.00 (US$ 280.00) accounting for 5.7% of forest income (Table 4). Another popular product collected by households is medicine (83.3%) with an average of about 50 kg per year. However, in terms of monetary value per household charcoal, honey and poles score high. The values of these products are KES 144,156.00 (US$ 1,601.00), 69,424.00 (US$ 771.00), and 32,959.00 (US$ 366.00), respectively (Table 3). Household who graze their livestock in public forest ranged from 57.1% (Kapsimbeiywo) and the highest of 77.9% of households in Mariashoni. Overall, 66.8% of the households reported using the forest as a source of fodder for their livestock. The monetary value of this use ranged from KES 11,983.00 (US$ 133.00) to 17,974.00 (US$ 200.00) per household per year. Wood fuel (firewood and charcoal) is the dominant source of forest income with a mean of 49.1% of forest income per household and this was followed by food products (26.5%) and structural

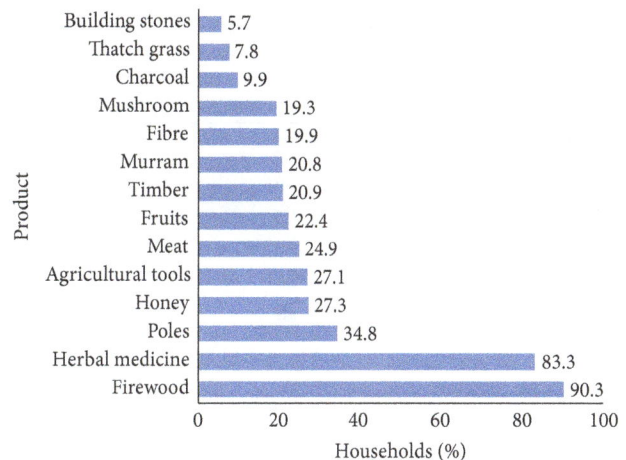

FIGURE 2: Proportion of households (%) collecting various forest products from East Mau forest.

and fibre products (17.4%). Though charcoal is not the most collected products (9.9%) of households yet its contribution was significant contributing 43.4% to household forest income due its high value. Other products which made significant contribution to household forest income were poles and honey each contributing 13.0% and 12.4%, respectively. The total forest income ranged from 28.8% to 36.5% with overall mean of 32.5% (Table 4).

3.4.3. Forest Dependence. The households in East Mau are dependent on East Mau forest for various products and services.

The net forest income and relative forest income are summarized in Table 5. The forest dependence was calculated as the ratio of total forest environmental income to the total household income and expressed as a percentage. The level of dependence was greater than 25% in all study locations, ranging from 28.8% to 36.5% with overall mean of 33.7% (Table 5). The absolute forest income and relative forest income were not significantly different between households in the five study locations.

Absolute forest income and relative forest income (%) were not significantly different across study locations ($F_{(4,309)} = 1.76$; $p > 0.05$) and between ethnic groups ($F_{(1,245)} = 0.307$, $p > 0.05$). However, absolute forest income and relative forest income (%) were significantly different among wealth classes ($p < 0.01$), meaning there is substantial difference in absolute forest income (Poor = $46,275.90 \pm 2,822.40$, Moderate household = $67,277.30 \pm 3,932.40$ and Rich household = $81,463.80 \pm 3,797.70$) and relative forest income (%) (Poor = 41.40 ± 2.13, Moderate household = 35.60 ± 2.03 and Rich household = 26.30 ± 1.30). The Poor households benefit less in absolute terms from the forest resources than the Moderate and the Rich (Table 5) (Poor < Moderate < Rich). However, in relative terms (% forest income) the poor derive more than the two categories (Poor > Moderate > Rich) (Table 5).

TABLE 3: Quantities and monetary value of forest products collected by households per year.

Product	Units	Quantities	Value (KES)	Value (US$)
Firewood	kg	4,070.45 ± 167.67	25,447.47 ± 1104.60	282.75 ± 12.27
Herbal Medicine	kg	48.78 ± 2.69	7,677.09 ± 1781.22	85.30 ± 19.79
Poles	Number	343.22 ± 17.62	32,959.22 ± 1855.49	366.21 ± 20.62
Honey	kg	102.39 ± 16.95	69,424.33 ± 5301.33	771.38 ± 58.90
Agricultural tools	Number	104.73 ± 17.50	1,053.82 ± 174.60	11.71 ± 1.94
Meat	kg	125.24 ± 12.84	12,919.20 ± 1502.18	143.55 ± 16.69
Fruits	Kg	256.68 ± 23.44	9,573.34 ± 552.13	106.37 ± 6.13
Timber	Running feet	171.38 ± 18.46	18,292.06 ± 1963.06	203.25 ± 21.81
Murram	Tons	120.22 ± 38.21	102.18 ± 32.48	1.14 ± 0.36
Fibre	kg	251.77 ± 38.98	4,227.20 ± 383.12	46.97 ± 4.26
Mushroom	kg	257.92 ± 45.98	3,021.28 ± 467.80	33.57 ± 5.20
Charcoal	kg	4,505.55 ± 1103.20	144,156.77 ± 22375.53	1,601.74 ± 248.62
Thatch grass	kg	179.08 ± 27.80	4,530.72 ± 7,142.99	50.34 ± 79.37
Building stones	Running feet	34.50 ± 4.20	1,000.00 ± 656.05	282.75 ± 7.29

Values are arranged as means, followed by standard error of means.

TABLE 4: Contribution of forest products category to forest income.

Product	Location Kapsimbeiywo	Silibwet	Kapkembu	Nessuit	Mariashoni	Mean
Fuel	**17.20**	**59.80**	**50.30**	**51.40**	**66.90**	**49.10**
Firewood	10.80	3.90	5.10	4.40	4.10	5.70
Charcoal	6.40	55.90	45.10	47.00	62.90	43.40
Food	**26.60**	**28.30**	**28.70**	**29.00**	**19.70**	**26.50**
Fruits	1.00	3.10	2.80	1.90	1.90	2.10
Honey	9.40	15.50	13.40	13.90	10.00	12.40
Mushroom	14.10	7.60	10.00	8.40	5.50	9.10
Meat	2.00	2.00	2.60	4.80	2.30	2.70
Structural and fibre	**46.00**	**7.10**	**14.40**	**11.80**	**7.40**	**17.40**
Timber	6.40	2.70	5.40	4.00	2.10	4.10
Poles	39.40	4.30	8.50	7.60	5.00	13.00
Agricultural tools	0.20	0.10	0.60	0.30	0.20	0.30
Grass	**7.60**	**3.80**	**4.60**	**5.10**	**4.60**	**5.10**
Thatch grass	1.10	1.10	0.80	2.00	2.80	1.60
Fodder	6.50	2.70	3.70	3.10	1.80	3.60
Herbal medicine	2.60	0.90	2.00	2.60	1.20	1.90
Others	0.00	0.10	0.00	0.00	0.20	0.10
Total	**100.00**	**100.00**	**100.00**	**100.00**	**100.00**	**100.00**
% of total household income	28.80	30.70	32.90	36.50	33.40	32.50
Absolute value (KES)	47,662.00	63,427.00	65,218.00	66,580.00	71,642.00	62,906.00
Absolute value (US$)	530.00	705.00	725.00	740.00	796.00	699.00

3.5. Discussions

3.5.1. Socioeconomic and Demographic Characteristics. The average family size in the study areas of (8.8 ± 3.2) is higher than national average of 5.3 persons per households [12]. However, households in Mariashoni showed lower family size. Male headed households were dominant in the study locations and this is consistent with customs of the local people where males are expected to be the heads of households and only females attain this role through bereavement. It was established that there was significant variation in asset endowment (land, physical assets, and livestock) between male and female headed households. Because crop farming and livestock are main livelihood activities in the study area ownership and access to land is one of the key determinants of livelihood options of the local people. On average, households

TABLE 5: Absolute forest income, relative forest income (%) by study location, wealth status, and ethnicity.

Variable	Absolute forest income (KES)	Relative forest income (%)
Location		
Kapsimbeiywo	$47,662.10 \pm 6,236.81^a$	28.85 ± 3.70^a
Silibwet	$63,427.11 \pm 6,470.64^a$	30.71 ± 3.34^a
Kapkembu	$65,217.56 \pm 4,801.03^a$	32.89 ± 2.18^a
Nessuit	$66,579.73 \pm 3,762.37^a$	36.46 ± 1.84^a
Mariashoni	$71,641.51 \pm 4,711.57^a$	33.42 ± 2.40^a
Overall mean	$65,836.28 \pm 2,232.06$	33.73 ± 1.10
	$(F_{(4,309)} = 1.76, p > 0.05)$	$(F_{(4,294)} = 1.18, p > 0.05)$
Wealth status		
Poor	$46,275.90 \pm 2,822.40^a$	41.40 ± 2.13^a
Moderate	$67,277.30 \pm 3,932.40^b$	35.60 ± 2.03^b
Rich	$81,463.80 \pm 3,797.70^c$	26.30 ± 1.30^c
	$(F_{(2,309)} = 23.87, p < 0.01)$	$(F_{(2,296)} = 18.35, p < 0.01)$
Ethnicity		
Indigenous	$63,536.12 \pm 3,961.22^a$	31.93 ± 1.75^a
Nonindigenous	$62,658.47 \pm 2,196.54^a$	33.15 ± 1.25^a
	$(F_{(1,241)} = 0.74, p > 0.05)$	$(F_{(1,245)} = 0.307, p > 0.05)$

Note. Means (column) with a common (letters) superscript imply the mean difference is not significant different at $p \leq 5\%$ level.

in Nessuit and Mariashoni have less land compared to households in other locations.

Most of the study areas (Mariashoni, Nessuit, and Kapkembu) were once part of East Mau forest. However, they were excised in 1990s and early 2000 for human settlement [7]. Each household in the settlement scheme was allocated 2.5 ha. The results showed that households in Nessuit and Mariashoni currently have smaller land size than originally allocated. This is most likely due to land transactions which might have occurred in the two locations. This finding was corroborated by key informants who reported increased number of new settlers due to high productivity of the land for food and cash crops. This fact was also reflected in the heterogeneity of the local population showed by household data which showed that most of the household heads (64.8%) were not indigenous to current place of residence. Households in Kapsimbeiywo have the highest access to land and this is reflected in the fact that about 78% of households have alternative access to land. This phenomenon of emigration from other areas in search of land and livelihood opportunities conforms to what has been established in other African societies where migration is influenced by demographic trends and the search for livelihood opportunities [16].

Households in the study area have adapted a diverse portfolio of livelihood activities such as farming, livestock keeping, forest product, small trade and remittance. The most common livelihood activity is farming and livestock keeping.

The local indigenous communities, the Ogiek have largely depended on livestock and forest resources. This is, however, changing due to the growing influence of immigrants from other counties. There is evidence of increasing diversification of income opportunities by the indigenous community. This is consistent with other studies on rural communities where livelihood diversification strategies is predominant [17–19]

because single livelihood strategy is insufficient for the needs of most rural households [20]. There was a strong association between educational attainment and ethnicity ($\chi^2 = 3.49$, df = 2, $p < 0.05$).

The household heads of nonindigenous group had higher postsecondary qualifications compared to indigenous households. Mariashoni and Nessuit dominated by Ogiek community had fewer schools. Livestock size (TLU) in the study area showed significant results pointing to the fact that the ownership of large herds is associated with access to alternative land. The households which had alternative land also showed large livestock size and lowest forest grazing incidence. Forest grazing is dependent on seasonal availability of fodder on the farms and forest grazing is an alternative resource. The implication is that alternative land ownership accounts for the additional livestock units owned.

3.5.2. Forest Dependence. The result from this study has shown that local people depend primarily on forest resources for subsistence needs and occasionally for sale. The highest contribution to household forest income is fuel wood (50%) and food products (27%). The high market value from fuel wood use category could be explained by the significantly high level of firewood collection by majority of households (90.3%) and the relatively high value of charcoal.

The study has revealed that forest income contributes between 25% and 36.5% of household income in the study area. This could be explained by low level of investment in tree growing and less retention of natural forests on individual farms and ease of access to public resources. The findings on forest reliance confirm what others have concluded in other parts of Africa, for example, Cavendish [21], found out that 35% of rural household income is derived from environmental products in Zimbabwe. Another study in Malawi showed that forest income contributes to 30% of household income [22]. Forest income contributes about 39% of the household income in Ethiopia highlands and nearly equaled combined livestock and agricultural incomes [23].

Another study by Kalaba et al. [24] in Miombo woodlands of Zambia showed that forest income contributed 43.9% to the average household income. In a compressive comparative analysis of environmental income, Angelsen et al. [25] revealed that environmental income accounted for 28% of household income in 24 developing countries. Therefore the findings of this study are in agreement with similar findings elsewhere and corroborate the importance of forest resources to households. In terms of who benefits more from forest resources, the moderately poor and higher income households derive higher absolute forest income than poor households. This is probably because the rich households extract high value products such as timber, poles which require large capital investments such as equipment which are inaccessible by poor households and therefore primarily engaged in low value and often labour intensive forest extractive activities [26]. Limited access to financial and social capital has been advanced by various authors [25, 27] to explain the inability of the poor households to benefit substantially from environmental resources. However, in relative forest income, poor households showed higher reliance on forest resources. These findings on the higher dependency on forest resources by poor households are consistent with findings of [16, 19, 23, 28–32].

4. Conclusion and Recommendations

The study has revealed the important role of forest resources in household income. It was found out that forest income share are higher for poor households. However, in absolute terms, the better off households are advantaged. Poor households showed high dependence on the forest resources despite most collection/usage being illegal. On average 33% of annual household income is generated by consumption and sale of forest products. With the increasing population in East Mau and surrounding areas, the demand on forest resources are likely to rise and this will exert pressure on the state of forest resources in East Mau. However, reflecting on the findings of this study, it would be imprudent to exclude local community from accessing forest resources because; it may lead to increased poverty. One way of managing the situation would be to allow low level extractive activities such as firewood collection and enforcing licensing procedures to allow for low extraction level, essentially for subsistence use and discourage commercial extraction. Another way to ease the pressure on East Mau is to promote intensification of tree growing on farms through support for agroforestry or farm forestry intervention. Another strategy is to lower the opportunity cost of engaging in forest resources by creating robust income opportunities independent of forest product extraction or improving the technical efficiency of agricultural and production systems in order to minimize illegal forest exploitation. These measures may improve rural livelihoods and conserve forest resources and biodiversity.

Appendix

Estimation of the Value Forest Grazing

According to the household data livestock data the mean livestock numbers 4.9 livestock units and 67% of households graze their animals inside the forest and forest fodder/browse make up to 40% of the fodder requirements. From literature, the dry fodder requirement for livestock is taken to be about 2-3% of the body weight per day [33] and a livestock unit (250 kg) requires a minimum quantity of fodder for maintenance of between 5.0 and 7.5 kg per day.

Step 1. Calculate the number of households who graze their animals = (43,527 ∗ 67)/100 = 29,163.00.

Step 2. Calculate the total number of livestock units grazing inside the forest = 29163 ∗ 4.9 = 142,898.00.

Step 3. Calculate the total dry matter requirements for the total livestock units for the whole year from the forest.

One TLU requires between 5.0 and 7.5 kg per day; therefore 365 days = 142,898 ∗ (5.0–7.5) ∗ 365.

The total dry matter requirements per year is between 260,788.85 and 391,183.28 kg.

40% of the total fodder requirements are obtained from the forest and therefore forest contributes between 104,315.54 and 156,473.31 kg.

Step 4. Convert the estimate quantities of dry matter into Hay equivalent.

One bale of hay weighs 30 kgs; the number of equivalent hay is between 3,477.20 and 5,215.80 bales.

Step 5. Calculate the monetary value of hay using the current market price. The current market price of 1 bale is KES 150.

The total value of forest grazing is KES 521,577.75 and 782,366.55 per year. This is equivalent to between KES 11,983.00 and 17,974.00 per household per year.

Ethical Approval

Ethical issues (including plagiarism, informed consent, misconduct, data fabrication and/or falsification, double publication, and redundancy) have been completely observed by the authors.

Conflict of Interests

The authors declare that there is no conflict of interests regarding the publication of this paper.

Acknowledgments

The authors express their gratitude to East Mau community for their willingness to share with us their information on forest use. Special thanks are due to research assistants and local administrative officials in the study areas for their support during data collection. Special thanks are due to Mr. Frank Mairura for assisting in data analysis. The authors are grateful to the Director Kenya Forestry Research Institute for financially supporting this study.

References

[1] S. Chao, *Forest People: Numbers across the World*, Forest Peoples Program, Moreton-in-Marsh, UK, 2012.

[2] World Bank, *Global Issues for Global Citizens: An Introduction to Key Development Challenges*, Edited by V. K. Bhargava, The World Bank Report, Washington, DC, USA, 2006.

[3] P. Vedeld, A. Angelsen, J. Bojö, E. Sjaastad, and G. K. Berg, "Forest environmental incomes and the rural poor," *Forest Policy and Economics*, vol. 9, no. 7, pp. 869–879, 2007.

[4] C. Shackleton and S. Shackleton, "The importance of non-timber forest products in rural livelihood security and as safety nets: a review of evidence from South Africa," *South African Journal of Science*, vol. 100, no. 11-12, pp. 658–664, 2004.

[5] C. M. Shackleton and S. E. Shackleton, "Household wealth status and natural resource use in the Kat River valley, South Africa," *Ecological Economics*, vol. 57, no. 2, pp. 306–317, 2006.

[6] J. K. Sang, "The Ogiek in Mau Forest: Case Study 3—Kenya—Forest Peoples Program, 2001," http://www.forestpeoples.org/sites/fpp/files/publication/2010/10/kenyaeng.pdf.

[7] UNEP, KFWG, DRSRS, and EU-BCP, *Eastern and South Western Mau Forests Reserves: Assessment and Way Forward*, UNEP, Nairobi, Kenya, 2006.

[8] W. G. Sombroek, H. M. Braun, and B. J. Van der Pouw, "The Exploratory soil map and agro-climatic map of Kenya," Report E 1, Kenya Soil Survey, Nairobi, Kenya, 1980.

[9] S. S. China, *Land use planning using GIS [Ph.D. thesis]*, University of Southampton, Southampton, UK, 1993.

[10] R. Jaetzold and H. Schmidt, *Farm Management Handbook for Kenya*, Ministry of Agriculture, Nairobi, Kenya, 1982.

[11] P. M. Kundu, *Application of remote sensing and GIS techniques to evaluate the impact of land use cover change on stream flows: the case of River Njoro in Eastern Mau-Kenya [Ph.D. thesis]*, Faculty of Environment and Resources Development, Egerton University, Njoro, Kenya, 2007.

[12] KNBS, *Kenya Population and Housing Census*, vol. 1A, Ministry of Planning and National Development, Nairobi, Kenya, 2010.

[13] O. Mugenda and A. Mugenda, *Research Methods Quantitative and Qualitative Approaches*, African Center for Technology Studies (ACTS), Nairobi, Kenya, 1999.

[14] B. M. Campbell and M. Luckert, "Towards understanding the role of forests in rural livelihoods," in *Uncovering the Hidden Harvest: Valuation Methods for Woodland and Forest Resources*, B. M. Campbell and M. K. Luckert, Eds., People and Plants Conservation Series, pp. 1–10, Earthscan, London, UK, 2002.

[15] Y. H. Chan, "Biostatistics 102: quantitative data-parametric and non-parametric tests," *Singapore Medical Journal*, vol. 44, no. 8, pp. 391–396, 2005.

[16] K. Heubach, *The socio-economic importance of non-timber forest products for rural livelihoods in West African savanna ecosystems: current status and future trends [Ph.D. thesis]*, Goethe Universität Frankfurt, Frankfurt, Germany, 2011.

[17] B. Belcher, M. Ruíz-Pérez, and R. Achdiawan, "Global patterns and trends in the use and management of commercial NTFPs: implications for livelihoods and conservation," *World Development*, vol. 33, no. 9, pp. 1435–1452, 2005.

[18] F. Ellis, *Rural Livelihoods and Diversity in Developing Countries*, Oxford University Press, Oxford, UK, 2000.

[19] P. Kamanga, P. Vedeld, and E. Sjaastad, "Forest incomes and rural livelihoods in Chiradzulu District, Malawi," *Ecological Economics*, vol. 68, no. 3, pp. 613–624, 2009.

[20] W. D. Sunderlin, A. Angelsen, B. Belcher et al., "Livelihoods, forests, and conservation in developing countries: an overview," *World Development*, vol. 33, no. 9, pp. 1383–1402, 2005.

[21] W. Cavendish, "Poverty, inequality and environmental resources: quantitative analysis of rural households," Centre for the Study of African Economies (CSAE) Paper Series, Paper 93, 1999.

[22] M. Fisher, "Household welfare and forest dependence," *Environment and Development Economics*, vol. 9, no. 2, pp. 135–154, 2004.

[23] G. Mamo, E. Sjaastad, and P. Vedeld, "Economic dependence on forest resources: a case from Dendi District, Ethiopia," *Forest Policy and Economics*, vol. 9, no. 8, pp. 916–927, 2007.

[24] F. K. Kalaba, C. H. Quinn, and A. J. Dougill, "Contribution of forest provisioning ecosystem services to rural livelihoods in the Miombo woodlands of Zambia," *Population and Environment*, vol. 35, no. 2, pp. 159–182, 2013.

[25] A. Angelsen, P. Jagger, R. Babigumira et al., "Environmental income and rural livelihoods: a global-comparative analysis," *World Development*, vol. 64, no. 1, pp. S12–S28, 2014.

[26] M. Arnold and I. Townson, "Assessing the potential of forest product activities to contribute to rural incomes in Africa," Natural Resources Perspectives 37, Overseas Development Institute, London, UK, 1998.

[27] S. Dewi, B. Belcher, and A. Puntodewo, "Village economic opportunity, forest dependence, and rural livelihoods in East Kalimantan, Indonesia," *World Development*, vol. 33, no. 9, pp. 1419–1434, 2005.

[28] A. Angelsen and S. Wunder, "Exploring the forestry-poverty linkage: key concepts, issues and research implications," CIFOR Occasional Paper 40, CIFOR, Bogor, Indonesia, 2003.

[29] B. Babulo, *Economic Valuation and Management of Common-Pool Resources: The Case of Enclosures in the Highlands of Tigray, Northern Ethiopia*, vol. 762 of *Doctoraatsproefschrift*, Faculteit Bio-Ingenieurswetenschappen Van de K.U. Leuven, 2007.

[30] W. Cavendish, "Empirical regularities in the poverty-environment relationship of rural households: evidence from Zimbabwe," *World Development*, vol. 28, no. 11, pp. 1979–2003, 2000.

[31] P. Illukpitiya and J. F. Yanagida, "Farming vs forests: trade-off between agriculture and the extraction of non-timber forest products," *Ecological Economics*, vol. 69, no. 10, pp. 1952–1963, 2010.

[32] J. K. Mariara and C. Gachoki, "Forest dependence and household welfare: empirical evidence from Kenya," CEEPA Discussion Paper 41, CEEPA, 2008.

[33] B. Ganesan, "Extraction of non-timber forest products, including fodder and fuelwood, in Mudumalai, India," *Economic Botany*, vol. 47, no. 3, pp. 268–274, 1993.

4

Effect of Disturbance Regimes on Spatial Patterns of Tree Species in Three Sites in a Tropical Evergreen Forest in Vietnam

Do Thi Ngoc Le,[1,2] Nguyen Van Thinh,[1,3] Nguyen The Dung,[2] and Ralph Mitlöhner[1]

[1]*Tropical Silviculture & Forest Ecology, Georg-August-Universität Göttingen, Büsgenweg 1, 37077 Göttingen, Germany*
[2]*Vietnam Forestry University, Xuan Mai Town, Chuong My District, Hanoi 100000, Vietnam*
[3]*Silvicultural Research Institute (SRI), Vietnamese Academy of Forest Sciences, Duc Thang, Bac Tu Liem District, Hanoi 100000, Vietnam*

Correspondence should be addressed to Do Thi Ngoc Le; dothingocle81@yahoo.com

Academic Editor: Scott D. Roberts

The effects of disturbance regimes on the spatial patterns of the five most abundant species were investigated in three sites in a tropical forest at Xuan Nha Nature Reserve, Vietnam. Three permanent one-ha plots were established in undisturbed forest (UDF), lightly disturbed forest (LDF), and highly disturbed forest (HDF). All trees ≥5 cm DBH were measured in twenty-five 20 m × 20 m subplots. A total of 57 tree species belonging to 26 families were identified in the three forest types. The UDF had the highest basal area (30 m^2 ha^{-1}), followed by the LDF (17 m^2 ha^{-1}) and the HDF (13.0 m^2 ha^{-1}). The UDF also had the highest tree density (751 individuals ha^{-1}) while the HDF held the lowest (478 individuals ha^{-1}). Across all species, there were 417 "juveniles," 267 "subadults," and 67 "adults" in the UDF, while 274 "juveniles," 230 "subadults," and 36 "adults" were recorded in the LDF. 238 "juveniles," 227 "subadults," and 13 "adults" were obtained in the HDF. The univariate and bivariate data with pair- and mark-correlation functions of intra- and interspecific interactions of the five most abundant species changed in the three forest types. Most species indicated clumping or regular distributions at small scale, but a high ratio of negative interspecific small-scale associations was recorded in both the LDF and HDF sites. These were, however, rare in the UDF.

1. Introduction

A forest stand comprises a set of trees characterized by their locations and sizes. Tree diameter distributions can provide information on tree sizes but cannot address tree locations. However, tree diameters are associated with tree positions, and growth is sensitive to both spatial interaction among trees [1] and local habitat characteristics [2]. The theory of marked point processes provides a formal framework for an analysis of the spatial characteristics of tree diameter distributions, in which the points indicate tree locations and the marks denote particular tree characteristics such as diameter at breast height, tree height, and growth during a given time span [3, 4]. Ecologists have become increasingly interested in studying spatial patterns in ecology [5–8]. In addition, tree species associations at different life stages or age classes have already received considerable attention [5, 9–11]. The spatial pattern of a particular species, especially the adult-juvenile

relationship, provides useful information on the species' regeneration process [10, 12, 13]. The spatial pattern of trees is an important indicator of stand history, population dynamics, and species interaction in forests [14, 15], all of which are helpful in understanding mechanistically the processes and patterns of plant communities [16].

Spatial distributions and the spatial patterns (both vertical and horizontal) of trees in forests are important elements for understanding forest ecosystem dynamics [17]; however, the potential for ecological understanding has not yet been fully recognized [18]. A spatial point pattern is a set of locations, distributed within a region of interest, which have been generated by some spatial process [19]. Many methods and indices have been developed in order to interpret and assess spatial distributions [20–23]. The spatial distribution pattern, within plant populations, is influenced by various ecological and evolutionary processes which take place during the life history of a plant, such as seed dispersal, intra- and interspecific competition, and environmental heterogeneity. The spatial structure of a forest is largely determined by the relationships within neighboring groups of trees [24]. The main factors in a forest structure are the spatial distribution, species diversity, and variations in tree dimensions [25]. In recent decades, several structural indices such as the Clark and Evans aggregation index [20], pair-correlation function [26], and species diversity indices [13] have been developed to quantify spatial forest structure.

Tropical rain forests have been used in recent studies to demonstrate the effects of high diversity on spatial distribution [5, 27]. This explains why interspecific species have functional similarity and may adapt to average environment conditions [28]. One approach to exploring spatial plant dynamics is to use a point pattern analysis of fully mapped plant locations [29]. Spatial statistics like Ripley's K function [23] and the pair-correlation function [26, 30] quantify the small-scale spatial correlation structure of a pattern which contains information on positive/negative interactions among plants. In addition, point pattern analysis is ideally suited to control for environmental heterogeneity and focus on neighborhood processes. Spatial patterns of individuals within populations are closely linked to ecological processes; consequently, ecological processes may be deduced from spatial patterns [31–33].

There is little published information on the spatial patterns of tree species in tropical evergreen forests as affected by different disturbance regimes in Vietnam. In the present study, univariate $g(r)$ statistics with the null model of complete spatial randomness and the mark-correlation function were used for point pattern analysis at different scales. We analyzed the intraspecific and interspecific association (relationship or correlation) of five abundant tree species in three forest types in a tropical evergreen forest in Xuan Nha Nature Reserve, Vietnam. This study will provide meaningful knowledge for predicting the spatial patterns of the most dominant tree species after disturbance and could have further silvicultural implications for management practice (e.g., selecting native tree species and choosing site conditions for enrichment planting) in restoration zones of the nature reserve.

2. Materials and Methods

2.1. Study Site. The study site (Figure 1) is located in Xuan Nha Nature Reserve ($20°36'–20°48'$N, $104°29'–104°50'$E). The reserve has a total area of 27,100 ha, of which 15,300 are tropical evergreen forests and 2,600 are tropical forests on limestone [34]. The nature reserve's elevation ranges from 300 to 2,000 m above sea level. The topography is composed of two shallow-sided valleys, which run across the reserve from the Laotian border. Xuan Nha Nature Reserve is located in a tropical monsoon climate where the mean temperature is 21.3°C. The rainy season lasts from May to October (with a maximum of 332.4 mm of rain in August), while the dry season goes from November to April (with a minimum of 32.4 mm in February). Mean annual precipitation is about 1,673 mm. Six major soils are found in the study site: (1) thick layers of dark yellow ferralsols at 700–1,700 m in altitude that develop on shale or metamorphic rocks with a medium

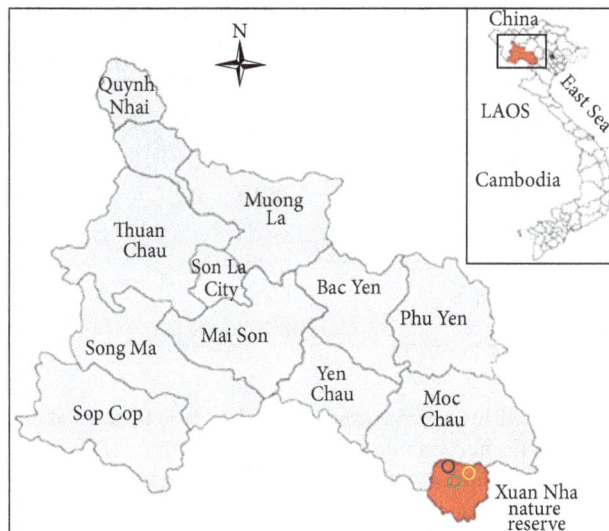

FIGURE 1: The location of Xuan Nha Nature Reserve in Vietnam. Dark blue, yellow, and green colors indicate the undisturbed forest (UDF), lightly disturbed forest (LDF), and highly disturbed forest (HDF), respectively.

texture; (2) thin layers of yellow-brown ferralsols at 700–1,700 m altitude which develop on limestone or metamorphic limestone and have coarse to medium texture; (3) thick layers of light yellow ferralsols at 700–1,700 m on shale or metamorphic rocks with medium texture; (4) thick layers of light/grey-yellow ferralsols that develop at 300 to 1,000 m in low hilly or mountainous regions on slate, siltstone, sandstone, and conglomerate with a coarse to medium texture; (5) grey-yellow ferralsols modified by paddy fields with medium texture that appear in the surrounding villages; and (6) alluvial soils deposited at the foot of mountains, river banks, and streams [35].

In accordance with the collected data, approximately 43% of the forest area (app. 7,872.20 hectares) is disturbed, while 42.8% (7,821.4 ha) is classified as undisturbed (Table 1). This disturbed area is divided into two categories: lightly disturbed and heavily disturbed. Undisturbed and disturbed forests refer to different disturbance regimes, where "disturbance" is defined as the impact level of human activities in the forest. Disturbance in this context refers to selective timber harvesting, felling small-sized trees for nontimber products, and past (pre-2003) illegal logging.

Undisturbed forests (primary or rich forests) are areas that do not show evidence of damage from human activities; these are relatively stable forests not yet (or less) influenced by humans or natural disasters. Because such forests are limited to inaccessible and protected areas, they are not used for production purposes. UDFs are extremely rich forests with a timber volume of standing trees between 201 and 300 $m^3 ha^{-1}$. LDFs (average forest) are influenced by humans or natural disasters, leading to changes in their structure. This type demonstrates low and minor damage from human beings and has a timber reserve of standing trees between 101 and 200 $m^3 ha^{-1}$. Forests may be classified as LDFs if past logging

TABLE 1: Stratification of vegetation types and characteristics of three main forest types at Xuan Nha Nature Reserve, Vietnam.

Forest type	Area (ha)	Species richness (N/ha)	Family richness (N/ha)	Tree density (N/ha)	Basal area (m^2/ha)
Undisturbed (rich forest)	7,821.4	49	25	751	29.8
Lightly disturbed (average forest)	2,288.5	42	21	540	16.9
Heavily disturbed (poor forest)	2,179.5	30	18	478	13.0
Wood and bamboo mixed forest	483.2	—	—	—	—
Bamboo forest	2,921.0	—	—	—	—
Limestone forest	1,549.4	—	—	—	—
Nonforested lands	1,024.5	—	—	—	—
Total	**18.267,5**				

has resulted in the removal of only 6 to 10% of trees above a minimum harvestable size (40–50 cm DBH). HDFs (postexploitation forests, poor forests) are degraded as a result of human activities that have severely impacted their canopy structure, productivity, and volume at various levels. Such forests have been exploited for their timber or other forest products. 10 to 30% of trees above a minimum harvestable size (40–50 cm DBH) have been removed, and they have a reserve of standing trees between 10 and 100 m^3 ha^{-1}.

The reserve has a high tree species diversity with 173 families and 1,074 species. The main vegetation observed at Xuan Nha Nature Reserve is monsoon evergreen broad-leaved forest; numerous rare species of flora have been found here, including 65 rare tree species that account for 6.1% of the total species listed in Vietnam's Red Book [35]. At the UDF, *Lithocarpus ducampii* is the most dominant species, followed by *Syzygium cuminii*, *Vatica odorata*, *Toxicodendron succedaneum*, and *Cinnamomum parthenoxylon*. *Lithocarpus ducampii*, *Vatica odorata*, and *Syzygium cuminii* dominated in the LDF, while *Engelhardtia roxburghiana* and *Nephelium melliferum* could be found in the HDF. In the UDF, the five most dominant and species-rich families were Lauraceae, Fagaceae, Myrtaceae, Dipterocarpaceae, and Anacardiaceae. Fagaceae was the most important family in the LDF, followed by Lauraceae, Sapindaceae, Dipterocarpaceae, and Myrtaceae. Anacardiaceae dominated only in the UDF; Juglandaceae was restricted to the HDF. The top family in the HDF was Lauraceae, followed by Dipterocarpaceae, Sapindaceae, Juglandaceae, and Fagaceae. Across the three forest types, Lauraceae was the most dominant, but Fagaceae, Myrtaceae, and Dipterocarpaceae were also common families. The tree density of the LDF was lower than that of the UDF, which had the highest density (751 trees ha^{-1}) of all trees ≥5 cm DBH among the three forest sites; the HDF held the lowest tree density (478 stems ha^{-1}).

2.2. Sampling Design.

Three permanent one-ha plots (100 m × 100 m each) were established in undisturbed, lightly disturbed, and highly disturbed forests (one plot each) in 2003. The corners of each plot were marked with a concrete post and a GPS device. Each one-ha plot was further divided into twenty-five 400 m^2 subplots (20 m × 20 m) (Figure 2).

2.3. Data Collection.

Data were collected in 2013. Within each plot, all living trees larger than 5 cm in diameter at breast

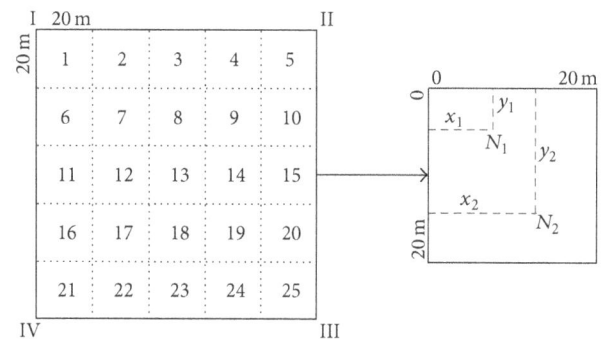

FIGURE 2: Sampling design (sample size: 1 ha). Symbols I, II, III, and IV are the four corners of the sampled plot. 25 subplots (20 m × 20 m) were set up in order to measure the coordinates of all trees with DBH ≥5 cm. N_1 (x_1, y_1); N_2 (x_2, y_2) are the tree coordinates sampled.

height (DBH 1.3 m) were marked with paint and their DBH and height were measured. All individuals were classified into three life history stages: "juveniles" (5 cm ≤ DBH < 10 cm), "subadults" (10 cm ≤ DBH < 30 cm), and "adults" (DBH ≥ 30 cm) [36]. Tapes were used to measure the coordinates of trees as follows: the starting point in each subplot (400 m^2) was fixed at the northwestern corner. The position of each tree (X, Y coordinates) was determined by measuring the distance to the two edges of the plot (Figure 2). All trees were recorded and labelled in order to avoid missing any ones. Tree species were identified in the field with specialists' help; unidentifiable specimens were taken to the Vietnam Forestry University Herbarium for identification.

2.4. Data Analysis.

Tree density (stem ha^{-1}) was calculated from the count of all individuals from the 25 subplots; tree basal area (BA, m^2 ha^{-1}) was calculated by using the following equation: BA = $(3.14 \times (DBH)^2)/4$ (m^2). The total basal area per ha was calculated by the sum of the BA of all trees in the 25 subplots. Species richness was taken by counting the number of species occurring in all subplots of each forest type [22]. The R-package software was used to measure nearest neighbor distance (NND) [37]. Only the five most abundant tree species in the three forest types were used to simulate spatial distributions and the relationships among them. Pair-correlation functions $g(r)$, mark-correlation functions, and null models were used to analyze spatial patterns in this study.

2.4.1. Pair-Correlation Functions. The pair-correlation function was used as a summary statistic to quantify the spatial structure of the uni- and bivariate patterns [8, 26]. Based on intertree distances, the univariate pair-correlation function $g_{11}(r)$ can be used to determine whether a point distribution is random, aggregated/clustered, or regular at distance r in which those patterns occur. The parameter $g(r)$ indicates whether a pattern is random (complete spatial randomness (CSR); $g(r) = 1$), clumped ($g(r) > 1$), or regular ($g(r) < 1$) at a given radius r.

The pair-correlation function $g_{11}(r)$ for the univariate pattern of species 1 can be defined based on the neighborhood density $O_{11}(r) = \lambda_1 g_{11}(r)$, that is, the mean density of trees belonging to species 1, surrounded by rings with radius r and width $d(r)$ [8], where λ_1 is the intensity (number of species 1's trees in the plot). Under the null model of complete spatial randomness (CSR), the points are independently and randomly distributed over the entire plot [8]. The values of $g_{11}(r)$ within the 95% confidence envelope indicate that the spatial structure of the given distances does not differ significantly from the CSR. Values of $g_{11}(r)$ above 95% confidence envelope indicate that a distance class is more aggregated than under CSR; values of $g_{11}(r)$ below the 95% confidence envelope indicate lower aggregation (approaching a more regular pattern) at that scale [38]. The pair-correlation function for bivariate patterns (composed of trees from species 1 and 2) follows, where the quantity $g_{12}(r)$ is the ratio of the observed mean density of species 2's trees in the rings around species 1's trees to the expected mean density of species 2's trees in those rings [8]. The corresponding neighborhood density function yields $O_{12}(r) = \lambda_2 g_{12}(r)$. $g_{12}(r) = 1$ shows independence (no interaction), $g_{12}(r) > 1$ indicates attraction between two point patterns at distance r, and $g_{12}(r) < 1$ indicates repulsion/inhibition between the two patterns at distance r.

2.4.2. Effect of Interspecific Competition on Tree Growth. A mark-correlation function (MCF) $K_{mm}(r)$ [26, 39] using DBH as marks was used to analyze the distance-dependent size correlation of trees for distances up to 50 m. The similarity or dissimilarity between the DBH marks of two trees at a distance r apart is quantified by the equation $f(m_1, m_2) = m_1 \times m_2$, where m_1 and m_2 are the DBH values of two neighboring trees. $K_{mm}(r)$ is defined as the normalized mean value of $f(m_1, m_2)$ for all marks at distance r. Marks are considered independent and positively or negatively correlated at distance r if $K_{mm}(r) = 1$, $K_{mm}(r) > 1$, or $K_{mm}(r) < 1$, respectively.

2.4.3. Null Models. The univariate statistic is used to analyze the spatial pattern of one object, while the bivariate statistic is used to analyze the spatial association of two objects (pattern 1 and pattern 2) [8, 40]. Based on a homogeneity test on the spatial pattern of all adult trees (DBH $\geq$ 10 cm), we applied complete spatial randomness (CSR) as the null model [8, 39]. A Monte-Carlo approach was used to test for significant departures from the null models. Each of the 199 simulations of a point process underlying the null model generates a summary statistic; simulation envelopes with $\alpha <$ 0.05 were calculated from the fifth highest and lowest values

of the 199 simulations [26]. All analyses were performed using the software Programita [8].

3. Results

3.1. Population Structure of Individuals in the Three Life Stages. The adults of all tree species were used to test for environmental heterogeneity in the three forest types. The five most abundant species (*Engelhardtia roxburghiana, Lithocarpus ducampii, Syzygium cuminii, Nephelium melliferum,* and *Cinnamomum parthenoxylon*; Table 2) were analyzed for all other associations (e.g., intra-and interspecific). Across all species, there was a higher proportion of "juveniles," "subadults," and "adults" in the UDF than in the LDF or HDF. All tree species across the three life stages are mapped in Figure 3.

The diameter distributions of the five most abundant species are shown in Figure 4. The frequency distribution of all tree species demonstrated an inverted J-shaped curve in which the number of individuals gradually decreased with increasing diameter classes, a trend found across all three forest sites. The distribution of *Engelhardtia roxburghiana* varied among the three forest types; a unimodal distribution was found in the HDF (Figure 4(c)), whereas the UDF and LDF had reverse J-shapes (Figures 4(a) and 4(b)). Conversely, *Syzygium cuminii* displayed a unimodal distribution in all three forest types. Both species had no individuals in the size class greater than 30 cm, and 88.4% of the total tree density came from the 5–20 cm diameter class. However, a very high abundance of 5–10 cm DBH individuals of *L. ducampii* and a high number of 10 cm *S. cuminii* trees were observed in the UDF.

3.2. Intraspecific Interactions. The spatial patterns of the five most abundant species in the three forest types are shown in Figure 5; the univariate $g(r)$ statistics with the null model of complete spatial randomness (CSR) displayed different spatial patterns for these species at various scales. There were no statistically significant departures from randomness for *Engelhardtia roxburghiana* in the HDF (Figure 6(a)), but, in the UDF, it was clumped at scales of 6 m and 9-10 m (Figure 5(a)). *Cinnamomum parthenoxylon* was aggregated at scales of 13 and 28 m in the UDF (Figure 5(d)) and randomly distributed at all distances in both the HDF and LDF sites (Figures 6(c) and 6(d)). *Engelhardtia roxburghiana* in the LDF and *Cinnamomum parthenoxylon* in the HDF both occurred in larger spatial clusters of trees (>40 m) and were thus evidence for clustering at larger spatial scales. *Lithocarpus ducampii* was significantly aggregated at a scale of 9 m in the HDF (Figure 5(b)), but predominantly random in the LDF (Figure 5(c)) and UDF (Figure 6(b)). *Syzygium cuminii* was significantly aggregated at a large scale of 39 m in the HDF (Figure 5(e)) but showed an independent distribution at all scales in the LDF (Figure 6(e)); no significant pattern was found at all scales in the UDF (Figure 6(f)). *Nephelium melliferum*, in contrast, was randomly distributed at all distances in the HDF (Figure 6(g)) but tended to have a regular distribution at a larger scale (>50 m) in the LDF (Figure 6(h)). This species displayed clumped distribution at two scales of 4 and 16 m in the UDF (Figure 5(f)). *Cinnamomum parthenoxylon*

TABLE 2: The five most abundant tree species of each forest site in Xuan Nha Nature Reserve, Vietnam.

Species	Tree density [N/ha]	Basal area [m²/ha]	Mean DBH [cm]	Maximum DBH [cm]	Median NN [m]
Undisturbed forest					
S. cuminii	103	2.17	15.5	28.2	12.33
L. ducampii	78	4.91	22.4	119.5	14.95
N. melliferum	53	1.75	17.7	60.3	11.94
C. parthenoxylon	50	2.73	23.1	61.8	13.68
E. roxburghiana	46	2.61	21.2	70.1	12.98
Lightly disturbed forest					
L. ducampii	96	2.80	15.4	55.4	10.99
S. cuminii	60	1.17	14.9	28.2	9.70
C. parthenoxylon	50	1.46	17.8	41.4	11.58
E. roxburghiana	44	1.28	16.9	42.7	13.29
N. melliferum	42	1.27	17.5	41.7	12.00
Heavily disturbed forest					
E. roxburghiana	61	1.80	17.3	57.0	5.66
L. ducampii	50	1.60	17.6	42.0	12.70
S. cuminii	49	0.89	14.2	42.0	11.34
N. melliferum	47	1.26	17.0	28.0	11.76
C. parthenoxylon	42	1.32	18.9	32.0	12.45

and *Nephelium melliferum* in both the HDF and LDF sites showed a random spatial distribution at all distances up to 50 m.

The mark-correlation function showed different spatial distributions of the five most abundant tree species. For *Engelhardtia roxburghiana*, only a negative correlation was detected at scales of up to 2 m in the HDF (data not shown), but there was a significant positive association in the range of 14-15 m in the LDF (Figure 7(a)); no statistically significant correlation was found in the UDF. *Lithocarpus ducampii* had a positive correlation at scales of 26 m in the HDF (Figure 7(b)) and 1 and 5-6 m in the LDF (Figure 7(c)); in the UDF, it exhibited a negative association at 16-17 and 48-49 m (Figure 7(d)). *Cinnamomum parthenoxylon* displayed a similar negative correlation at different scales of 22-23 m in the HDF (Figure 7(e)), 3–6 m in the LDF (Figure 7(f)), and 8 m in the UDF (Figure 7(g)). Only two instances of relationship were observed for *Syzygium cuminii*: one was positive at 7 and 9 m; the other was negative at a distance of 12 m in the HDF (Figure 7(h)). *Nephelium melliferum* demonstrated positive trends at a scale of 4 m and a negative correlation at a distance of 8 m in the HDF (Figure 7(i)). In the LDF, it tended to have a negative correlation at the particular scales of 24 and 47 m (Figure 7(j)) and a positive association in the UDF at the scales of 28-29 m (Figure 7(k)).

3.3. Interspecific Interactions. The results of the bivariate spatial pattern analysis using pair- and mark-correlation functions for the five abundant species in the three forest types are shown in Figure 8. A total of 20 (5 × 4) bivariate point pattern analyses were simulated for pairs of the five most abundant tree species. *Engelhardtia roxburghiana* and *Lithocarpus ducampii* showed a trend for positive association (statistical attraction) at scales of 17 and 42 m in the HDF

(Figure 8(a)); in the UDF, they demonstrated a positive correlation at scales of 6 and 15 m and a negative association at a distance of 2-3 m (Figure 8(b)). However, this relationship was independently distributed at all scales in the LDF. There was a significant positive association (attraction) between *Engelhardtia roxburghiana* and *Cinnamomum parthenoxylon* at distances of 6 and 28 m in the LDF (Figure 8(c)) and 39 m (Figure 8(d)) in the HDF, but no such association was found at all scales in the UDF. *Engelhardtia roxburghiana* showed no interaction with *Syzygium cuminii* at all distances for both the LDF and HDF; however, a significant positive association was detected between the two at the specific scales of 4, 7, 10, 16, and 21 m in the UDF (Figure 8(e)). Between *Engelhardtia roxburghiana* and *Nephelium melliferum*, there was only a negatively associated trend at scales of 33 and 8 m in the HDF and UDF (Figures 8(f) and 8(g)). Two instances of statistically significant departures from randomness were obtained between *Lithocarpus ducampii* and *Cinnamomum parthenoxylon* in the HDF: one was towards repulsion distribution at a small distance of 5 m; the other towards positive association at a scale of 22 m (Figure 8(h)). In the LDF, no spatial interaction of *Lithocarpus ducampii* and *Cinnamomum parthenoxylon* was recorded at all distances. *Lithocarpus ducampii* and *Syzygium cuminii* were significantly segregated (attraction) at a small distance of 7 m and negatively associated (repulsion) at 11 m in the HDF (Figure 8(i)); in comparison, this relationship in the LDF showed significant repulsion at the scales of 4 and 33 m (Figure 8(j)). The only clear significant aggregation obtained was for the interaction between *Lithocarpus ducampii* and *Syzygium cuminii* at the small scales of 4 and 7 m as well as the larger distances of 16 and 28 m in the UDF (Figure 8(k)). A similar result was recorded for *Lithocarpus ducampii* and *Nephelium melliferum* at different distances; in the HDF, the relationship exhibited a

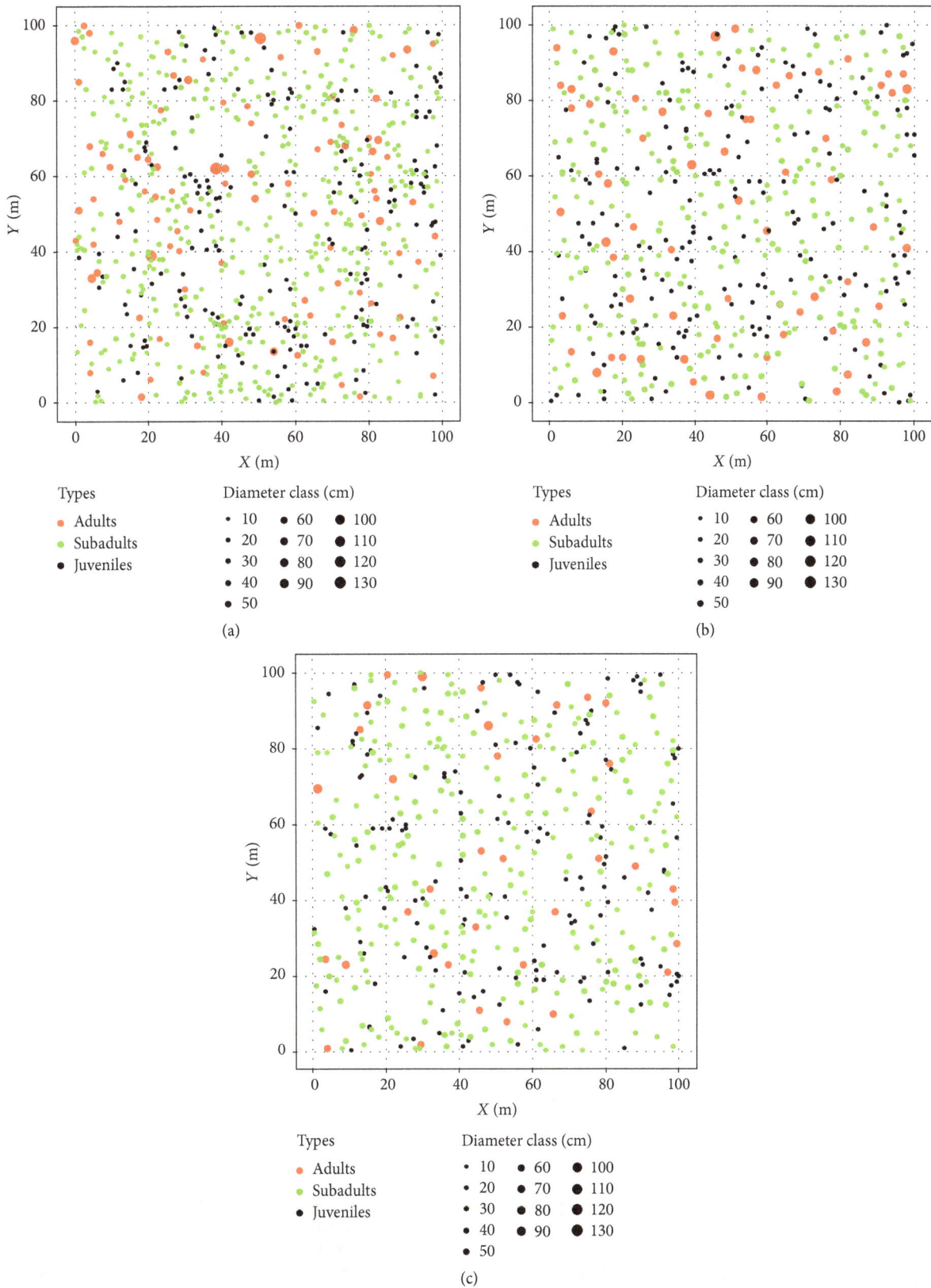

FIGURE 3: Distribution of all tree individuals in the three life stages: "juveniles," "subadults," and "adults." Symbol size is proportional to the DBH of the individuals with an actual size from 6.0 to 129.2 cm. The unit of (X, Y) axes is meter, where (a) is the undisturbed forest, (b) the lightly disturbed forest, and (c) the highly disturbed forest.

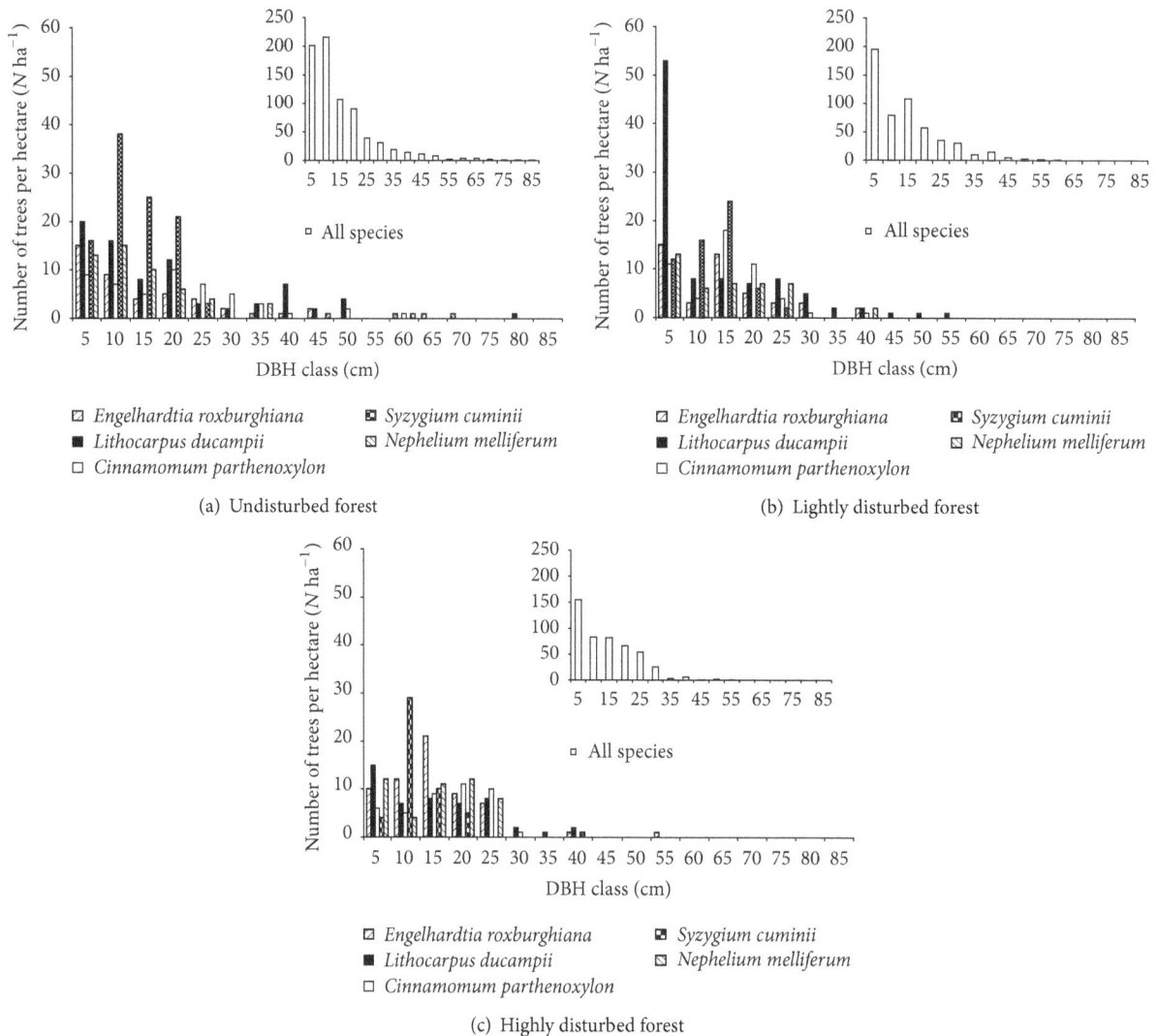

(a) Undisturbed forest

(b) Lightly disturbed forest

(c) Highly disturbed forest

FIGURE 4: Diameter distributions of all living trees and the five major tree species in each forest type in Xuan Nha Nature Reserve, Vietnam.

significant positive spatial association at scales of 5 and 31 m with a significant negative spatial association at 7, 9, and 30 m (Figure 8(l)).

The complementary analyses using the mark-correlation function are shown in Figure 9. *Engelhardtia roxburghiana* and *Lithocarpus ducampii* were spatially uncorrelated at all scales in the three forest types, but the former displayed a negative spatial interaction with *Cinnamomum parthenoxylon* at scales of 18–20 m in the HDF (Figure 9(a)); this interaction was strongly positive at 7–9, 21–24, and 36 m in the UDF (Figure 9(b)). A significant negative association was observed between *Lithocarpus ducampii* and *Cinnamomum parthenoxylon* at small scales of 3 m and greater than 32-33 m in the HDF (Figure 9(c)), whereas no spatial interaction was recorded in the UDF. In the HDF, however, two tendencies were obtained between *Lithocarpus ducampii* and *Nephelium melliferum*: a positive association at scales of 6 and 8-9 m and a significant repulsion at a scale greater than 30 m (Figure 9(d)). *Cinnamomum parthenoxylon* showed a negative

spatial interaction with *Syzygium cuminii* at the range of 24-25 m in the UDF (Figure 9(e)), whereas in the HDF and LDF sites there was only a spatially independent interaction between them. A similar result was obtained for *Syzygium cuminii* and *Nephelium melliferum* in different scales; they showed a significant repulsion at spatial distances of 6, 10–12, and 15-16 m in the UDF (Figure 9(f)).

4. Discussion

4.1. Intraspecific Competition. Environmental heterogeneity and dispersal limitation, which usually influence the large-scale spatial patterns of trees, are important processes that lead to an aggregation pattern [41, 42]. Our investigation revealed that there was change in the spatial distribution from aggregation to randomness (and even to regularity) for the five most abundant species; these displayed a significant degree of spatial aggregation at several small distances of 10, 30, and 40 m (Figure 4), suggesting the importance of

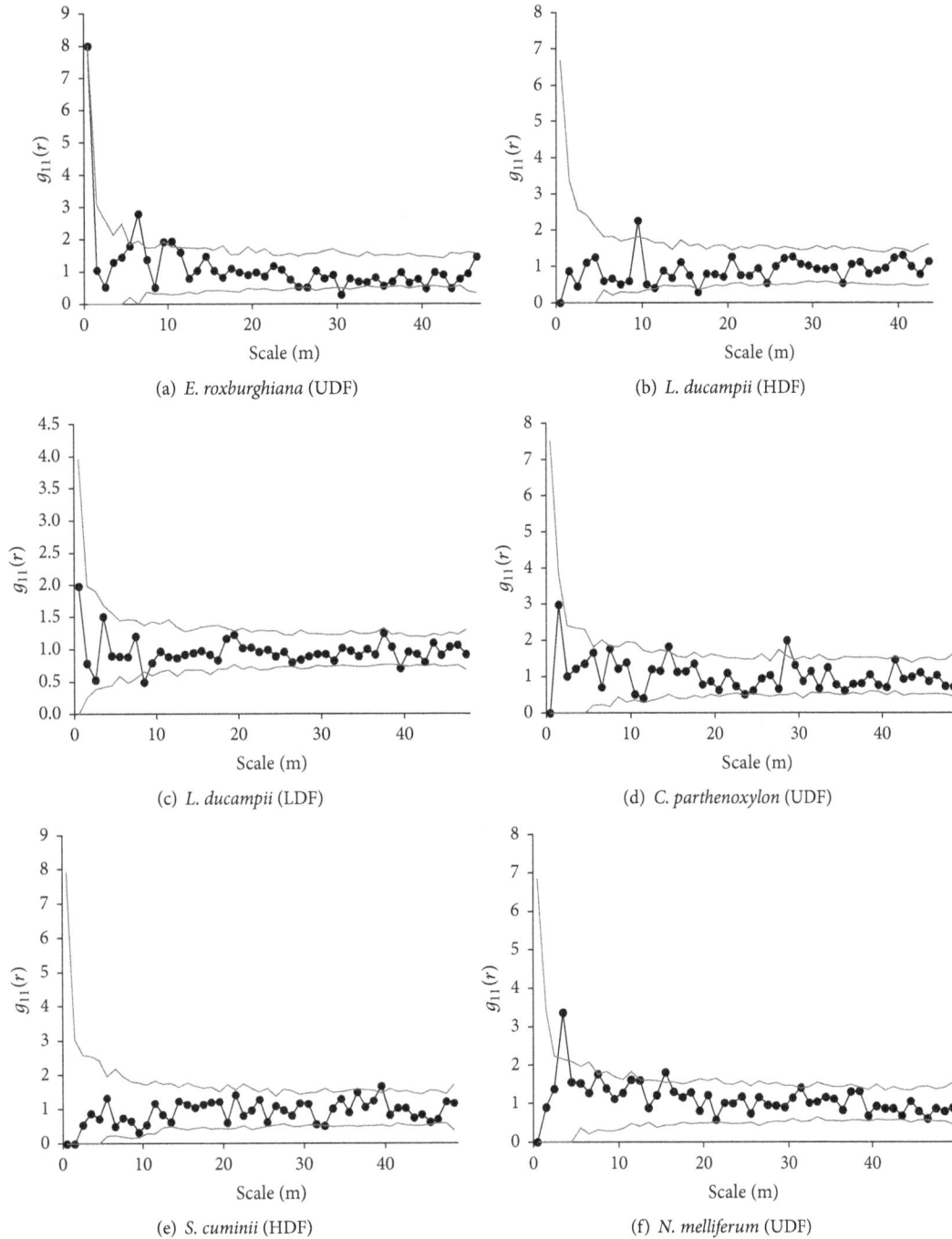

FIGURE 5: Spatial patterns of the five most abundant tree species shown by the pair-correlation function $g(r)$ with the null model of CSR. Black lines are observed patterns and grey lines are approximate 95% confidence envelopes.

intraspecific competition in governing the spatial distribution and traits of tree populations in these forests. Of the five most abundant species, *Engelhardtia roxburghiana* and *Syzygium cuminii* had a widely distributed aggregation in this area. *Lithocarpus ducampii* was only found to have a regular distribution as succession progressed, providing further evidence of competition in the development of an evergreen forest structure. Hubbell [12] and Condit et al. [43] suggested that poor seed dispersal is related to greater clumping

intensities of populations; however, our results showed that aggregations of abundant species were relatively loose [44]. At a given scale of observation, it has been found that the aggregation pattern occurs more frequently than random and regular patterns in natural tropical forests, where large trees organize themselves in a regular way [5]. However, this general rule did not apply to the Xuan Nha forests.

The question of whether trees compete in larger size classes [5, 45] is important for our general understanding of

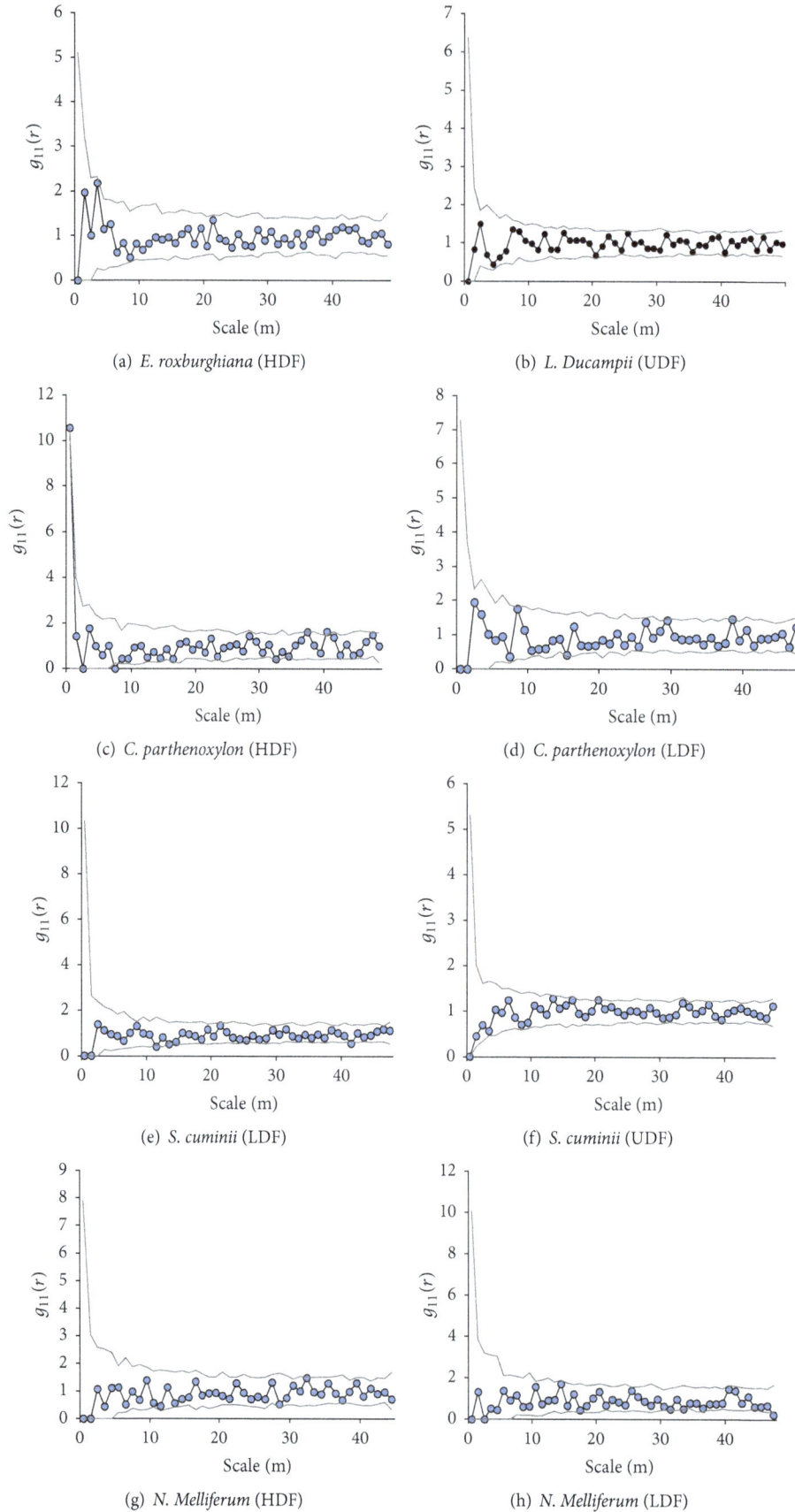

FIGURE 6: Spatial patterns of the five most abundant tree species shown by the pair-correlation function $g(r)$ with the null model of CSR. Black lines are observed patterns and grey lines are approximate 95% confidence envelopes.

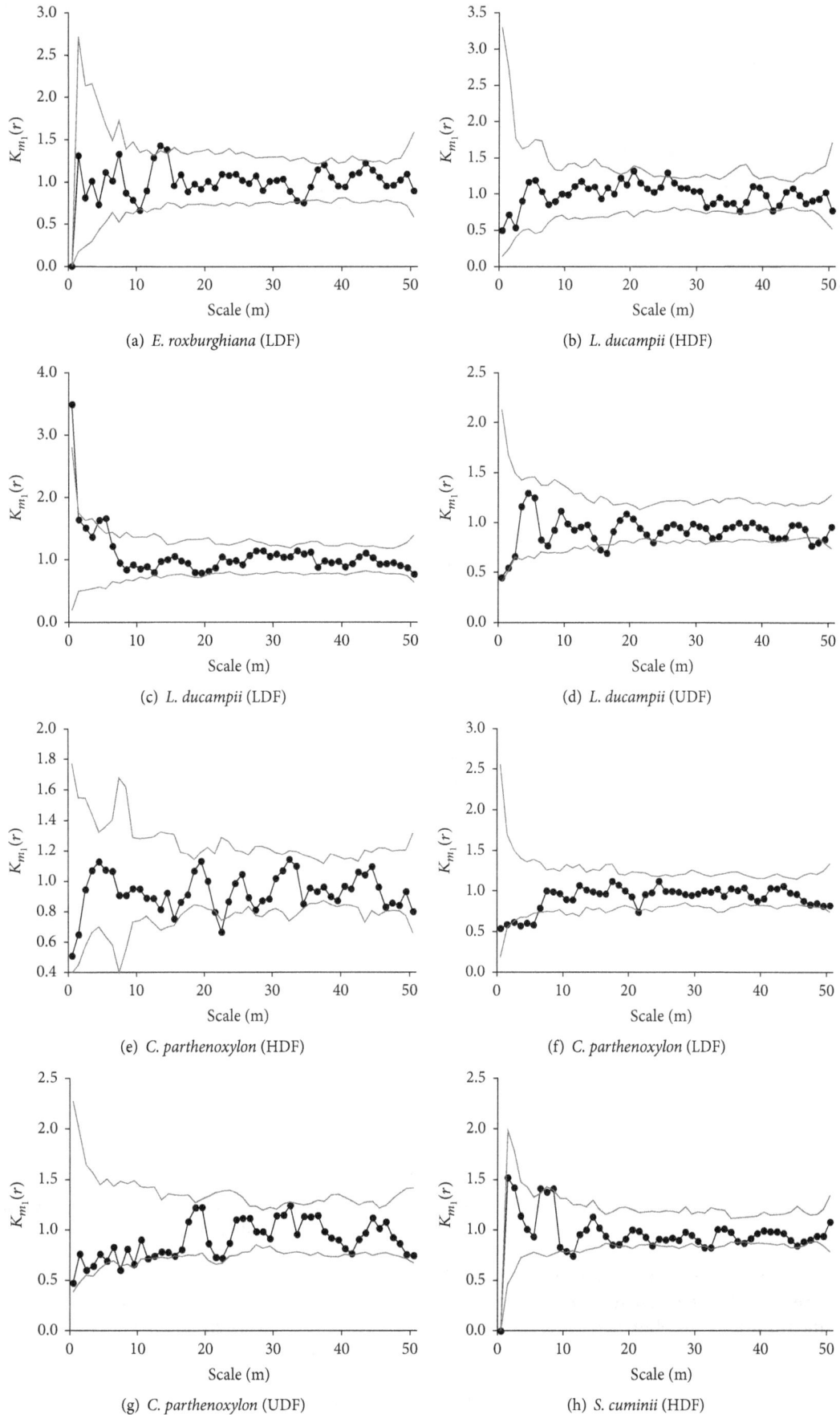

(a) *E. roxburghiana* (LDF)

(b) *L. ducampii* (HDF)

(c) *L. ducampii* (LDF)

(d) *L. ducampii* (UDF)

(e) *C. parthenoxylon* (HDF)

(f) *C. parthenoxylon* (LDF)

(g) *C. parthenoxylon* (UDF)

(h) *S. cuminii* (HDF)

FIGURE 7: Continued.

(i) *N. melliferum* (HDF)

(j) *N. melliferum* (LDF)

(k) *N. melliferum* (UDF)

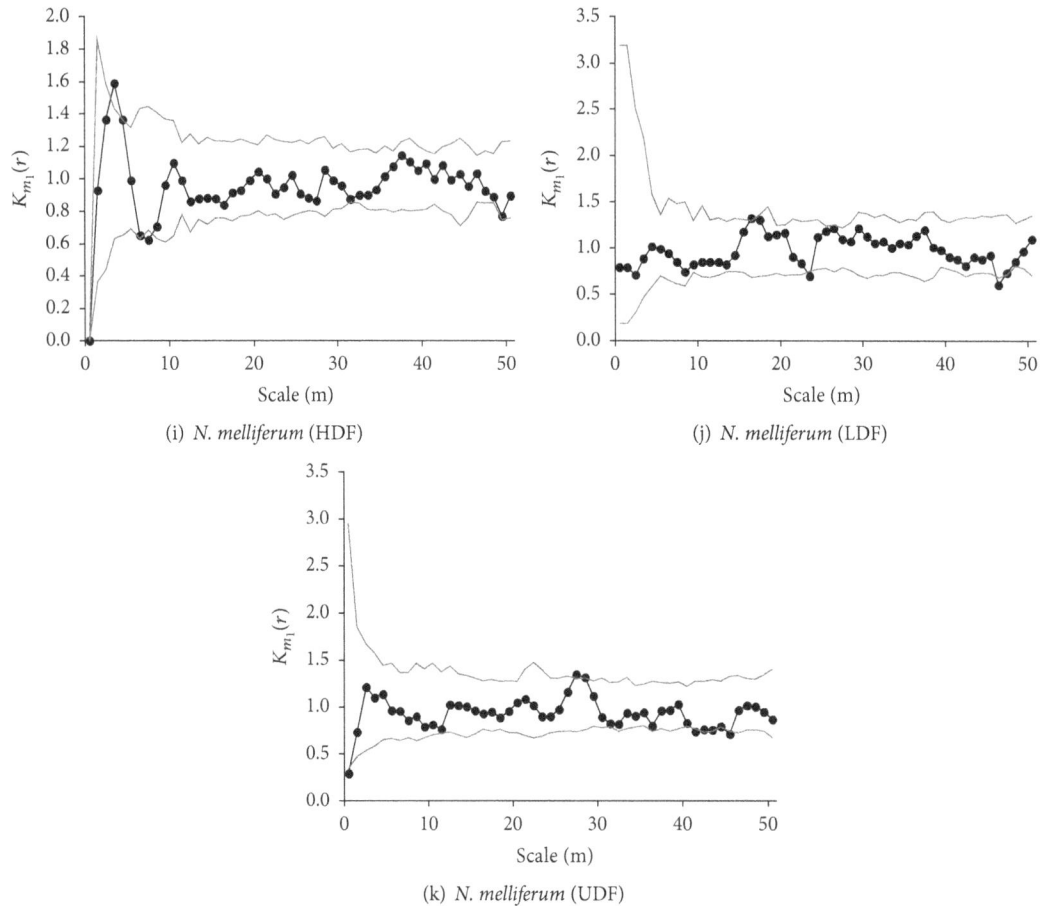

FIGURE 7: Spatial patterns of the five most abundant tree species shown by mark-correlation function with the null model of uncorrelated marks. Black lines are observed patterns and grey lines are approximate 95% confidence envelopes.

tropical forest ecology. Three different patterns (regularity, randomness, and clumping with a tendency towards regularity) were found by Pélissier [46] among large adult trees in three different plots in tropical India. This was also recorded in our results, but these patterns tended toward randomness. We found evidence for competition among trees within the smaller size classes of the three forest types, but this competition was lower than expected given their relatively high densities; indeed, there was only slight regularity or aggregation at a radius of less than 10 m. Clustered spatial distributions are often typical in naturally regenerated stands [47]. In very old forests at Wind River on the Pacific coast of the USA, North et al. [48] found that trees were clustered at all distances. Clustering also appeared to be present at shorter distances, although this was not seen as differing from a random spatial process [49].

4.2. Interspecific Competition. In this study, we conducted a comprehensive spatial pattern analysis to assess species associations among the five most abundant tree species across three different tropical forest types in Northern Vietnam. These species comprised 43.9%, 54.1%, and 52.1% of all trees in the UDF, LDF, and HDF sites, respectively. Analogous analyses of data from tropical forests in Sinharaja (Sri Lanka)

and Barro Colorado Island (Panama) also revealed significant small-scale associations [27]. However, we cannot exclude the possibility that these differences may owe as much to variation in other factors such as environmental heterogeneity, range of conditions, or species richness. Further comparisons across sites would help to assess the effects of these possibilities. Our analyses therefore indicate that there were some small differences between the spatial structures of the Changbaishan (CBS) and Xuan Nha forests; for example, we found that most species cooccurred in small neighborhoods less often than expected. Only 8% of all species pairs shared roughly the same plot areas in the CBS forests [50], whereas most Xuan Nha plot species pairs interacted both at small and large scales, or else had no association. Individuals of different species thus showed a clear tendency towards independence of one another at all distances.

The analyses of spatial association patterns among the five abundant species in the three different forest types present in Xuan Nha Nature Reserve revealed a balance between the attraction and repulsion interaction of spatial structures. The selective analysis of small-scale effects found surprising results, especially because more than one-third of all species pairs had independent association at all scales. We also found that the positive small-scale association pattern was caused by

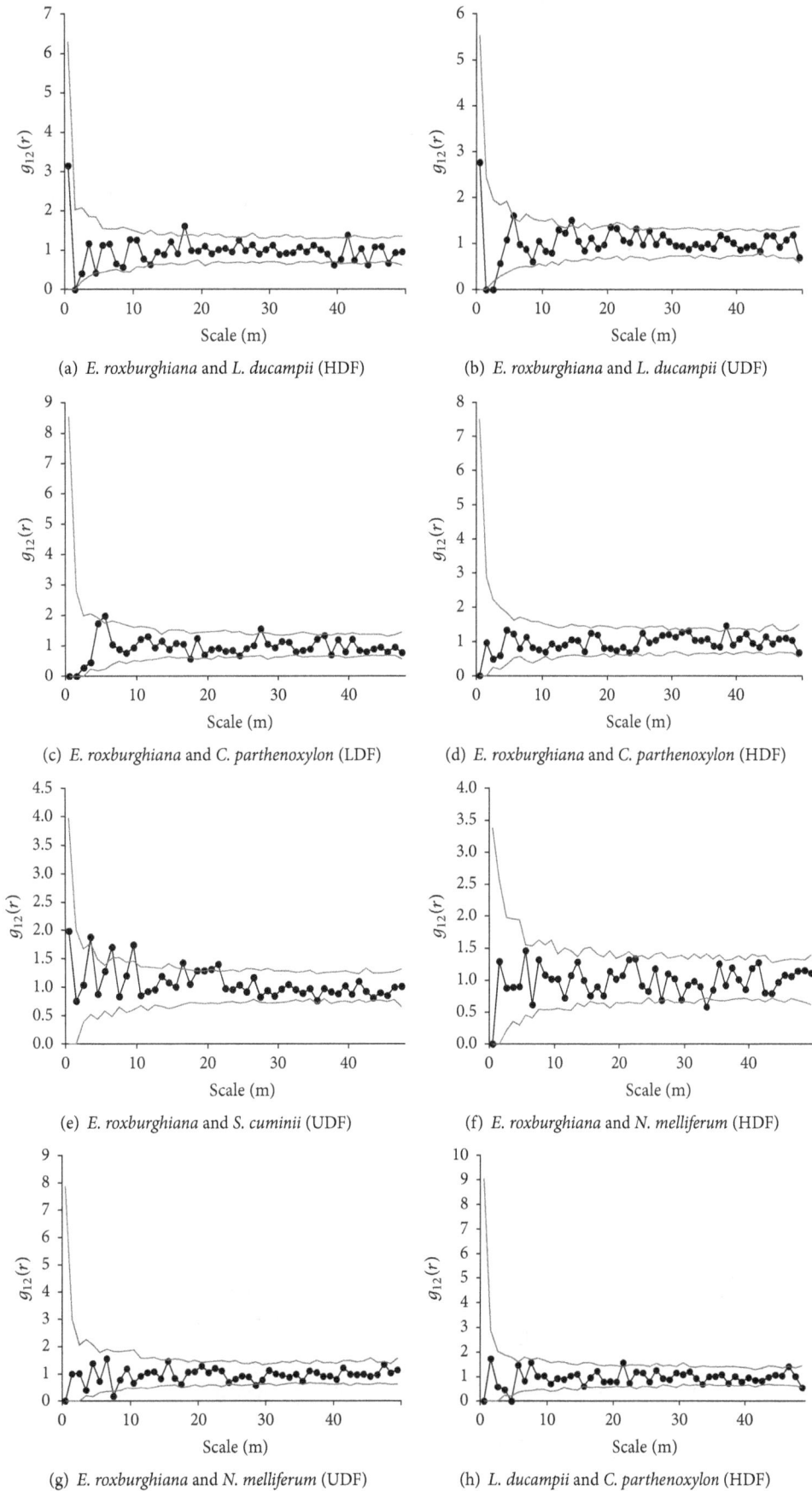

(a) *E. roxburghiana* and *L. ducampii* (HDF)

(b) *E. roxburghiana* and *L. ducampii* (UDF)

(c) *E. roxburghiana* and *C. parthenoxylon* (LDF)

(d) *E. roxburghiana* and *C. parthenoxylon* (HDF)

(e) *E. roxburghiana* and *S. cuminii* (UDF)

(f) *E. roxburghiana* and *N. melliferum* (HDF)

(g) *E. roxburghiana* and *N. melliferum* (UDF)

(h) *L. ducampii* and *C. parthenoxylon* (HDF)

FIGURE 8: Continued.

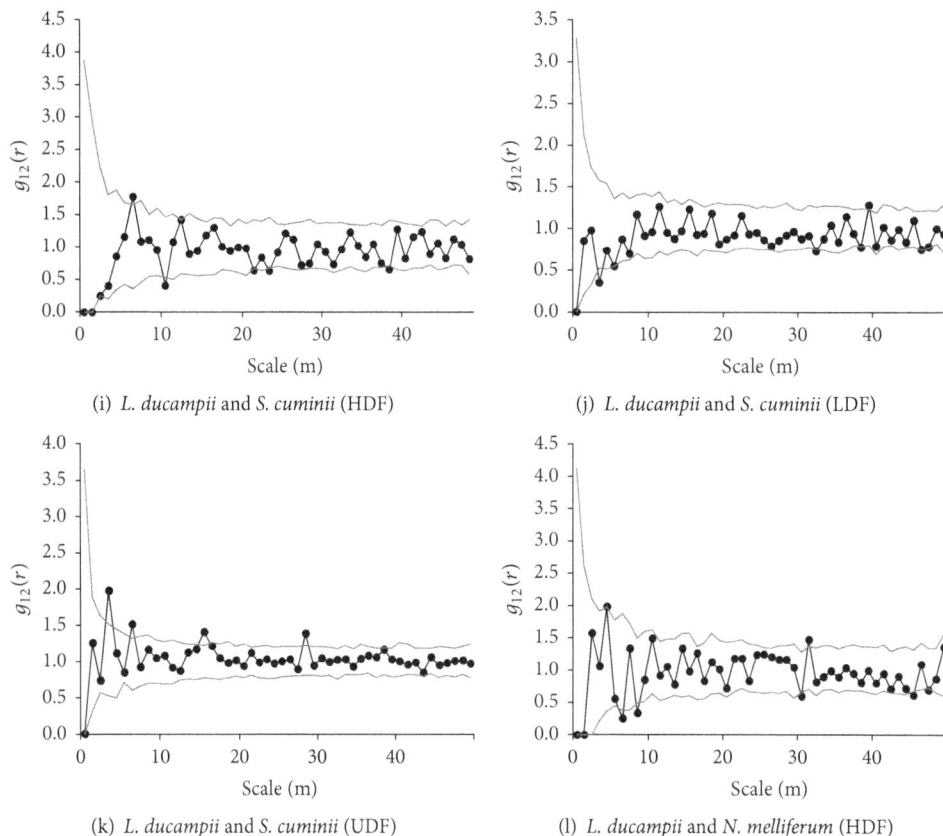

(i) *L. ducampii* and *S. cuminii* (HDF)

(j) *L. ducampii* and *S. cuminii* (LDF)

(k) *L. ducampii* and *S. cuminii* (UDF)

(l) *L. ducampii* and *N. melliferum* (HDF)

FIGURE 8: Spatial correlations among the five most abundant tree species are shown by the bivariate pair-correlation. The observed patterns (dark line) that lie beyond the confidence envelopes (grey lines) indicate significant departures from the null models of toroidal shift (spatial independence).

the aggregation of neither small individuals nor larger trees. Small-scale attraction rather than competition appeared to be a critical factor in shaping spatial distribution patterns; positive associations at small distances can thus be expected when species have similar requirements for establishment. Likewise, there are positive interspecific neighborhood effects on tree survival (facilitation), and/or species are dispersed across the same microhabitats [51].

5. Conclusion

Our analysis based on a systematic design may have been constrained by a lack of observation (pseudoreplication). We used a single one-ha permanent sample plot per forest type, and while this may reflect current conditions, it does not facilitate a detailed statistical analysis. Other, more efficient methods are recommended for a better understanding of tree species' distribution/density and disturbance regimes. Our results indicate that high intensity human disturbance adversely affected floristic composition, tree species density, and diversity among the three forest types. The basal areas recorded in the three forest types are quite low; indicating that these forests should be conserved and protected and all timber cutting and logging activities (legal or illegal) should be avoided until the forest recovers in terms of quantity and quality. Enrichment planting for species with low densities in

the disturbed forests should be carried out; the promotion of natural regeneration by tending, weeding, and making bare soil in this area will also accelerate the recovery process.

The main goals of the present study were to detect the effects of disturbance degrees on the spatial patterns and distributions of tree species. Through spatial point-pattern analyses, our results clearly show the impacts of disturbance degrees on horizontal structure, and intra- and interspecific interactions. The univariate and bivariate data with pair- and mark-correlation functions of intra- and interspecific interactions within the five most abundant tree species indicate that most species demonstrated clumping and regular distributions at small scales. A high proportion of negative interspecific small-scale association (inhibition) and positive association (attraction) were recorded in both the LDF and HDF sites but were rare in the UDF. Based on the present results, silvicultural implications such as the site selection of enrichment planting and the promotion of natural regeneration could be applied in lightly and highly disturbed forests. We suggest that some species characterized by a negative interspecific association, for example, *Lithocarpus ducampii* and *Cinnamomum parthenoxylon,* should not be planted in homogeneous environments. This competition and difference in habitat existed in the LDF and HDF sites, while other habitat associations or dispersal limitations existed in the UDF. The results obtained from this study could

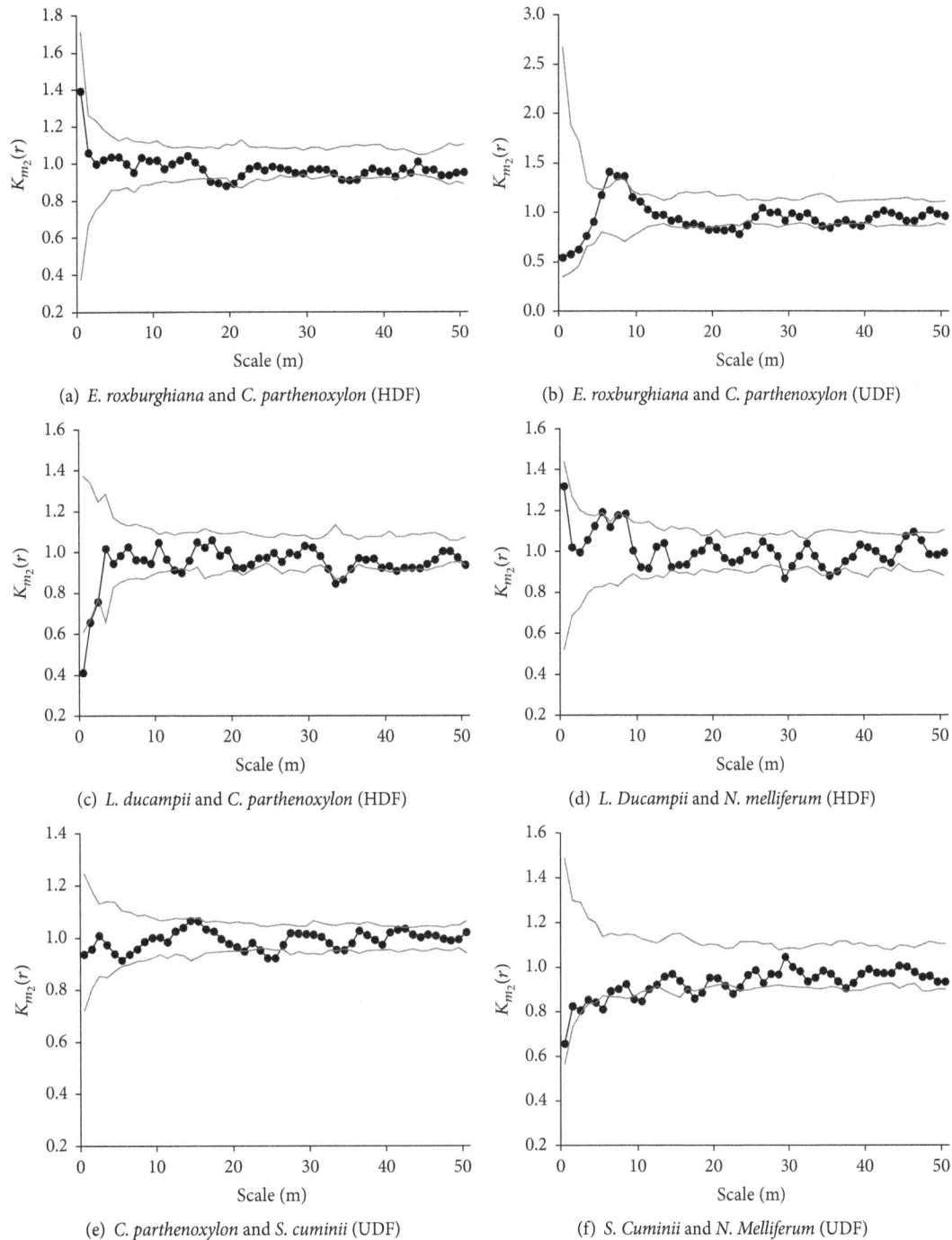

(a) *E. roxburghiana* and *C. parthenoxylon* (HDF)

(b) *E. roxburghiana* and *C. parthenoxylon* (UDF)

(c) *L. ducampii* and *C. parthenoxylon* (HDF)

(d) *L. Ducampii* and *N. melliferum* (HDF)

(e) *C. parthenoxylon* and *S. cuminii* (UDF)

(f) *S. Cuminii* and *N. Melliferum* (UDF)

FIGURE 9: Spatial correlations among the five most abundant tree species are shown by the bivariate mark-correlation functions. The observed patterns (dark line) that lie beyond the confidence envelopes (grey lines) indicate significant departures from the null models of independent marks.

be improved if the series data regarding recruitment and mortality were available in order to explain the underlying dynamic processes.

Conflict of Interests

The authors declare that there is no conflict of interests regarding the publication of this paper.

Acknowledgments

The authors are grateful to the Forest Inventory and Planning Institute (FIPI) for providing data of the permanent sample plots. They wish to thank all the people who contributed to this work, especially Mr. Dung and Mr. Huong who identified tree species for them in the field. They profoundly thank the forest ranger, Mr. Phung Huu Thu, who worked with them under difficult conditions in the field. Thanks also go to Professor Dr. J. Saborowski who provided them with statistical advice at times of critical need. They thank Ms. Tina Marie Joaquim who corrected and checked grammatical errors in the paper. They acknowledge support by the German Research Foundation and the Open Access Publication Funds of the Göttingen University.

References

[1] G. Kunstler, C. H. Albert, B. Courbaud et al., "Effects of competition on tree radial-growth vary in importance but not in intensity along climatic gradients," *Journal of Ecology*, vol. 99, no. 1, pp. 300–312, 2011.

[2] M. Kariuki, M. Rolfe, R. G. B. Smith, J. K. Vanclay, and R. M. Kooyman, "Diameter growth performance varies with species functional-group and habitat characteristics in subtropical rainforests," *Forest Ecology and Management*, vol. 225, no. 1–3, pp. 1–14, 2006.

[3] J. Lancaster, "Using neutral landscapes to identify patterns of aggregation across resource points," *Ecography*, vol. 29, no. 3, pp. 385–395, 2006.

[4] L. Parrott and H. Lange, "Use of interactive forest growth simulation to characterise spatial stand structure," *Forest Ecology and Management*, vol. 194, no. 1–3, pp. 29–47, 2004.

[5] F. He, P. Legendre, and J. V. LaFrankie, "Distribution patterns of tree species in a Malaysian tropical rain forest," *Journal of Vegetation Science*, vol. 8, no. 1, pp. 105–114, 1997.

[6] J. A. Logan, P. White, B. J. Bentz, and J. A. Powell, "Model analysis of spatial patterns in mountain pine beetle outbreaks," *Theoretical Population Biology*, vol. 53, no. 3, pp. 236–255, 1998.

[7] C. Skarpe, "Spatial patterns and dynamics of woody vegetation in an arid savanna," *Journal of Vegetation Science*, vol. 2, no. 4, pp. 565–572, 1991.

[8] T. Wiegand and K. A. Moloney, "Rings, circles, and null-models for point pattern analysis in ecology," *Oikos*, vol. 104, no. 2, pp. 209–229, 2004.

[9] S. Barot, J. Gignoux, and J.-C. Menaut, "Demography of a savanna palm tree: predictions from comprehensive spatial pattern analyses," *Ecology*, vol. 80, no. 6, pp. 1987–2005, 1999.

[10] D. N. Hamill and S. J. Wright, "Testing the dispersion of juveniles relative to adults: a new analytic method," *Ecology*, vol. 67, no. 4, pp. 952–957, 1986.

[11] D. A. King, S. J. Wright, and J. H. Connell, "The contribution of interspecific variation in maximum tree height to tropical and temperate diversity," *Journal of Tropical Ecology*, vol. 22, no. 1, pp. 11–24, 2006.

[12] S. P. Hubbell, "Tree dispersion, abundance, and diversity in a tropical dry forest," *Science*, vol. 203, no. 4387, pp. 1299–1309, 1979.

[13] R. W. Sterner, C. A. Ribic, and G. E. Schatz, "Testing for life historical changes in spatial patterns of four tropical tree species," *Journal of Ecology*, vol. 74, no. 3, pp. 621–633, 1986.

[14] P. Haase, "Spatial pattern analysis in ecology based on Ripley's K-function: introduction and methods of edge correction," *Journal of Vegetation Science*, vol. 6, no. 4, pp. 575–582, 1995.

[15] S. Vacek and J. Lepš, "Spatial dynamics of forest decline: the role of neighbouring trees," *Journal of Vegetation Science*, vol. 7, no. 6, pp. 789–798, 1996.

[16] J. H. Hou, X. C. Mi, C. R. Liu, and K. P. Ma, "Spatial patterns and associations in a Quercus-Betula forest in northern China," *Journal of Vegetation Science*, vol. 15, no. 3, pp. 407–414, 2004.

[17] T. T. Veblen, D. H. Ashton, and F. M. Schlegel, "Tree regeneration strategies in a lowland Nothofagus-dominated forest in south-central Chile," *Journal of Biogeography*, vol. 6, no. 4, pp. 329–340, 1979.

[18] J. F. Franklin, T. A. Spies, R. V. Pelt et al., "Disturbances and structural development of natural forest ecosystems with silvicultural implications, using Douglas-fir forests as an example," *Forest Ecology and Management*, vol. 155, no. 1–3, pp. 399–423, 2002.

[19] P. J. Diggle, *Statistical Analysis of Spatial Point Patterns*, Academic Press, London, UK, 1983.

[20] P. J. Clark and F. C. Evans, "Distance to nearest neighbor as a measure of spatial relationships in populations," *Ecology*, vol. 35, no. 4, pp. 445–453, 1954.

[21] P. Greig-smith, "The use of random and contiguous quadrats in the study of the structure of plant communities," *Annals of Botany*, vol. 16, no. 2, pp. 293–316, 1952.

[22] E. C. Pielou, "The use of point-to-plant distances in the study of the pattern of plant populations," *Journal of Ecology*, vol. 47, no. 3, pp. 607–613, 1959.

[23] B. D. Ripley, "Tests of 'randomness' for spatial point patterns," *Journal of the Royal Statistical Society Series B: Methodological*, vol. 41, no. 3, pp. 368–374, 1979.

[24] C. Hui, L. C. Foxcroft, D. M. Richardson, and S. MacFadyen, "Defining optimal sampling effort for large-scale monitoring of invasive alien plants: a Bayesian method for estimating abundance and distribution," *Journal of Applied Ecology*, vol. 48, no. 3, pp. 768–776, 2011.

[25] A. Pommerening, "Approaches to quantifying forest structures," *Forestry*, vol. 75, no. 3, pp. 305–324, 2002.

[26] D. Stoyan and H. Stoyan, *Fractals, Random Shapes, and Point Fields: Methods of Geometrical Statistics*, Wiley, Chichester, UK, 1994.

[27] T. Wiegand, S. Gunatilleke, and N. Gunatilleke, "Species associations in a heterogeneous Sri Lankan dipterocarp forest," *The American Naturalist*, vol. 170, no. 4, pp. 77–95, 2007.

[28] S. P. Hubbell, "Neutral theory and the evolution of ecological equivalence," *Ecology*, vol. 87, no. 6, pp. 1387–1398, 2006.

[29] S. Getzin and K. Wiegand, "Asymmetric tree growth at the stand level: random crown patterns and the response to slope," *Forest Ecology and Management*, vol. 242, no. 2-3, pp. 165–174, 2007.

[30] J. Illian, A. Penttinen, H. Stoyan, and D. Stoyan, *Statistical Analysis and Modelling of Spatial Point Patterns*, John Wiley & Sons, 2008.

[31] P. Greig-Smith, "Pattern in vegetation," *The Journal of Ecology*, vol. 67, no. 3, pp. 755–779, 1979.

[32] J. Leps, "Can underlying mechanisms be deduced from observed patterns," in *Spatial Processes in Plant Communities*, F. Krahulec, A. D. Q. Agnew, and J. H. Willems, Eds., pp. 1–11, SPB Academic Publishing, The Hague, The Netherlands, 1990.

[33] A. S. Watt, "Pattern and process in the plant community," *The Journal of Ecology*, vol. 35, no. 1/2, pp. 1–22, 1947.

[34] A. W. Tordoff, B. Q. Tran, T. D. Nguyen, and H. M. Le, *Sourcebook of Existing and Proposed Protected Areas in Vietnam*, Birdlife International in Indochina and Ministry of Agriculture and Rural Development, Hanoi, Vietnam, 2004.

[35] Plan of conservation andsustainable development Xuan Nha Nature Reserve until 2020, Son La Protection of Department, 2013.

[36] N. H. Hai, K. Wiegand, and S. Getzin, "Spatial distributions of tropical tree species in northern Vietnam under environmentally variable site conditions," *Journal of Forestry Research*, vol. 25, no. 2, pp. 257–268, 2014.

[37] B. Adrian and T. Rolf, "Spatstat: an R package for analyzing spatial point patterns," *Journal of Statistical Software*, vol. 12, no. 6, pp. 1–42, 2005.

[38] M. J. Lawes, M. E. Griffiths, J. J. Midgley, S. Boudreau, H. A. C. Eeley, and C. A. Chapman, "Tree spacing and area of competitive influence do not scale with tree size in an African rain forest," *Journal of Vegetation Science*, vol. 19, no. 5, pp. 729–738, 2008.

[39] S. Getzin, T. Wiegand, K. Wiegand, and F. He, "Heterogeneity influences spatial patterns and demographics in forest stands," *Journal of Ecology*, vol. 96, no. 4, pp. 807–820, 2008.

[40] F. Goreaud and R. Pélissier, "Avoiding misinterpretation of biotic interactions with the intertype K 12-function: population independence vs. random labelling hypotheses," *Journal of Vegetation Science*, vol. 14, no. 5, pp. 681–692, 2003.

[41] S. Getzin, C. Dean, F. He, J. A. Trofymow, K. Wiegand, and T. Wiegand, "Spatial patterns and competition of tree species in a Douglas-fir chronosequence on Vancouver Island," *Ecography*, vol. 29, no. 5, pp. 671–682, 2006.

[42] G. Shen, M. Yu, X.-S. Hu et al., "Species-area relationships explained by the joint effects of dispersal limitation and habitat heterogeneity," *Ecology*, vol. 90, no. 11, pp. 3033–3041, 2009.

[43] R. Condit, P. S. Ashton, P. Baker et al., "Spatial patterns in the distribution of tropical tree species," *Science*, vol. 288, no. 5470, pp. 1414–1418, 2000.

[44] Z. Luo, B. Ding, X. Mi, J. Yu, and Y. Wu, "Distribution patterns of tree species in an evergreen broadleaved forest in eastern China," *Frontiers of Biology in China*, vol. 4, no. 4, pp. 531–538, 2009.

[45] H. A. Peters, "Neighbour-regulated mortality: the influence of positive and negative density dependence on tree populations in species-rich tropical forests," *Ecology Letters*, vol. 6, no. 8, pp. 757–765, 2003.

[46] R. Pélissier, "Tree spatial patterns in three contrasting plots of a southern Indian tropical moist evergreen forest," *Journal of Tropical Ecology*, vol. 14, no. 1, pp. 1–16, 1998.

[47] A. Van Laar and A. Akça, *Forest Mensuration*, Managing Forest Ecosystems, Springer Science & Business Media, 1997.

[48] M. North, J. Chen, B. Oakley et al., "Forest stand structure and pattern of old-growth western hemlock/Douglas-fir and mixed-conifer forests," *Forest Science*, vol. 50, no. 3, pp. 299–311, 2004.

[49] R. P. Duncan and G. H. Stewart, "The temporal and spatial analysis of tree age distributions," *Canadian Journal of Forest Research*, vol. 21, no. 12, pp. 1703–1710, 1991.

[50] X. Wang, T. Wiegand, Z. Hao, B. Li, J. Ye, and F. Lin, "Species associations in an old-growth temperate forest in north-eastern China," *Journal of Ecology*, vol. 98, no. 3, pp. 674–686, 2010.

[51] I. Martínez, T. Wiegand, F. González-Taboada, and J. R. Obeso, "Spatial associations among tree species in a temperate forest community in North-western Spain," *Forest Ecology and Management*, vol. 260, no. 4, pp. 456–465, 2010.

5

Economic Contribution to Local Livelihoods and Households Dependency on Dry Land Forest Products in Hammer District, Southeastern Ethiopia

Dagm Fikir,[1] Wubalem Tadesse,[2] and Abdella Gure[3]

[1]*University of Gondar, P.O. Box 196, Gondar, Ethiopia*
[2]*Ethiopian Environment and Forest Research Institute, P.O. Box 24536, 1000 Addis Ababa, Ethiopia*
[3]*Hawassa University Wondo Genet College of Forestry and Natural Resources, P.O. Box 128, Shashemene, Ethiopia*

Correspondence should be addressed to Dagm Fikir; dagtfsm@gmail.com

Academic Editor: Piermaria Corona

The study was conducted in Hammer district, Southern Ethiopia, to provide empirical evidence on economic contribution to local livelihoods and households dependency on dry forest products. One agropastoral and two pastoral kebeles were purposively selected, and data was collected through household survey, group discussions, market assessments, and field observation. A total of 164 households, selected based on a random sampling procedure, were interviewed using structured questionnaire. The study found that income from forest products contributes 21.4% of the total annual household income. The major dry forest products include honey, fuel wood, gum and resin, and crafts and construction materials, contributing 49%, 39%, 6%, and 6% of the forest income, respectively. Households of the pastoral site earned more forest income and were relatively more dependent on forest products income than those in the agropastoral study site. Significant variation was also found among income groups: households with higher total annual income obtain more forest income than those with lower income, but they are relatively less dependent on forest products than the lower counterpart. Besides, various socioeconomic and contextual factors were found to influence forest income and dependency. The findings of the study provide valuable information up on which important implications for dry land forest development and management strategies can be drawn.

1. Introduction

Dry forests play an important role in the livelihoods of approximately one billion people worldwide [1]. In Africa, where 60% of rural dwellers are poor [2], dry forests represent an important resource base for livelihoods and economic development [3, 4]. About 320 million people in the continent depend on dry forest resources to meet many of their basic needs [5]. In Sub-Saharan Africa, where poverty and underdevelopment, in terms of infrastructure, government services, markets, and jobs, are characteristics features for most of its parts, about a quarter of a billion people live in or around dry forests [6] and depend on these resources for several products such as building materials, food, cropland, fuel wood, and nonwood products.

Over the past years, the recognition of the widespread reliance of rural people on forest products and the poverty-forest use relationships have spawned a growing scientific interest in demonstrating the economic dependence on forest products and understanding its determinants [7–10] The most important argument behind such scientific studies is that empirically based knowledge on people-forests interactions is an important tool in devising alternative strategies for livelihood security, poverty reduction, and dry forest conservation [11, 12]. Moreover, knowledge about the significance of environmental income is also important to the environmental conservation debate and the tradeoffs and synergies that exist between use and protection.

Economic valuation studies undertaken in different dry forest and woodland countries showed forest related income contributions ranging from 6% to 45% [9, 11, 13–18]. Studies do also suggest that such contribution, or extent of people's dependency, varies substantially across households due to different factors like wealth status, education, livestock and land holdings, awareness levels, age, gender, household sizes, access to forests and markets, and nonfarm activities [13, 19–21].

Of widely discussed factors in forest dependency literature is that of the relationship between wealth and forest products extraction. While some studies reported that poorer households extracted more forest products and are more dependent on forest products than wealthier one [8, 22], others found that households with higher total annual income extracted more forest products than those with lower income, but they are relatively less dependent on forest products than the lower counterpart [15, 23, 24]. This implies that forest-wealth relationship may vary from place to place, and it suggests that there is still a need to increase our understanding in different contexts and geographical areas.

It has also been suggested that the role of forest products may vary with the type of the traditional source of livelihood: due to the reason that farmers, agro/semipastoral, sedentary pastoral, and pastoralist differ in their social and cultural backgrounds, spatial provenance, and their history of close association with nature, their different resource use patterns might result in differences in the economic importance of forest products among them [10, 23, 25]. This obviously implies that site and livelihood specific investigations on forest dependence are necessary so that policy and management interventions targeted to support rural livelihoods and promote sustainable resource use can be tailor-made to suit intercommunity heterogeneity.

While existing studies from different dry forest areas have widen our knowledge on various aspects of forest-people interaction, there are still many other dry forest areas where empirical researches on the level of people's dependence on forests and determinant factors of forest dependence are still essential areas of research to be addressed. Ethiopia is no exception with respect to such aforementioned research needs. About 55 million ha of its land is covered by dry forests, which is its largest remaining forest vegetation [26]. These forests are largely found in the arid and semiarid lowland areas of the country, which are inhabited by rural peoples whose major livelihood strategies are traditional pastoralism and agropastoralism. Dry forests in these areas have long remained as integral components of such pastoral and agropastoral livelihood systems. However, only a few studies have so far tried to unveil the economic importance of dry forests in the country, and even such studies covered some parts of the lowland areas in three regional states, Amhara [21, 27], Oromia [28, 29], and Tigray region [15, 30] whereas there is a dearth of information for the dry forests of most of the areas in the Southern Nations, Nationality, and Peoples Regional State, one of which is Hammer district, the current study area.

The objectives of this study were to provide empirical evidence on the contribution of dry forest product income to the annual income of rural households. The study addressed such specific questions as the following: (i) what are the major

dry forest products utilized in the study area? (ii) What is the absolute and relative contribution of dry forest income to the annual household income? (iii) Does the income contribution differ between pastoral and agropastoral livelihood types and across income groups? (iv) What factors affect household's forest income and forest income dependence?

2. Materials and Methods

2.1. Study Area. The study was conducted in Hammer district, found in South Omo zone of Southern Nations, Nationalities, and Peoples Regional State. Located at 839 km south of Addis Ababa, the country's capital, Hammer district lies between $4°25'-5°30'$N and $36°5'-36°59'$E (Figure 1). It covers an area of 5742 km^2 and has a total population of 59,160. The altitude in the district varies between 371 m.a.s.l. and 2084 m.a.s.l. [31]. The climate is dominantly a semiarid type, which covers about 95% of the area. The district has an erratic, variable rainfall and high ambient temperature ranging in 26–35°C [32]. The rainfall pattern is bimodal, with a primary rainy season between March and May and occasional rain between October and December. The average annual rainfall varies from 581 mm in the lowlands to 796 mm in the highland parts of the district [33].

The major land cover types in the district are forest and shrub land (39.2%), bush land (26%), savanna and grass land (25.7%), and settlement and farm land (9.1%), among which the woody vegetation accounts for significant proportion, 65.2% of the total land area of the district [33]. The major livelihood activities in the area include pastoralism (dominated by livestock production) and agropastoralism (crop and livestock production). However, the former is the dominant one, which covers approximately 75% of the area, and is commonly practiced in the southern and eastern lowland regions of the district, at altitudes below 1000 m.a.s.l. The latter, which covers small proportion of the district, is common in the highland territories in the north and northeast, with an elevation of above 1000 m.a.s.l [31, 33].

2.2. Sampling and Data Collection. Prior to the actual survey, visits were made to the district and secondary information relevant to the study was gathered from all possible sources. Based on the baseline information documented by zonal and district agricultural offices and by previous studies [33], the district was first stratified into two: pastoral and agropastoral *kebeles* (administrative units of the district). Then, after a thorough discussion with experts (from South Omo Zone Agriculture and Rural Development Department, Jinka Agricultural Research Center and HWPDO), one agropastoral (*Shanko-lala*) and two pastoral kebeles (*Angude* and *Mirsha-Bitagelefa*) were purposively selected based on dry forest/woodland cover, representativeness, and accessibility (Figure 1). Hereafter, agropastoral and pastoral are referred to as AGPAS and PAS, respectively.

Field data collection at the selected kebeles was carried out from November 2012 to February 2013 using such methods as household survey, group discussions, market assessments, and field observation [34]. For household survey, a total of 164 households (61 from AGPAS kebele and 103 from

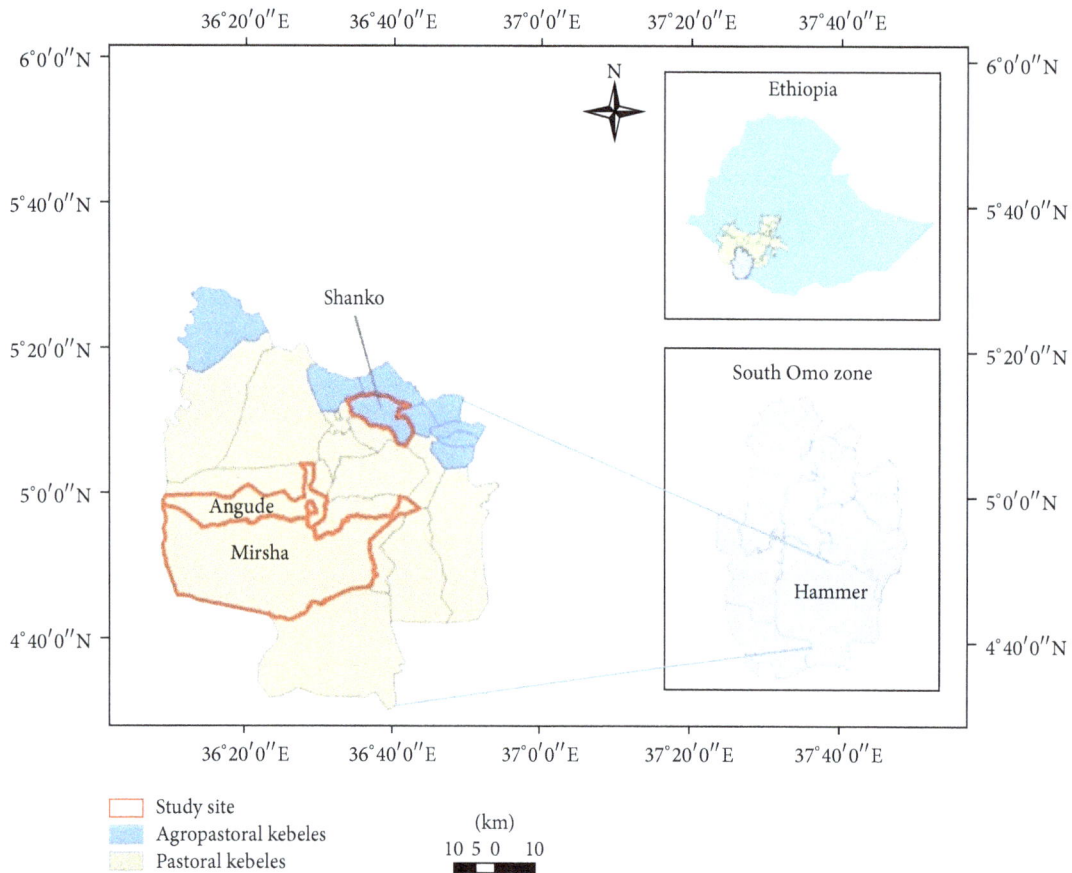

FIGURE 1: Location map of the study area.

PAS kebele, which was 10% of total households in each kebele) were sampled based on a random sampling procedure, with kebele registers used as sampling frames. The survey was carried out using a structured questionnaire aimed at capturing both qualitative and quantitative information. The questionnaire was comprised of such major issues as sociodemographic characteristics (such as sex, age, family size, and education level), major assets such as land and livestock, livelihood activities, and forest product extraction. Local enumerators were recruited from the respective sample kebeles. All enumerators were fluent speakers of the respective local languages. They were trained on data collection procedures, interviewing techniques, and the detailed contents of the questionnaire. The questionnaire was pretested to check its appropriateness for gathering all the required data.

Three group discussions, one at each selected kebele, were undertaken. Eight to twelve members were involved in the discussion, where such members include local elders, informal and formal institutional committee members, village leaders, and experts like development agents. Major points raised for discussion included several issues regarding livelihood activities; forest/woodland benefits; type and utilization of major forest products; and their socioeconomic role. During the entire discussion process, the lead author served as facilitator while insiders fully participated in the dialogue. In addition to the above methods, a rough market inventory was also conducted in some nearby local markets so as to identify the marketable forest products and record market prices of different products. Finally, field observation was carried out as a supporting data collection approach, to observe the major products collected in the forest and the respective species, particularly gum and resin bearing tree species, and also bee hives and honey products of the forest. This was done at the nearby forests in each of the three sample kebeles.

2.3. Household Income Accounting. Valuation of income from dry forests and other livelihood activities was done following Cavendish [34]. Forest products identified and included in forest income accounting were honey, fuel wood, gums and resins, hand crafts, and construction materials. The values of wild foods and fodder (grazing) were not included due to methodological difficulties. Income in this study includes both subsistence and cash incomes, calculated based on the annual (December 2011 up to December 2012) estimate of the amount produced, consumed, and sold from the different income sources. Subsistence income (the value of a product directly consumed by the household or given away to friends and relatives as gifts) was estimated using the actual market prices when available and, otherwise, surrogate market prices were used. Wherever bartering system was reported, the market price of the taken-on item was used to determine the value of the sold item. All costs for material inputs, hired labor, transportation, and marketing were deducted from the gross income of each income source. Family labor was not considered as a cost.

2.4. Data Analysis. The data collected from the survey was first checked. During the data checking, 17 of the 161 questionnaires were found incomplete and removed from data processing and analysis. Qualitative data were summarized by way of text analyses, while quantitative data were analyzed by descriptive and inferential statistics. Descriptive statistics, including proportions or percentages, averages, and others, were used for describing socioeconomic characteristics of sample respondents. Inferential statistics was used to study relationships between variables using different statistical tests.

To test statistical differences between AGPAS and PAS sites in the mean annual absolute and relative income, independent sample t-test and nonparametric Mann–Whitney test were employed, respectively. To distinguish between absolute and relative forest income among the different wealth groups, sample households were divided into three income (total annual household income) terciles, operationally categorized as first, second, and third terciles, respectively. And one-way ANOVA and Kruskal-Wallis test were run to test the difference among income groups in absolute and relative forest income, respectively.

To identify factors influencing household income from the forest, multiple linear regression analysis using Ordinary Least Square (OLS) method was employed. Two separate OLS models were used in this study: one for absolute forest income and the other for forest income dependence (measured as the relative share of forest income in the total annual household income). Different explanatory variables, hypothesized to affect forest income and income dependence, were included in the regression model [20]. Presence of multicollinearity and heteroskedasticity problems was first checked prior to running the model, and as the later problem was detected, a robust OLS model was used.

3. Results and Discussion

3.1. Sociodemographic Characteristics of Sample Respondents. The majority of the sample respondents (77%) were male, while the remaining (23%) were female. The household size of respondents ranges from two to 13 with the average of seven members. Specific to the two study sites, average household size was seven at AGPAS and six at PAS. The sample respondents were, on average, 38 and 42 years old, respectively, at AGPAS and PAS. Education level was very low in both study sites: a significant proportion of the sample respondents (88%) do not have formal education.

Land and livestock are important fixed assets of households in the study area. On average, a household owned 2.2 ha of land (ranges from 0.25 to 6 ha) and 16.5 tropical livestock units (TLU) (ranges from 1.6 to 102.8 TLU). Looking into variations in the land and livestock asset holdings between the two study sites, average landholding size was 1.54 ha at AGPAS (ranges from 0.45 to 3.4 ha), whereas it was 2.64 ha at

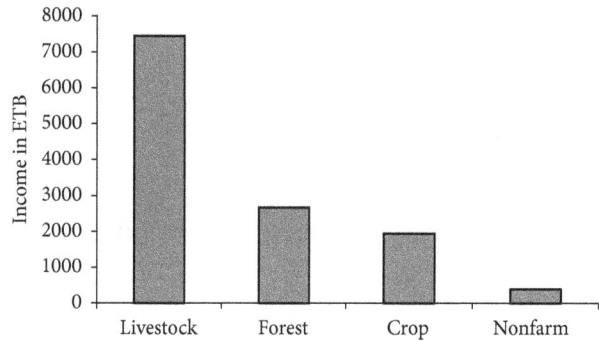

FIGURE 2: Major sources of annual household income and their contributions (in ETB) in the study area.

PAS (ranges from 0.25 to 6 ha), and the average livestock asset was 7.6 TLU at AGPAS (ranges from 1.6 to 25.3) and 21.5 TLU at PAS (ranges from 1.8 to 102.8).

3.2. Household Income and Its Sources. Households in the study areas rely on a wide range of economic activities mainly related to natural resource extraction. The major sources of income identified from the household survey were livestock, crop, forest, and nonfarm activities (such as daily labor and petty trade). Although not common, aid and remittance were mentioned as sources of livelihoods. The contribution of the major income sources to the total annual household income (the combined income generated from the main household economic activities) is presented in Figure 2. Our data showed that the average total annual income of the sample households was Ethiopian Birr (ETB) 12,450. Among the four income sources, income from livestock sector contributed the largest share (59.7%), followed by income from forest products (21.4%). Income from crop production and nonfarm activities accounted for 15.7% and 3.2% of the total annual household income, respectively.

3.3. Household Income from Forest Products. In the current study, livestock production dominates the livelihoods of households in the study area. In fact, as the sole source of feed for livestock was the surrounding forest which provides grasses and/or woody plants year round, this seems to signify that the forest is used as rangelands. However, the current study revealed that provision of grazing or browsing feed was not the only attribute of the dry forests, but the forest is also a source of many other products like honey, fuel wood, gums and resins, hand crafts, and construction materials, that, through their subsistence and cash income, are of importance for household livelihoods. Forest income contributes 21.4% to the total income, which was second to livestock income, even without inclusion of income from fodder and grazing. If the income from fodder and grazing was accounted, total forest income would have been higher, illustrating further the importance of forests for household income in the study area.

The 21.4% forest income share found in the current study is comparable to other studies. The share of forest income in the total annual household income was 15% in Chiradzulu District, Malawi [35], 23% in Gore District, Southwestern

TABLE 1: Mean annual household income (in ETB) from different forest products.

Forest products	PAS (N = 94)		AGPAS (N = 53)		Total (N = 147)	
	Mean	SE	Mean	SE	Mean	SE
Honey	1448.5	116.6	1117.92	228.13	1329.32	111.3
Fuel wood	1005.5	108.8	1128.08	178.01	1049.72	94.4
Gum-resin	237.4	25.1	—	—	151.83	18.6
Other products	136.1	10.50	146.54	12.99	139.83	8.2
Total forest	2827.5	124.6	2392.5	289.5	2670.7	132

Ethiopia [36], and up to 22% in Africa, Asia, and Latin America [37]. However, it is also relatively lower than the share reported from other studies conducted in other arid and semiarid areas of Ethiopia. For instance, Worku et al. [38] found a 34% annual income contribution from dry forests in Somali region, Southeastern Ethiopia. In Wenbera district, Northwestern Ethiopia, income from dry forests contributed 39% of the total annual household income [39]. 32% of the annual household income in Somali region, Eastern Ethiopia, was obtained from the collection and sale of one forest product, Gum-resin [28]. The lower income share in the current study compared with the aforementioned studies could likely be attributed to the low engagement of households in extraction activities, and most importantly to the limited enabling production and marketing environment in the current study area, particularly for some economically valuable products such as gum and resin.

3.3.1. Major Forest Products: Use, Collection, and Income. As indicated earlier, the forest provides different products that, through their subsistence and cash income, are of importance for household livelihoods. The major products include honey, fuel wood, gum and resin, hand crafts, and construction materials, and the first three were the top three important products in terms of their contribution to household income (Table 1). The description of the collection and production of these major products and respective income contribution are presented below.

Honey. In the current study, about 61 and 68% of sample households at the AGPAS and PAS site indicated the production of honey as one of their forest based livelihood activities. According to the household survey, honey production in the study areas was mainly carried out by placing hives hanged in a forest. Honey from the forest is produced/harvested three to five times annually; and three to five Kilograms of honey can be produced in one bee hive in one harvest. As reported during the survey, households delivered the raw honey to the nearby market, without product processing or any other value adding activity. The raw honey was reported to be sold at an average price of ETB 40 per kg.

Income from honey contributed a significant proportion of the annual forest income of households at both study sites: of the total forest income of 2390 ETB at AGPAS and

2827 ETB at PAS, honey accounted for 47 and 51% (Table 1). In fact, being a much-valued product from forested areas, collection and production of honey from the forest have been reported as one of the most important forest based livelihood activities around the world [40]. In Ethiopia, rural households engagement in honey production from hives hanged in a forest has been widely reported for humid and subhumid forest areas of the country, particularly the southern and southwestern areas endowed with natural high forest [41–43], whereas it has seldom been reported for arid and semiarid lowland areas. For instance, no engagement in honey production was reported from studies conducted in the dry forests of Liben and Afdher districts, south eastern dry lands [38], and in Quara (northwestern), Asgede-Tsimbla (northern), and Yabelo (southern) lowlands [22]. Likewise, Abebaw et al. [21] found few households engaged in honey production from the dry forest of Metema district, northwestern region. Though it is implicit to justify the variation between the finding of the current study and the aforementioned ones, the current study confirmed that the dry forests have the potential to provide honey as the humid forests do.

Fuel Wood. Income from fuel wood collection was the second most important forest income; it accounted for 47 and 36% of the annual forest income at AGPAS and PAS, respectively (Table 1). Given that fuel wood has remained as the major energy sources for most of the rural as well as urban households, coupled with the relentless population growth and the subsequent increase in wood demand, it might not be uncommon to find fuel wood harvesting as a major forest based livelihood activity. During group discussion, participants stated that the activity of fuel wood collection for income generation has become common and that the emerging and growing market demand, which was the result of expansion of nearby towns and increment of urban population, was one of the major responsible factors for such activity. Extraction of fuel wood from the forest was reported as a major source of forest related incomes for rural households in different parts of Ethiopia [21, 22, 44]. In Malawi, Chilongo [45] found that fuel wood harvest is the dominant forest related activities and attached this to the fact that fuel wood is the main source of cooking energy in the country. A global assessment also shows the dominant role of fuel wood in forest environmental incomes for the rural poor [37].

Gums and Resins (GR). Gums and resins (GR) are one of the most economically valuable products of dry forests of several regions. In the current study, collection and sale of GR products as a source of income were observed only at PAS study site. In this study site, households extracted GR from different species. The most commonly known GR product was frankincense collected from *Boswellia neglecta*, which is locally named as *"tikuretan"* (meaning: black incense). Next to frankincense, gum-resins from four *Commiphora* species, *Commiphora africana, C. myrrh, C. habessinica,* and *C. schimperi* were the major products collected and utilized in the study area, and products from all species are locally

TABLE 2: Mean annual income (in ETB) of pastoral and agropastoral households from different sources.

Sources	AGPAS (N = 53)		PAS (N = 94)		Comparison test	
	Mean ± SE	%	Mean ± SE	%	[a]t value (df)	[b]Z value
(1) Livestock income	6969.75 ± 681.37	54.5	7716.99 ± 597.3	62.73	0.79 (145)	−1.85
(2) Forest income	2392.55 ± 289.51	18.4	2827.53 ± 124.6	22.99	1.38 (72)	−3.24*
(3) Crop income	2981.21 ± 195.66	23.4	1371.94 ± 49.2	11.15	7.98 (59)*	−5.25*
(4) Nonfarm income	424.4 ± 125.11	3.5	385.09 ± 71.6	3.13	0.29 (145)	−1.11
Total	12767.9 ± 838.49		12301.5 ± 666.9		0.43 (145)	

a: Independent sample t-test for difference in absolute income; the test for forest income and crop income was based on unequal variance. b: Mann–Whitney test for difference in relative income (percentage). *Significant at $p < 0.001$; df: degree of freedom.

named as "*nechetan*" (meaning: white incense). Production and/or harvesting of the product is mainly done by collecting the naturally oozed product from the trunk and branches of the tree, as well as from the ground. Artificial tapping was not common in the study area. No specialized production techniques, product handling cares, and product processing were observed in the study. Collectors simply collected the product, made the fresh product have a ball-like structure (sometimes they also use dry ash so as to avoid cracking), and delivered it to the nearby market. The average annual income from GR was estimated to be 152 ETB (Table 1).

GR extraction as local income source is in fact reported in different parts of Ethiopia's lowland forest areas, particularly in the country's northern, northwestern, southern, and southeastern areas [21, 38, 41, 46]; however, the level of people's participation in GR production and the estimated GR income in the current study is relatively lower than that reported in most of these previous studies. One major attributable factor for this can be (though differences in resource base might exist) the lack of enabling production and marketing environment in the current study area as compared to others. For instance, in most of the northern forested areas, there are many private and state companies engaged in the production and marketing of GR, and they create seasonal employment for the tapping, collection, and grading of frankincense [41, 46, 47]. On the other hand, Ethiopian Natural Gum Production and Marketing Enterprise (ENGPME), which has been the main actor in the gum and resin business in Ethiopia since its origin, is playing a major role in GR production and marketing through establishing its own production areas in the different localities in the northern and northwestern parts of the country, as well as a major buyer of GRs from individual rural collectors and cooperatives in the south and southeastern parts of the country [39]. In the current study area, information from group discussion showed that ENGPME has established one of its own warehouse sites at the study area in 1989 and started buying GR from individual pastoralists. However, soon after pastoral households had increasingly engaged in the collection and sale of GR, the company, ENGPME, has left the area. Besides, with the last few years' effort to organize rural households in the form of cooperatives and engage them as new actors in the frankincense supply chain, households in Northern and Southern Ethiopia formed cooperatives and earned direct income from the sale of GR [29, 39, 46,

47]. In our study area, until the time when data collection was conducted, no intervention has been made regarding the initiation and mobilization of pastoralists to establish cooperatives.

3.4. Forest Income and Its Contribution: Comparison between Pastoral and Agropastorals. Important finding in the current study was that the absolute and relative forest and all other three income components differed between the households of AGPAS and PAS sites (Table 2). While income from livestock production took the lion share in both cases, its contribution was slightly higher at PAS (63%) than at AGPAS (54%), which confirms the fact that pastoralism is a livestock dominated livelihood. Likewise, while income from crop production was the second share (23%) at AGPAS, its importance was relatively lower at PAS (11%). With regard to forest income, with average share of 18% and 23% income from forest was second to livestock income at PAS, whereas it was third after livestock and crops at AGPAS, respectively. Though differences in the amount and share of income between the two sites were observed for all income sources, statistically significant differences were found only in the cases of absolute crop income, relative crop income, and relative forest income, all at $p < 0.01$ (Table 2).

The higher dependency on forest income of pastoralists as compared to agropastoralists could be attributed by different factors. For instance, it might be related to time and labor resource: the engagement of agropastoralists in crop farming may reduce time as well as labor resources for forest product collection, whereas pastoralists may not face such scarcity of time and labor as livestock production is relatively less labor intensive and/or as forest product collection could be carried out parallel to livestock rearing activities. On the other hand, the increasing sedentarization in agropastoral areas and the subsequent forest clearance for crop cultivation may reduce the forest resource base and hence reduce the amount that could be extracted from the forest. A study in Liben and Afdher, Southeastern Ethiopia, found that pastoral households earned more dry forest income than agropastoralists, and it mentioned labor scarcity due to competing agriculture and distance to the forest as factors hindering agropastoralists' dry forest income [38]. The same study also reported that the mobility of pastoralists also allowed them to cover wider areas of forests with species that bear valuable products such as gums and resins. Similarly, a study conducted in arid and

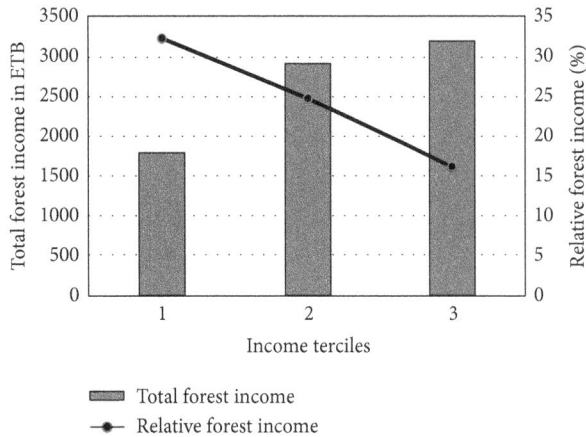

FIGURE 3: Total and relative forest income (in ETB) across income tercile groups.

semiarid areas of Afar, Eastern Ethiopia, revealed that since pastoralism is more compatible with forest based activities in terms of labor allocation, pastoral households earn more forest income than agropastorals [48].

3.5. Forest Income and Its Contribution: Comparison among Income Groups. Splitting the sample respondents into three income terciles based on total annual household income, the study examined how important the forest is for different wealth classes in terms of its absolute as well as relative income. As shown in Figure 3, it is the upper income group that obtained higher dry forest income (3232 ± 283), followed by middle group (2955 ± 181). The lowest wealth group earned the least income (1825 ± 151). Such variation in dry forest income among income groups was statistically significant ($p < 0.001$). In contrast to the trend observed in mean dry forest income which increased with total household income, the share of forest income decreased with higher total annual household income (Figure 3). The lower income group generated 32.4%, middle 24.8%, and upper 16% of their income from forest. The Kruskal-Wallis test revealed that such variation in the share of forest income among the three income groups was statistically significant (Chi-square: 31.1, 2, $p < 0.000$). Looking into product specific income, the study found no clear trends between income groups and the collected dry forest product type.

The finding implies that households with higher total annual income extracted more forest products than those with lower income, but they are relatively less dependent on forest products than the lower counterpart. This may be due to the fact that poor households have fewer assets in terms of land, livestock, and cash to generate more income from agriculture, or in other words, since they have only minimal access to farmland and livestock assets, they may have no chance to replace extraction of forest products by adequate production of agricultural products. Thus they tend to depend more on forests. Similar trends were reported in other studies [15, 23, 24, 38] while there are also contrasting results showing that poorer households extracted more forest

products and are more dependent on forest products than wealthier one [8, 22].

Another important inference regarding the wealth-forest income link can also be given here. For instance, the greater average number of forest product collectors in all income groups in this study indicates that forest products in the study area are used in support of current basic consumptions. On the other hand, the results that lower income groups obtained low income but had high dependency on forest products imply that forest product extraction primarily serves poorer households as a gap-filling activity or as a safety net in times of crisis while remaining a low return activity. Bognetteau and Wirtu [49] stated that "... for the poorer households, forest products usually contribute less in financial terms, but they are an important safety net, especially through the open access situation of some of these products."

3.6. Determinants of Forest Income and Forest Income Dependence. Two separate multiple linear regression models were carried out to identify factors influencing (i) households' annual forest income and (ii) households' level of dependency on forest income. The result is presented in Table 3. All explanatory variables, except household head's age, were found to have a statistically significant effect in one or both of the two (forest income and income dependency) models.

To start with site factor, its effect on forest income was not statistically significant, whereas it had a significant and positive influence ($p < 0.01$) on dependency, which implied that the level of dependency on forest income increased with the likely of the households to be pastoral and this supported the mean comparison test indicated previously. Sex of the household head (being male) had a significant and negative influence ($p < 0.05$) on forest dependency, while its effect on forest income was not statistically significant. According to the analysis result, a female headed household is more likely to be forest-dependent than a male headed household. As suggested in the forest dependence literature, women are likely to participate more in common property resources than men and may be more involved in gathering activities than men [50]. The finding of the current study regarding sex is in line with other studies [30, 44, 51].

Family size significantly influenced both the forest income and level of dependency on forest income ($p < 0.05$) where an increase in family size resulted in an increased income and dependency. This may be related to the need to have additional income (in addition to main livelihood activities) so as to support larger number families, and on the other hand, the quantity of production may rise as more number of household members is engaged in product collection, and hence the income may increase. This finding is in line with others such as [9, 38, 52].

The two important household assets, land and livestock size, were found to influence forest income and dependency negatively; however, their effect on income was not significant. According to the regression result, larger landholding size and livestock asset significantly decreased (both at $p < 0.01$) the level of dependency on forest income. Given that larger land and livestock resources may reflect higher total income, the negative sign would seem to contradict our

TABLE 3: OLS regression of household forest income (A) and forest income dependence (B) against socioeconomic characteristics.

Factor	(A) Forest income				(B) Forest income dependence			
	Coefficient	SE	t-value	p value	Coefficient	SE	t-value	p value
Constant	1463.596	1576.25	0.929	0.355	19.99	13.603	1.47	0.144
Site (dummy, 1 if pastoral site)	494.499	534.478	0.929	0.357	13.083	5.074	2.578	0.011*
Sex of household head (dummy, 1 if male)	−77.247	271.202	−0.285	0.776	−7.276	3.175	−2.292	0.023*
Age of household head (yrs)	25.948	17.918	1.448	0.15	0.014	0.178	0.079	0.94
Household size (number)	258.303	106.476	2.426	0.017	1.409	0.672	2.097	0.038*
Land size (in hectare)	−156.805	113.184	−1.385	0.168	−2.358	0.7	−3.369	0.001**
TLU (tropical livestock unit)	−10.95	17.621	−0.621	0.535	−0.655	0.171	−3.83	0
TLU squared	0.066	0.137	0.482	0.632	0.005	0.001	3.25	0.001**
Access to extension (dummy, 1 if yes)	949.616	303.252	3.131	0.002**	6.268	2.611	2.401	0.018*
Nonfarm activity (dummy, 1 if yes)	−672.183	224.207	−2.998	0.003**	−5.623	2.019	−2.785	0.006**
Distance to market (in km)	−152.298	49.471	−3.079	0.003**	−0.728	0.503	−1.447	0.15
Distance to forest (in km)	−315.238	121.513	−2.594	0.011*	−1.328	1.154	−1.151	0.252
R-square	0.41				0.37			
F-value	12.27				5.42			
p value	0.000				0.000			

*Significant at 5%. **Significant at 1%.

earlier findings that high income households extracted more forest products. However, since the relation is insignificant, it cannot possibly be a contradiction. The significant negative effect of these assets on dependency confirmed the earlier finding that rich households are less dependent on forest income than the poor ones. Similar finding was reported by Fisher [13] and McElwee [53].

On the other hand, the regression result for nonfarm activity supported the hypothesis that availability of nonfarm activity reduces the need for environmental income: households with nonfarm activity were found to earn higher forest income (p < 0.01) and to be more dependent on forest income (p < 0.01) than those without. This can possibly be because engagement in other income generating activities like nonfarm may reduce quantity of extraction and hence income, by competing for and taking over labor as well as time that would otherwise be invested for forest related activities.

Extension contact was the other important factor that significantly affected both absolute forest income and income dependence (p < 0.01). Households having an extension contact are more likely to earn higher forest income and to be more dependent on forest income. Even though extension service was not so far pronounced in the study area, the positive influence of these services on forest income can be an indication of how the presence of encouraging and enabling environment is important for rural dwellers to better obtain substantial returns from forest product extraction. Similar to this finding, Gebrelibanos [54] found that frequency of extension contact has a positive and significant impact in adopting specific technology or getting new information.

Forest income showed significant reduction with increasing distance to the forest (p < 0.05) and to the market (p < 0.01). The larger the distance from the forest the lower the income from forest: obviously, this is related to less or limited access to the resource and associated time

and energy costs, both of which became significant with increasing distance from the forest. Particular to the study area: large distance from the forest coupled with the poor infrastructural situation and the unsuitable terrain of the area limit the quantity of extraction. The negative relationship between income and distance to forest is also in agreement with the report of Abebaw et al. [21] for Metema district in northwestern Ethiopia and Asfaw et al. [44] for Jelo Afromontane forest of Eastern Ethiopia. Regarding market distance, forest income is higher for those who are near to the market place than those far away, which may be related to the incurrence of time and transportation costs, as well as to the lack of or limited access to market information. Such negative influence of market distance in the current study has also been observed in different studies [9, 38, 55].

4. Conclusion and Recommendations

The livelihood of the people in the study area depends on livestock rearing, crop production, forest product collection, and nonfarm activities. Livestock sector plays the dominant role in livelihoods of both pastoral and agropastoral households, while income from forest products supplements livestock dominated livelihoods. Pastoral households are relatively more dependent on forest related incomes than agropastoral households. Honey, fuel wood, GR, hand crafts, and construction materials are the five major forest income sources, where the top three products in terms of contribution to household income are honey, fuel wood, and GR. Forest income and dependency vary with household characteristics. Households with higher total annual income obtain more forest income than those with lower income, but they are relatively less dependent on forest products than the lower counterpart. Being pastoral and female headed household increases the level of dependency on forest income, while having larger land and livestock size reduces dependency.

Distance to forest and market reduces total forest income. Larger family size, no nonfarm activity, and access to extension contact increase both forest income and forest income dependency.

Based on the above findings, some important implications can be drawn. First, our results illustrate that forest products continue to play important role in the livelihood of pastoral and agropastoral communities of the dry land areas. Such important role of dry forests should be given a due focus in development and land use policies, because making and implementing policies that do not consider the role of forests may result in policy outcomes that do not harmonize development and conservation objectives.

Second, as far as improving household livelihoods and alleviating poverty in the study site are concerned, the present result on wealth-forest income link suggests that enhancing forest product collection and marketing for lower income households should certainly be encouraged as part of an overall strategy. Likewise, results from the regression analysis of determinants of forest income and income dependency equally offer a host of policy options open to forest conservation and management in the study area.

Third, dry forest of the study area is home to high value products such as honey and GR for which there are potential opportunities that could be explored further and might contribute in greater extents to household incomes. For this, awareness creation and capacity building for households on the production, harvesting, handling, and value-addition of these economically valuable products should be the major future intervention areas. Moreover, given that Ethiopia is known as being one of the major producers and sellers of honey and GR products and these products are enjoying fortunate market and technological opportunities, creating and promoting better commercialization environment in the study area would further lend additional weight to the existing contribution of these products to the national economy.

Finally, future attempts should be made to estimate and value the contribution of other nonmarketable products that have not been included in the current study (such as fodder and grazing) to enhance the economic contributions of dry forests to the local livelihoods and to the national economy at large.

Competing Interests

The authors declare that there is no conflict of interests regarding the publication of this paper.

Acknowledgments

The authors are grateful for the financial support from The Spanish Agency for International Development Cooperation (AECID). They also would like to express their gratitude to Ethiopian Environment and Forestry Research Institute (formerly, Forestry Research Center) for its institutional support. They are thankful to the local people of Shankolala, Angude, and Mirsha kebeles for granting them their valuable time and support during the study. Special thanks are also due to experts and officials from Jinka Agricultural Research Center, South Omo Zone Agriculture Department, Hammer District Pastoral Development Office, and Hammer District Administration Office, for their support during data collection. Finally, they are very pleased to thank Ms. Iria Soto for her invaluable technical support for this research.

References

[1] E. Mwangi and S. Dohrn, *Biting the Bullet: How to Secure Access to Dry Lands Resources for Multiple Users*, International Food Policy Research Institute, Washington, DC, USA, 2006.

[2] T. Oksanen, B. Pajari, and T. Tuomasjukka, Eds., *Proceedings from Forests in Poverty Reduction Strategies: Capturing the Potential*, Proceedings no. 47, European Forest Institute, Joensuu, Finland, 2003.

[3] F. Paumgarten and C. M. Shackleton, "Wealth differentiation in household use and trade in non-timber forest products in South Africa," *Ecological Economics*, vol. 68, no. 12, pp. 2950–2959, 2009.

[4] S. Shackleton, B. Campbell, H. Lotz-Sisitka, and C. Shackleton, "Links between the local trade in natural products, livelihoods and poverty alleviation in a semi-arid region of South Africa," *World Development*, vol. 36, no. 3, pp. 505–526, 2008.

[5] L. Petheram, B. Campbell, C. Marunda, D. Tiveau, and S. Shackleton, "The wealth of the dry forests. Can sound forest management contribute to the Millennium Development Goals in sub-Saharan Africa?" Forest Livelihood Briefs 4, CIFOR, Bogor, Indonesia, 2006.

[6] CIFOR, *Thinking beyond the Canopy. CIFOR Annual Report*, CIFOR, Bongor, Chad, 2008.

[7] W. Cavendish and B. M. Campbell, *Poverty, Environmental Income and Rural Inequality: A Case Study from Zimbabwe*, Centre for International Forestry Research, Bogor, Indonesia, 2002.

[8] C. Shackleton and S. Shackleton, "The importance of non-timber forest products in rural livelihood security and as safety nets: a review of evidence from South Africa," *South African Journal of Science*, vol. 100, no. 11-12, pp. 658–664, 2004.

[9] G. Mamo, E. Sjaastad, and P. Vedeld, "Economic dependence on forest resources: a case from Dendi District, Ethiopia," *Forest Policy and Economics*, vol. 9, no. 8, pp. 916–927, 2007.

[10] G. Thondhlana, P. Vedeld, and S. Shackleton, "Natural resource use, income and dependence among San and Mier communities bordering Kgalagadi Transfrontier Park, southern Kalahari, South Africa," *International Journal of Sustainable Development and World Ecology*, vol. 19, no. 5, pp. 460–470, 2012.

[11] W. Cavendish, "Empirical regularities in the poverty-environment relationship of rural households: evidence from Zimbabwe," *World Development*, vol. 28, no. 11, pp. 1979–2003, 2000.

[12] E. Sjaastad, A. Angelsen, P. Vedeld, and J. Bojö, "What is environmental income?" *Ecological Economics*, vol. 55, no. 1, pp. 37–46, 2005.

[13] M. Fisher, "Household welfare and forest dependence in southern Malawi," *Environment and Development Economics*, vol. 9, no. 1, pp. 135–154, 2004.

[14] C. M. Shackleton, S. E. Shackleton, E. Buiten, and N. Bird, "The importance of dry woodlands and forests in rural livelihoods and poverty alleviation in South Africa," *Forest Policy and Economics*, vol. 9, no. 5, pp. 558–577, 2007.

[15] B. Babulo, B. Muys, F. Nega et al., "Household livelihood strategies and forest dependence in the highlands of Tigray, Northern Ethiopia," *Agricultural Systems*, vol. 98, no. 2, pp. 147–155, 2008.

[16] M. Appiah, D. Blay, L. Damnyag, F. K. Dwomoh, A. Pappinen, and O. Luukkanen, "Dependence on forest resources and tropical deforestation in Ghana," *Environment, Development and Sustainability*, vol. 11, no. 3, pp. 471–487, 2009.

[17] J. C. Tieguhong and E. M. Nkamgnia, "Household dependence on forests around Lobeke National Park, Cameroon," *International Forestry Review*, vol. 14, no. 2, pp. 196–212, 2012.

[18] T. Yemiru, A. Roos, B. M. Campbell, and F. Bohlin, "Forest incomes and poverty alleviation under participatory forest management in the bale highlands, Southern Ethiopia," *International Forestry Review*, vol. 12, no. 1, pp. 66–77, 2010.

[19] J. A. Timko, P. O. Waeber, and R. A. Kozak, "The socio-economic contribution of non-timber forest products to rural livelihoods in Sub-Saharan Africa: knowledge gaps and new directions," *International Forestry Review*, vol. 12, no. 3, pp. 284–294, 2010.

[20] S. P. Kar and M. G. Jacobson, "NTFP income contribution to household economy and related socio-economic factors: lessons from Bangladesh," *Forest Policy and Economics*, vol. 14, no. 1, pp. 136–142, 2012.

[21] D. Abebaw, H. Kassa, G. T. Kassie, M. Lemenih, B. Campbell, and W. Teka, "Dry forest based livelihoods in resettlement areas of Northwestern Ethiopia," *Forest Policy and Economics*, vol. 20, pp. 72–77, 2012.

[22] B. Teshome, H. Kassa, Z. Mohammed, and C. Padoch, "Contribution of dry forest products to household income and determinants of forest income levels in the Northwestern and Southern Lowlands of Ethiopia," *Natural Resources*, vol. 06, no. 05, pp. 331–338, 2015.

[23] K. Heubach, R. Wittig, E.-A. Nuppenau, and K. Hahn, "The economic importance of non-timber forest products (NTFPs) for livelihood maintenance of rural west African communities: a case study from northern Benin," *Ecological Economics*, vol. 70, no. 11, pp. 1991–2001, 2011.

[24] M. R. Nielsen, M. Pouliot, and R. K. Bakkegaard, "Combining income and assets measures to include the transitory nature of poverty in assessments of forest dependence: evidence from the Democratic Republic of Congo," *Ecological Economics*, vol. 78, pp. 37–46, 2012.

[25] G. Thondhlana, S. Shackleton, and E. Muchapondwa, "Kgalagadi Transfrontier Park and its land claimants: a pre-and post-land claim conservation and development history," *Environmental Research Letters*, vol. 6, no. 2, Article ID 024009, pp. 1–12, 2011.

[26] M. Lemenih, T. Abebe, and M. Olsson, "Gum and resin resources from some *Acacia, Boswellia* and *Commiphora* species and their economic contribution in Liben, Southeastern Ethiopia," *Journal of Arid Environment*, vol. 56, pp. 146–166, 2003.

[27] A. Eshete, D. Teketay, and H. Hakan, "The socio-economic importance and status of populations of *Boswellia papyrifera* (Del.) Hochst. In Northern Ethiopia: the case of north gonder zone," *Forests, Trees and Livelihoods*, vol. 15, no. 1, pp. 55–74, 2005.

[28] M. Lemenih and F. Bonger, "Dry forests of Ethiopia and their silviculture," in *Silviculture in the Tropics, Tropical Forestry*, S. Günter, M. Weber, B. Stimm, and R. Mosandi, Eds., pp. 261–272, Springer, Berlin, Germany, 2011.

[29] A. Worku, M. Lemenih, M. Fetene, and D. Teketay, "Socio-economic importance of gum and resin resources in the dry woodlands of Borana, southern Ethiopia," *Forests Trees and Livelihoods*, vol. 20, no. 2-3, pp. 137–156, 2011.

[30] B. Babulo, B. Muys, F. Nega et al., "The economic contribution of forest resource use to rural livelihoods in Tigray, Northern Ethiopia," *Forest Policy and Economics*, vol. 11, no. 2, pp. 109–117, 2009.

[31] HWPDO, *Hammer Woreda Pastoralist Development Office. Different Reports and Archives in the Office*, Dimeka, Hammer Woreda, Ethiopia, 2012.

[32] A. Terefe, A. Ebro, and T. Zewedu, "Rangeland dynamics in South Omo zone of Southern Ethiopia: assessment of rangeland condition in relation to altitude and grazing types," *Livestock Research for Rural Development*, vol. 22, no. 10, pp. 1–12, 2010.

[33] T. A. Belay, Ø. Totland, and S. R. Moe, "Woody vegetation dynamics in the rangelands of lower Omo region, Southwestern Ethiopia," *Journal of Arid Environments*, vol. 89, pp. 94–102, 2013.

[34] W. Cavendish, "Qualitative methods for estimating the economic value of resource use to rural households," in *Uncovering the Hidden Harvest*, B. M. Campbell and M. K. Luckert, Eds., pp. 17–63, Earthscan, London, UK, 2002.

[35] P. Kamanga, P. Vedeld, and E. Sjaastad, "Forest incomes and rural livelihoods in Chiradzulu District, Malawi," *Ecological Economics*, vol. 68, no. 3, pp. 613–624, 2009.

[36] B. Debela, *Contribution of Non-timber forest products to the rural household economy: gore District, Southwestern Ethiopia [M.S. thesis]*, Hawassa University Wondo Genet College Of Forestry and Natural Resources, Ethiopia, Ethiopia, 2004.

[37] P. Vedeld, A. Angelsen, E. Sjaastad, and G. K. Berg, *Counting on the Environment: Forest Incomes and the Rural Poor*, Environmental Economics Series Paper No. 98, The World Bank, Washington, DC, USA, 2004.

[38] A. Worku, J. Pretzsch, H. Kassa, and E. Auch, "The significance of dry forest income for livelihood resilience: the case of the pastoralists and agro-pastoralists in the drylands of southeastern Ethiopia," *Forest Policy and Economics*, vol. 41, pp. 51–59, 2014.

[39] Z. Mekonnen, A. Worku, T. Yohannes, T. Bahru, T. Mebratu, and D. Teketay, "Economic contribution of gum and resin resources to household livelihoods in selected regions and the national economy of Ethiopia," *Ethnobotany Research & Applications*, vol. 11, pp. 273–288, 2013.

[40] A. Mohammed, W. Tadesse, and A. Yadessa, "Counting on forests: non-timber forest products and their role in the households and national economy in Ethiopia," in *Commercialization of Ethiopian Agriculture. Proceedings of the 8th Annual Conference of Agricultural Economics Society of Ethiopia, February 24–26, 2005*, pp. 179–196, Agricultural Economics Society of Ethiopia (AESE), Addis Ababa, Ethiopia, 2006.

[41] B. Desalegn, "Assessment of the effect of ant (Dorylus fulvus) on honeybee colony (A. mellifera) and their products in west and South-West Shewa Zones, Ethiopia," *Ethiopian Journal of Animal Production*, vol. 7, no. 1, pp. 12–26, 2007.

[42] A. Shiferaw, M. Jaleta, B. Gebremedhin, and D. Hoekstra, *Increasing Economic Benefit from Apiculture through Value Chain Development Approach: The Case of Alaba Special District, Southern Ethiopia*, IPMS Ethiopia, 2010.

[43] C. Kinati, T. Tolemariam, K. Debele, and T. Tolosa, "Opportunities and challenges of honey production in Gomma district of Jimma zone, South-West Ethiopia," *Journal of Agricultural Extension and Rural Development*, vol. 4, no. 4, pp. 85–91, 2012.

[44] A. Asfaw, M. Lemenih, H. Kassa, and Z. Ewnetu, "Importance, determinants and gender dimensions of forest income in eastern highlands of Ethiopia: the case of communities around Jelo Afromontane forest," *Forest Policy and Economics*, vol. 28, pp. 1–7, 2013.

[45] T. Chilongo, "Livelihood strategies and forest reliance in Malawi," *Forests Trees and Livelihoods*, vol. 23, no. 3, pp. 188–210, 2014.

[46] M. Tilahun, L. Vranken, B. Muys et al., *Rural Households' Demand for Frankincense Forest Conservation in Tigray: A Contingent Valuation Analysis*, Bioeconomics Working Paper Series, Working Paper No. 2, Division of Bioeconmics Department of Earth and Environmental Sciences University of Leuven Geo-Institute, Leuven, Belgium, 2012.

[47] M. Lemenih, H. Kassa, G. T. Kassie, D. Abebaw, and W. Teka, "Resettlement and woodland management problems and options: a case study from North-Western Ethiopia," *Land Degradation and Development*, vol. 25, no. 4, pp. 305–318, 2014.

[48] D. Tsegaye, P. Vedeld, and S. R. Moe, "Pastoralists and livelihoods: a case study from northern Afar, Ethiopia," *Journal of Arid Environments*, vol. 91, pp. 138–146, 2013.

[49] E. Bognetteau and O. Wirtu, "Certification and market development for NTFPs: an option for sustainability," NTFP Research & Development in South-West Ethiopia Project, Policy Briefing Note 3, 2007.

[50] U. Narain, S. Gupta, and K. van't Veld, "Poverty and the environment: exploring the relationship between household incomes, private assets, and natural assets," *Land Economics*, vol. 84, no. 1, pp. 148–167, 2008.

[51] A. Mohamed, *The contribution of non-timber forest products to rural livelihood in southwest Ethiopia [M.S. thesis]*, Wageningen University and Research Center, Wageningen, The Netherlands, 2007.

[52] W. M. Fonta, H. E. Ichoku, and E. Ayuk, "The distributional impacts of forest income on household welfare in rural Nigeria," *Journal of Economics and Sustainable Development*, vol. 2, no. 2, pp. 1–13, 2011.

[53] P. D. McElwee, "Forest environmental income in Vietnam: household socioeconomic factors influencing forest use," *Environmental Conservation*, vol. 35, no. 2, pp. 147–159, 2008.

[54] A. Gebrelibanos, *Farmer's perception and adoption of integrated striga management technology in tahatay adiabo woreda, Tigray, Ethiopia [M.S. thesis]*, Haramaya University, 2006.

[55] E. Melaku, Z. Ewnetu, and D. Teketay, "Non-timber forest products and household incomes in Bonga forest area, southwestern Ethiopia," *Journal of Forestry Research*, vol. 25, no. 1, pp. 215–223, 2014.

6

Bark Thickness Equations for Mixed-Conifer Forest Type in Klamath and Sierra Nevada Mountains of California

Nickolas E. Zeibig-Kichas, Christopher W. Ardis, John-Pascal Berrill, and Joseph P. King

Department of Forestry and Wildland Resources, Humboldt State University, 1 Harpst St. Arcata, CA 95521, USA

Correspondence should be addressed to Nickolas E. Zeibig-Kichas; nek75@humboldt.edu

Academic Editor: Mark Finney

We studied bark thickness in the mixed-conifer forest type throughout California. Sampling included eight conifer species and covered latitude and elevation gradients. The thickness of tree bark at 1.37 m correlated with diameter at breast height (DBH) and varied among species. Trees exhibiting more rapid growth had slightly thinner bark for a given DBH. Variability in bark thickness obscured differences between sample locations. Model predictions for 50 cm DBH trees of each species indicated that bark thickness was ranked *Calocedrus decurrens* > *Pinus jeffreyi* > *Pinus lambertiana* > *Abies concolor* > *Pseudotsuga menziesii* > *Abies magnifica* > *Pinus monticola* > *Pinus contorta*. We failed to find reasonable agreement between our bark thickness data and existing bark thickness regressions used in models predicting fire-induced mortality in the mixed-conifer forest type in California. The fire effects software systems generally underpredicted bark thickness for most species, which could lead to an overprediction in fire-caused tree mortality in California. A model for conifers in Oregon predicted that bark was 49% thinner in *Abies concolor* and 37% thicker in *Pseudotsuga menziesii* than our samples from across California, suggesting that more data are needed to validate and refine bark thickness equations within existing fire effects models.

1. Introduction

There is interest in predicting fire-caused tree mortality in places where prescribed fire or wildfires are common [1–4]. Heat from flames or smoldering duff at a tree's base can kill trees, especially those with thinner bark [5–8]. Species-specific bark thickness equations are central to fire effects models such as FOFEM (First Order Fire Effects Model) [9, 10] and FFE-FVS (Fire and Fuels Extension for the Forest Vegetation Simulator) [1, 11, 12].

Tree bark plays a critical role in reducing mortality from fire. Bark protects living cambial tissues from external biotic and abiotic forces [14–17]. Different tree species exhibit distinct strategies in growth and the development of defense features with some allocating proportionally more resources to bark development than others [18–21]. However, there are many factors that can influence the formation of bark and little information exists comparing this trait across geographic gradients [22, 23]. The properties and function of bark are a result of complex evolutionary strategies by these organisms to perform more efficiently and competitively within their native ranges [21, 22].

Bark is comprised of various tissues covering the stem, branches, and roots of woody plants. It is found outside the secondary xylem and includes the inner living phloem and dead outer tissue [5, 24]. Inner bark is produced directly by the secondary cambium and consists of secondary phloem tissues [18]. Outer bark, also known as the rhytidome, is composed of periderm, cortical, and phloem tissue [25]. Bark plays an important physiological role in protecting trees from the environment and infectious microorganisms as well as containing mechanical injuries [18, 26]. Bark thickness (BT) is the most important characteristic for cambial protection from fire, more so than other bark properties like density, moisture content, or structure [14, 20, 27]. Trees with thicker bark are more likely to survive wildfire events. Thick bark provides an insulating layer of protection from heat for the underlying vascular tissues which can prevent cambial girdling [6, 28–31].

Standard fire-caused tree mortality models use BT, derived from species-specific BT equations, along with percent crown scorch as predictors of tree mortality [1, 9, 11, 12]. However, despite its importance in fire-induced tree mortality modeling, there are few studies assessing BT across a range of species and locations [32]. Studies analyzing external factors influencing BT are limited [32]. Site quality and soil fertility have been considered [33], but site quality cannot be easily altered by forest management. Measures of tree vigor, such as annual radial growth rates, can be associated with reduced likelihood of tree mortality during disturbances such as wildfire [29]. Competition can be reduced to enhance tree vigor but with an unknown influence on BT.

While there are studies correlating BT to tree diameter [34–37], we did not find any that tested for the influence of tree growth and vigor on BT. Therefore, we sought to test whether vigorous rapidly growing trees (in terms of crown ratio or recent growth rate) might allocate more or less resources to bark production. Also unknown is whether BT is an adaptation to fire that differs among areas with different climates and fire regimes. Therefore, we used regression analysis to test for such effects and compared our BT data and best BT models against existing BT models implemented within fire effects models. Our objectives were to

(i) examine how BT relates to measures of tree size [diameter at breast height (DBH)], recent growth rates (GR), vigor [crown ratio (CR)], and crown position [crown class (CC)];

(ii) quantify BT variation among species along a latitudinal gradient;

(iii) develop BT prediction models and compare their predictions against published BT models for the mixed-conifer forest type.

2. Materials and Methods

2.1. Study Sites. Data for the mixed-conifer forest type were collected from Klamath, Tahoe, and Sequoia National Forests and the Stanislaus-Tuolumne Experimental Forest in California. Sampling at these locations provided a latitudinal gradient across the range of mixed-conifer species (Figure 1). Eight species were sampled for BT: white fir (*Abies concolor*) [ABCO], red fir (*Abies magnifica*) [ABMA], incense-cedar (*Calocedrus decurrens*) [CADE], lodgepole pine (*Pinus contorta*) [PICO], Jeffrey pine (*Pinus jeffreyi*) [PIJE], sugar pine (*Pinus lambertiana*) [PILA], western white pine (*Pinus monticola*) [PIMO], and Douglas-fir (*Pseudotsuga menziesii*) [PSME]. Climate and geology varied among the study sites (Table 1). The Klamath National Forest (KNF) is recognized as one of America's most biologically diverse regions [38, 39]. It is situated in a transitional region between hotter and drier areas to the south and colder, wetter climate to the north [40]. Unlike our other study sites, KNF has the Shasta red fir variety (*Abies magnifica* var. *shastensis*). The Tahoe National Forest (TNF) data were collected in the Blackwood Creek watershed of the Lake Tahoe Basin Management Unit. This area has a Mediterranean continental climate with warm, dry

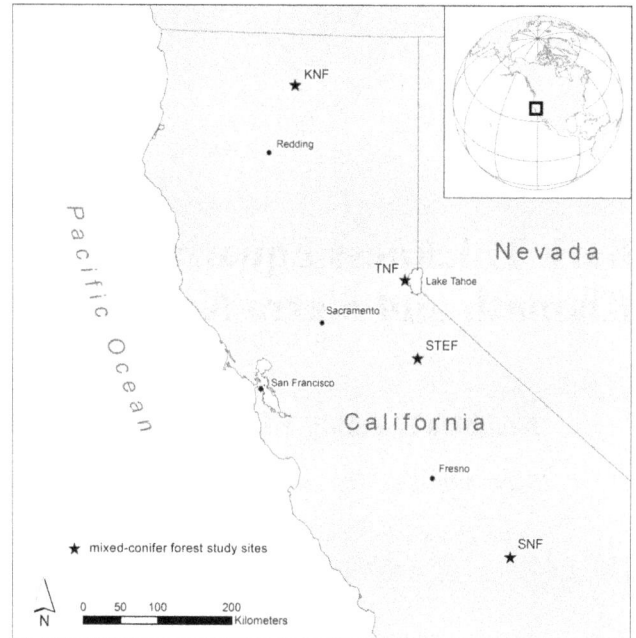

FIGURE 1: Bark thickness sampling in California. Stars show sample locations at Klamath National Forest (KNF), Tahoe National Forest (TNF), Stanislaus-Tuolumne Experimental Forest (STEF), and Sequoia National Forest (SNF). Major cities are shown for reference.

summers and cold winters with most precipitation falling as snow. The Stanislaus-Tuolumne Experimental Forest (STEF) is located near the town of Pinecrest, California. It is a mixed-conifer forest of high site quality on the western slope of the Sierra Nevada, at lower elevation than the other two Sierra Nevada sites (TNF and SNF). Climate of the region is characterized by warm, dry summers and cold, wet winters, with over half of annual precipitation falling as snow between December and March. The Sequoia National Forest (SNF) sampling was conducted in the Bull Run Creek watershed which has characteristic expansive areas of exposed rock, particularly at higher elevations. The area experiences warm to hot, dry summers and cool to cold, wet winters. Summer brings occasional thunderstorms but most precipitation is in the form of snow falling from October through April.

2.1.1. Field Data Collection. At KNF, TNF, and SNF, trees were sampled along transects spanning an elevation gradient. Every 100 m the closest tree was sampled, followed by one tree of each species in understory and overstory positions, giving data for a range of tree sizes and stand densities. Transects were not straight; they climbed and traversed the slope from bottom to top of each watershed, running approximately parallel to the main creek and avoiding road corridors. Trees below five cm DBH (diameter at breast height), noticeably unhealthy trees, and malformed trees were not sampled. Bark thickness, recent radial growth, and diameter measurements were taken at a height of 1.37 m on each sample tree. Total height and live crown base height were also measured to calculate live crown ratio (CR). Crown class was recorded for each sample tree.

TABLE 1: Description of study sites at Klamath National Forest (KNF), Tahoe National Forest (TNF), Stanislaus-Tuolumne Experimental Forest (STEF), and Sequoia National Forest (SNF). Average temperature is reported as the mean temperature during winter and summer months, respectively. Soil types include parent material (PM). Species sampled: Douglas-fir (PSME), red fir (ABMA), white fir (ABCO), incense-cedar (CADE), sugar pine (PILA), western white pine (PIMO), Jeffrey pine (PIJE), and lodgepole pine (PICO).

Site	Location	Average temp. (°C)	Annual precip. (mm)	Soil type and parent materials	Species sampled and sample size (number of trees)	Elevation range of sampling (m)
KNF	41.5003°N 123.3333°W	24–41	559	Gravely clay loam PM: basic igneous, metamorphic, and altered sedimentary rocks	ABCO (53), ABMA (48), CADE (29), PILA (17), PIMO (6), PSME (30)	1508–1868
TNF	39.5625°N 120.5625°W	6–23	1,397	Gravely loam PM: andesite and volcanic rock	ABCO (39), ABMA (53), PICO (29), PIJE (20), PIMO (23)	2013–2369
STEF	38.1677°N 120.0°W	0–17	940	Sandy to fine sandy loam PM: granite and diorite	ABCO (365), CADE (221), PIJE (51), PILA (189)	1820–1948
SNF	37.4167°N 119.1667°W	7–26	660	Sandy clay loam PM: sedimentary, granite, and granodiorite	ABCO (40), ABMA (37), CADE (16), PIJE (10), PILA (6)	2066–2499

Two BT measurements were taken using a handheld Swedish bark gauge at approximately 90 degrees apart around the tree circumference. Bark thickness was measured from the wood surface to the contour of the diameter tape wrapped snugly around the tree [41]. One shallow increment core was collected at the site of BT measurements. The collective width of the most recent five complete rings measured to the nearest 0.1 mm gave tree growth (GR) in terms of a five-year periodic average annual radial increment (although the series was not crossdated so it is possible that increment was over/underestimated due to missing/false tree rings).

We obtained independently collected data for BT and DBH in ABCO, CADE, PIJE, and PILA at the STEF site in the central Sierra Nevada (Andrew Slack, Humboldt State University, personal communication). Here, sample trees were nearest neighbors of individual PILA trees randomly selected throughout the forest for a different study. Growth and crown ratio were not measured at STEF.

2.2. Analysis. We used multiple linear regression and non-linear regression to model BT for each species at three study sites with growth and crown ratio data (KNF, TNF, and SNF). Individual variables were either square root- or log-transformed to reduce skewness in data distributions. We compared regression models with and without candidate predictor variables representing tree size (DBH), growth (GR, converted from radial increment to a basal area increment), and tree vigor (CR). The y-intercept was forced through the origin because a tree with zero DBH is exactly 1.37 m tall and essentially has zero BT at that height (the tip) but begins to develop bark at breast height as the tree grows taller. We used a second-order correction for Akaike information criterion (AICc) in model selection as this takes into account sample size by increasing the relative penalty for model complexity with small data sets [42]. Models with delta AICc < 2 were treated as similar, among which the most parsimonious

model was favored. We also calculated average BT prediction error (mm) in terms of root-mean-square error (RMSE) as an indicator of model performance.

The large sample size for ABCO and ABMA allowed for investigation of geographic location (north, central, and southern latitudes), as well as crown class (dominant, codominant, intermediate, and suppressed) influence on BT. Dummy variables were included in nonlinear regressions to test for differences in BT between categorical variables of site and crown class.

We examined performance of existing, widely used BT models by comparing predictions from these BT models against our California BT models and data. Specifically, we compared our California BT data against predictions from diameter inside bark equations for ABCO, CADE, PILA, and PSME in southwest Oregon [13] as well as those embedded in the Fire and Fuels Extension (FFE) of the Forest Vegetation Simulator (FVS) [11, 12]. We did not validate another commonly used fire mortality model, FOFEM (First Order Fire Effects Model) version 6.3.1, because the same bark thickness equations were also embedded in the Fire and Fuels Extension (FFE) version 2.0. Prediction errors were calculated in percent terms for each tree in our dataset (percent error = 100 × (predicted-actual)/predicted). We used R [43] and SPSS [44] to analyze data.

3. Results and Discussion

Sample trees covered a broad range of tree sizes, crown ratio, and growth (Supplementary File, Table S1, in Supplementary Material available online at http://dx.doi.org/10.1155/2016/1864039). Sample size across the KNF, TNF, and SNF sites differed among species. ABCO and ABMA were encountered most frequently along the sample transects. The independent dataset from STEF added records for four species, including very large PILA, as well as CADE, PIJE, and ABCO (Supplementary File, Table S2).

TABLE 2: Comparison of bark thickness (BT; mm) as a function of DBH (cm) across latitudinal gradient (north, central, south) and among crown classes (dominant, codominant, intermediate, and suppressed) for red fir (ABMA) and white fir (ABCO) at Klamath National Forest (KNF), Tahoe National Forest (TNF), and Sequoia National Forest (SNF). Coefficients and fit statistics for region and crown class dummy variable (d) in nonlinear regression.

Species	Region/crown class	Coefficient (d)	s.e.	Appr. 95% confidence limits	
				Lower	Upper
ABMA ($n = 138$)					
$\sqrt{BT} = 0.900\sqrt{DBH}^{d}$ (RMSE = 6.22 mm)	North (KNF)	0.875	0.032	0.812	0.938
	Central (TNF)	0.941	0.031	0.880	1.003
	South (SNF)	0.917	0.030	0.858	0.976
$\sqrt{BT} = 0.801\sqrt{DBH}^{d}$ (RMSE = 6.75 mm)	Dominant	0.964	0.042	0.881	1.047
	Codominant	0.973	0.051	0.872	1.074
	Intermediate	1.012	0.059	0.896	1.128
	Suppressed	1.041	0.068	0.906	1.176
ABCO ($n = 132$)					
$\sqrt{BT} = 1.005\sqrt{DBH}^{d}$ (RMSE = 6.15 mm)	North (KNF)	0.856	0.056	0.745	0.967
	Central (TNF)	0.888	0.029	0.831	0.945
	South (SNF)	0.865	0.028	0.810	0.921
$\sqrt{BT} = 0.878\sqrt{DBH}^{d}$ (RMSE = 6.20 mm)	Dominant	0.924	0.036	0.853	0.995
	Codominant	0.947	0.043	0.861	1.033
	Intermediate	0.964	0.051	0.863	1.065
	Suppressed	0.996	0.060	0.877	1.116

There was a positive trend of increasing tree size (DBH) and BT, although the BT of some conifers varied widely for any given tree size. In general, nonlinear relationships best explained our empirical data with the exception of PSME and CADE where simpler linear models were adopted. Incorporating tree vigor (in terms of recent growth, GR) improved model predictions of bark thickness for four mixed-conifer species, ABCO, ABMA, PICO, and PIJE, indicating that faster growth came at the expense of BT (Supplementary Files, Tables S3 and S4). However, in practice the small differences in prediction errors (8% overall average reduction in RMSE; range 4%–16% reduction by species) indicated that including GR as a predictor of BT only gave marginal improvements over the simplified models with only DBH as a parameter (Supplementary File, Table S3).

We found no significant difference in BT for ABMA or ABCO along the latitude gradient of northern (KNF), central (TNF), and southern (SNF) sample locations. On average, the *shastensis* variety of ABMA had thinner bark for a given tree size than ABMA along the Sierra Nevada. Bark thickness was slightly but not significantly greater at TNF than further south at SNF (Table 2) where mean GR for ABMA was 3.18 mm yr^{-1} (i.e., 31% faster). ABCO exhibited a similar but less pronounced trend of thicker bark at TNF and thinner bark at KNF but variability in BT at each site prevented detection of significant differences. Variability in BT was greater among ABMA than ABCO sample trees. For a given tree size in ABMA or ABCO, average BT by crown class was ranked as follows: suppressed > intermediate > codominant > dominant; however, these differences were

not statistically significant but were consistent with negative coefficients for GR indicating that faster-growing trees had thinner bark (Supplementary File, Table S3). It should be noted that there are several other factors beyond latitudinal variation which could contribute to the observed differences in BT, such as past management and disturbance regimes, as well as variations in climate and site quality between each of the study locations.

Independently collected BT data from STEF exhibited greater BT for a given DBH than BT at the three National Forest sites. In comparison with STEF data, predictions from our simple BT-DBH models for the KNF, TNF, and SNF locations (models shown in Supplementary File, Table S3) revealed that predicted BT was 10.5% less for ABCO, 20.9% less for CADE, 15.1% less for PIJE, and 21.2% less for PILA than the BT data for the STEF site. To increase the geographic range of applicability of our BT-DBH models, we merged the data from KNF, TNF, SNF, and STEF sites and fitted final models to this expanded dataset (Table 3 and Figure 2). The STEF dataset included BT for larger CADE, PIJE, and PILA than the other sites. Our final models indicated that BT among large-sized (100 cm DBH) California mixed-conifers was thickest for CADE and thinnest for PIMO and PICO (Figure 3). Modeled averages for BT in sample trees above 150 cm DBH ranked PSME > ABCO > PILA. Expected bark thickness (i.e., modeled average) for a 50 cm DBH tree fell into three groupings, where CADE, PIJE, and PILA had relatively thick bark, ABCO, PSME, and ABMA had intermediate BT, and PIMO and PICO had relatively thin bark. Our California models can be used to estimate BT

TABLE 3: Mixed-conifer forest type bark thickness (BT) models for red fir (ABMA), lodgepole pine (PICO), western white pine (PIMO), and Douglas-fir (PSME) at three sites: Klamath National Forest (KNF), Tahoe National Forest (TNF), and Sequoia National Forest (SNF), and models for white fir (ABCO), Jeffrey pine (PIJE), sugar pine (PILA), and incense-cedar (CADE) fitted to data from four sites including Stanislaus-Tuolumne Experimental Forest (STEF). Models predict square root of bark thickness in mm, as a function of DBH (cm).

Data	Model		Coefficient	s.e.	Pr > |t|	RMSE (mm)
ABCO $(n = 497)$	$\sqrt{BT} = a * \sqrt{DBH}^{b}$	a	1.005	0.031	<0.0001	7.46
		b	0.888	0.016	<0.0001	
ABMA $(n = 138)$	$\sqrt{BT} = a * \sqrt{DBH}^{b}$	a	0.886	0.060	<0.0001	6.88
		b	0.919	0.034	<0.0001	
PIJE $(n = 81)$	$\sqrt{BT} = a * \sqrt{DBH}^{b}$	a	1.298	0.109	<0.0001	10.29
		b	0.802	0.041	<0.0001	
PICO $(n = 29)$	$\sqrt{BT} = a * \sqrt{DBH}^{b}$	a	1.027	0.104	<0.0001	1.49
		b	0.603	0.057	<0.0001	
PILA $(n = 212)$	$\sqrt{BT} = a * \sqrt{DBH}^{b}$	a	1.521	0.116	<0.0001	14.50
		b	0.718	0.034	<0.0001	
PIMO $(n = 29)$	$\sqrt{BT} = a * \sqrt{DBH}^{b}$	a	1.299	0.156	<0.0001	4.24
		b	0.609	0.059	<0.0001	
PSME $(n = 30)$	$\sqrt{BT} = a * \sqrt{DBH}$	a	0.785	0.015	<0.0001	7.16
CADE $(n = 266)$	$\sqrt{BT} = a * \sqrt{DBH}$	a	0.946	0.009	<0.0001	14.18

TABLE 4: Performance of bark thickness (BT) models in the Fire and Fuels Extension for the Forest Vegetation Simulator (FFE-FVS) [11, 12] and Larsen & Hann [13] models for BT in Oregon applied to BT and DBH data for California mixed-conifers. Comparing BT data for Klamath National Forest (KNF), Tahoe National Forest (TNF), Stanislaus-Tuolumne Experimental Forest (STEF), and Sequoia National Forest (SNF) in California against FFE-FVS and Oregon BT model predictions in terms of percent difference between predicted BT and actual BT data calculated as 100 × (predicted-actual)/predicted. Negative percentage indicates underprediction by the FFE-FVS or Oregon models.

Species	FFE-FVS BT models [11, 12]	Oregon BT models [13]
White fir (ABCO)	−47.7%	−49.2%
Incense-cedar (CADE)	−55.2%	−14.4%
Jeffrey pine (PIJE)	−20.1%	—
Sugar pine (PILA)	−0.5%	11.2%
Douglas-fir (PSME)	6.9%	37.2%
Western white pine (PIMO)	−17.3%	—
Lodgepole pine (PICO)	−5.4%	—
Red fir (ABMA)	−53%	—

using forest inventory data and quantify diameter inside bark at breast height. The minimum and maximum DBH for each species define the range of application of BT models (Supplementary File, Tables S1 and S2).

The percent differences between our data and predicted BT from the FFE-FVS [11, 12] or Oregon [13] BT models indicated that these models generally underpredicted BT in California mixed-conifers, most noticeably for ABCO, ABMA, CADE, and PIJE (Table 4, Figure 4). Prediction errors were greatest among smaller trees, where underprediction was common. When compared against California BT data, the FFE-FVS model underpredicted BT for CADE, ABMA, and ABCO of all sizes. For PICO, PILA, and PIMO trees in California, the FFE-FVS model underpredicted BT for smaller trees and overpredicted BT for larger trees. The Oregon models overpredicted BT for PILA and PSME, especially among larger PILA and smaller PSME trees (Figure 4).

Underpredicted bark thickness for most species suggested that the FFE-FVS fire effects model may overestimate

fire severity in California. This finding is consistent with validation of the postfire tree mortality models in Hood et al. [1] where mortality was overpredicted for many of the same species that we studied. The models for conifers in Oregon indicated that bark was 49% thinner in *Abies concolor* and 37% thicker in *Pseudotsuga menziesii* than our sample from across California. These important differences and subtler differences according to location and growth rate suggested that more BT data are needed to validate and, if needed, refine equations within existing fire effects models or develop local or regional model variants. Overall, our findings suggest that further study and revision of BT models implemented within fire models are warranted for mixed-conifer forests in California.

4. Conclusion

Bark thickness correlated with DBH but varied among conifer species. Our modeling indicated that, among 50 cm DBH

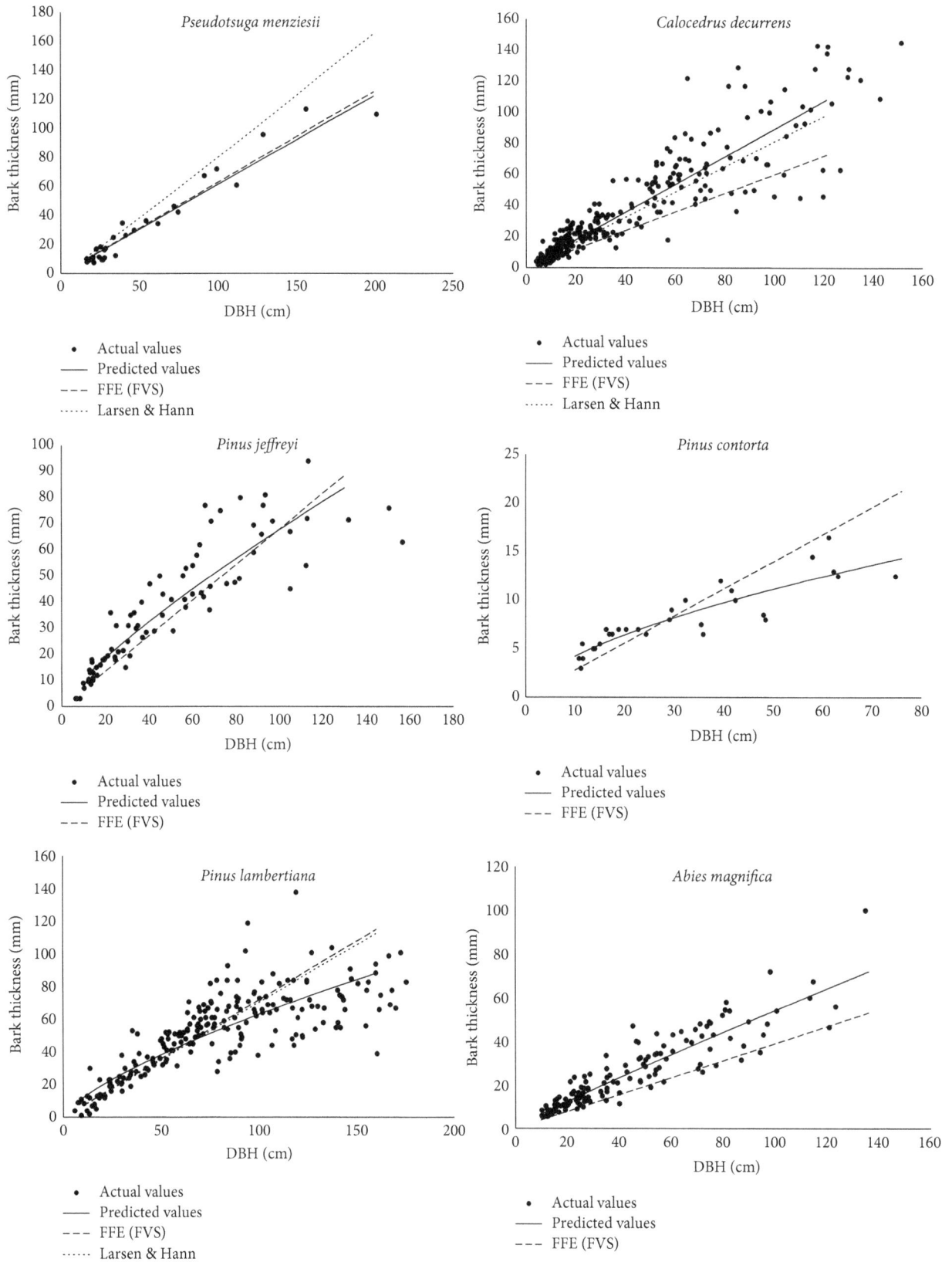

FIGURE 2: Continued.

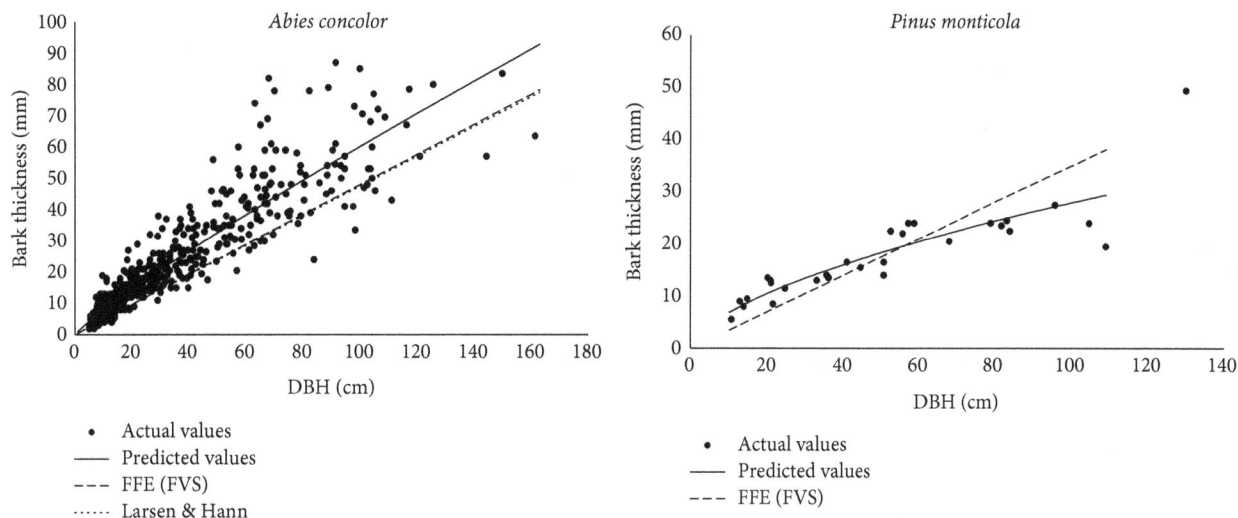

FIGURE 2: Bark thickness data and models (predicted values) for eight species within the mixed-conifer forests of California. Data were collected across a latitudinal gradient: north (Klamath National Forest), central (Tahoe National Forest and Stanislaus-Tuolumne Experimental Forest), and south (Sequoia National Forest). Dashed lines denote predictions from Fire and Fuels Extension for the Forest Vegetation Simulator (FFE-FVS) [11, 12] or Larsen & Hann [13] models for BT in Oregon. Note scale differences in x- and y-axes for each species.

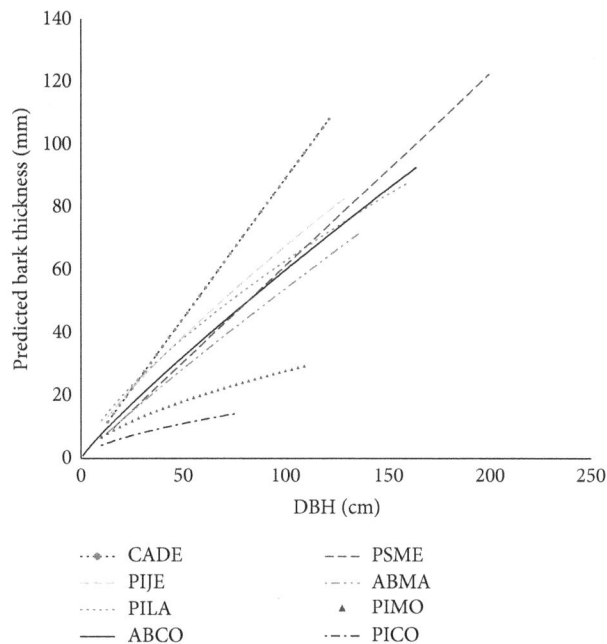

FIGURE 3: Comparing California mixed-conifer forest bark thickness model predictions for eight conifer species at Klamath National Forest (KNF), Tahoe National Forest (TNF), Stanislaus-Tuolumne Experimental Forest (STEF), and Sequoia National Forest (SNF) in California. Species (and codes): Douglas-fir (PSME), red fir (ABMA), white fir (ABCO), incense-cedar (CADE), sugar pine (PILA), western white pine (PIMO), Jeffrey pine (PIJE), and lodgepole pine (PICO).

conifers, BT ranked CADE > PIJE > PILA > ABCO > PSME > ABMA > PIMO > PICO. We did not detect regional differences in BT nor differences between crown class and only slight differences according to recent tree radial growth. These findings suggest that our linear and nonlinear models of BT-DBH have general application within California. We failed to find reasonable agreement between our newly developed BT model predictions and most existing BT models currently used to model fire-induced mortality in the mixed-conifer forest type in California.

Competing Interests

The authors declare that they have no competing interests.

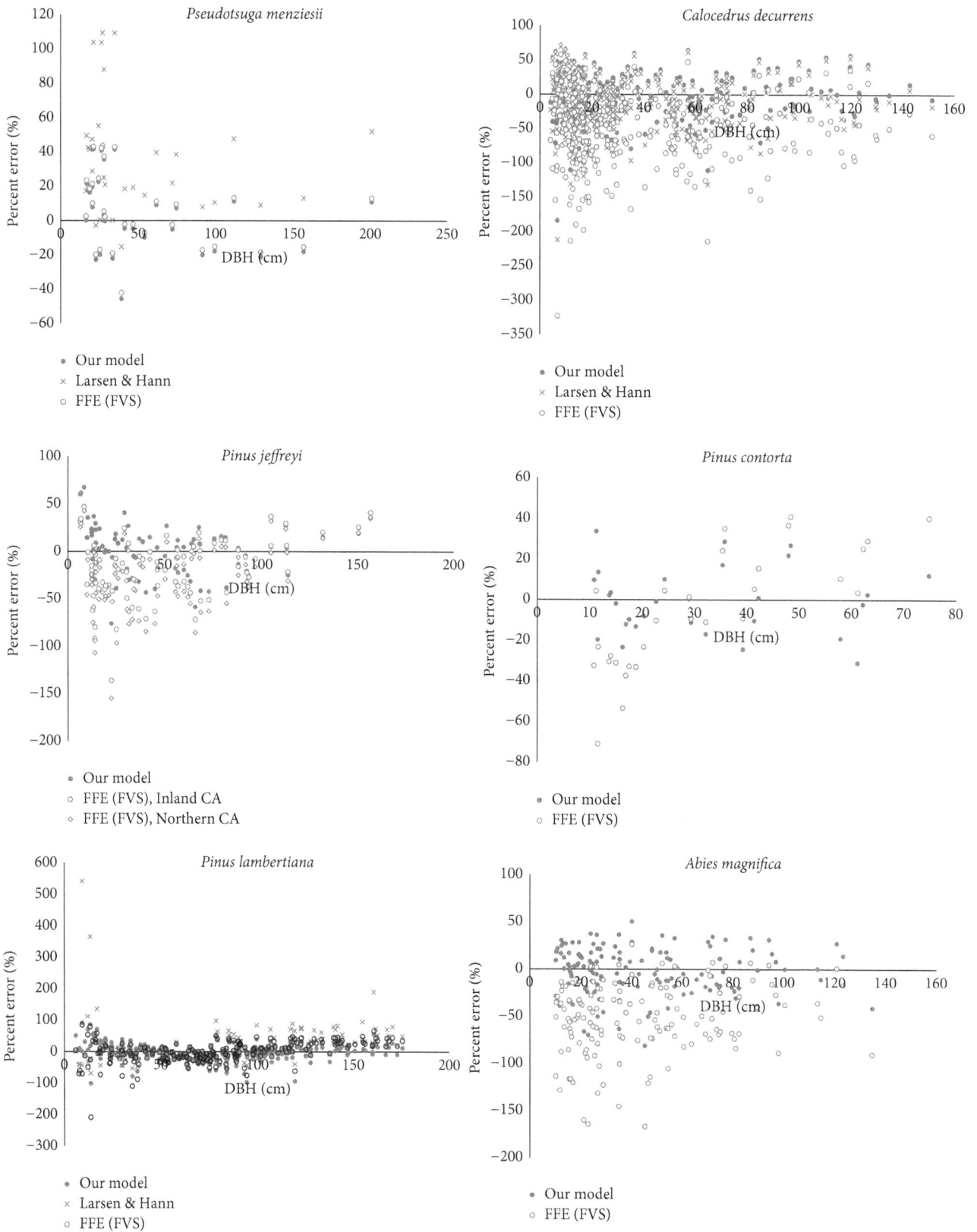

FIGURE 4: Continued.

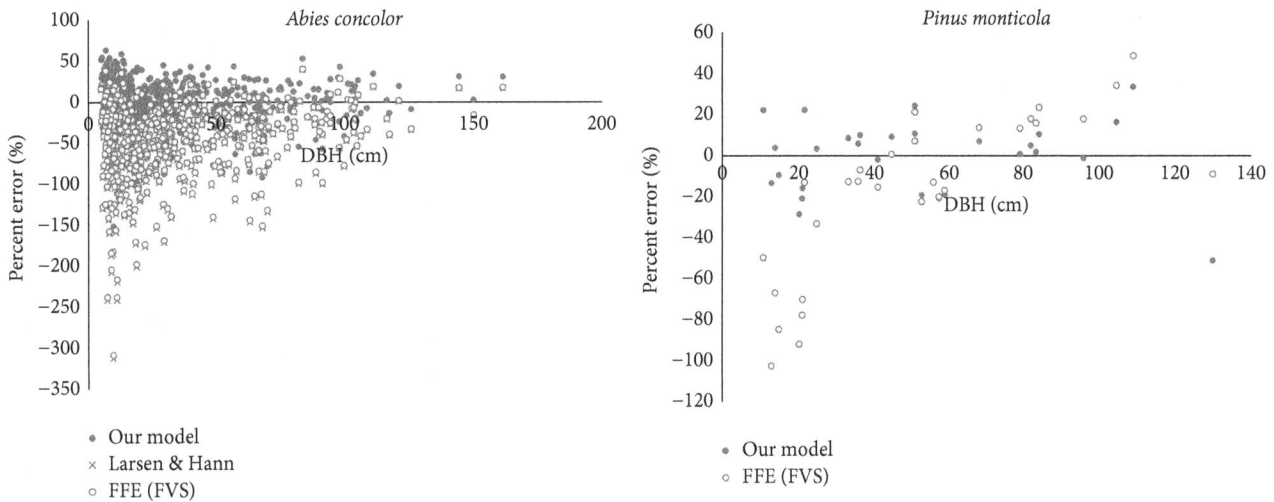

FIGURE 4: Bark thickness model prediction errors according to tree size (DBH) for eight species within the mixed-conifer forests of California. Prediction errors calculated in percent terms for our California models, for FFE-FVS (Fire and Fuels Extension for the Forest Vegetation Simulator) [11, 12], and for the Larsen & Hann [13] bark thickness models for Oregon, where percent error is the difference between predicted BT and actual BT data calculated as $100 \times$ (predicted-actual)/predicted.

Authors' Contributions

NEZ-K summarized and analyzed data and prepared figures. CWA assisted with data summary and performed validations. J-PB designed sampling, assisted with data analysis, and obtained funding. JPK assisted with data summary and validation. All authors contributed to writing of the manuscript.

Acknowledgments

Andrew Slack provided the independent bark thickness dataset and valuable advice. Scott Burdette collected field data. This work was supported in part by the USDA National Institute of Food and Agriculture, McIntire Stennis Cooperative Research Program.

References

[1] S. M. Hood, C. W. McHugh, K. C. Ryan, E. Reinhardt, and S. L. Smith, "Evaluation of a post-fire tree mortality model for western USA conifers," *International Journal of Wildland Fire*, vol. 16, no. 6, pp. 679–689, 2007.

[2] C. Hull Sieg, J. D. McMillin, J. F. Fowler et al., "Best predictors for postfire mortality of ponderosa pine trees in the Intermountain West," *Forest Science*, vol. 52, no. 6, pp. 718–728, 2006.

[3] J. L. Jones, B. W. Webb, B. W. Butler et al., "Prediction and measurement of thermally induced cambial tissue necrosis in tree stems," *International Journal of Wildland Fire*, vol. 15, no. 1, pp. 3–17, 2006.

[4] P. E. Dennison, S. C. Brewer, J. D. Arnold, and M. A. Moritz, "Large wildfire trends in the western United States, 1984–2011," *Geophysical Research Letters*, vol. 41, no. 8, pp. 2928–2933, 2014.

[5] K. W. Spalt and W. E. Reifsnyder, *Bark Characteristics and Fire Resistance: A Literature Survey*, Southern Forest Experiment Station, USDA Forest Service in cooperation with School of Forestry, Yale University, New Haven, Conn, USA, 1962.

[6] S. T. Michaletz and E. A. Johnson, "How forest fires kill trees: a review of the fundamental biophysical processes," *Scandinavian Journal of Forest Research*, vol. 22, no. 6, pp. 500–515, 2007.

[7] J. M. Varner, F. E. Putz, J. J. O'Brien, J. K. Hiers, R. J. Mitchell, and D. R. Gordon, "Post-fire tree stress and growth following smoldering duff fires," *Forest Ecology and Management*, vol. 258, no. 11, pp. 2467–2474, 2009.

[8] D. H. Hammond, J. M. Varner, J. S. Kush, and Z. Fan, "Contrasting sapling bark allocation of five southeastern USA hardwood tree species in a fire prone ecosystem," *Ecosphere*, vol. 6, no. 7, pp. 1–13, 2015.

[9] E. D. Reinhardt, R. E. Keane, and J. K. Brown, "First Order Fire Effects Model: FOFEM 4.0, User's Guide," Tech. Rep. INT-GTR-344, USDA Forest Service, Ogden, Utah, 1997.

[10] D. C. Lutes and R. E. Keane, *First Order Fire Effects Model: FOFEM 6.3, User's Guide*, USDA Forest Service, Missoula, Mont, USA, 2016.

[11] E. D. Reinhardt, N. L. Crookston, S. J. Beukema et al., "The fire and fuels extension to the forest vegetation simulator," Tech. Rep. RMRS-GTR-116, USDA Forest Service, Fort Collins, Colo, USA, 2003.

[12] E. D. Reinhardt, N. L. Crookston, S. J. Beukema et al., "Addendum to the fire and fuels extension to the forest vegetation simulator," Tech. Rep. RMRS-GTR-116, UDA Forest Service, Fort Collins, Colo, USA, 2009.

[13] D. R. Larsen and D. W. Hann, *Equations for Predicting Diameter and Squared Diameter Inside Bark at Breast Height for Six Major Conifers of Southwest Oregon*, Oregon State University College of Forestry, Corvallis, Ore, USA, 1985.

[14] B. Odhiambo, M. Meincken, and T. Seifert, "The protective role of bark against fire damage: a comparative study on selected introduced and indigenous tree species in the Western Cape, South Africa," *Trees*, vol. 28, no. 2, pp. 555–565, 2014.

[15] I. Roth, *Structural Patterns of Tropical Barks*, Schweizerbart Borntraeger, Stuttgart, Germany, 1981.

[16] J. L. Jones, B. W. Webb, B. W. Butler et al., "Prediction and measurement of thermally induced cambial tissue necrosis in

tree stems," *International Journal of Wildland Fire*, vol. 15, no. 1, pp. 3–17, 2006.

[17] R. L. Edmonds, J. K. Agee, and R. I. Gara, *Forest Health and Protection*, Waveland Press, Long Grove, Ill, USA, 2011.

[18] A. R. Biggs, W. Merrill, and D. D. Davis, "Discussion: response of bark tissues to injury and infection," *Canadian Journal of Forest Research*, vol. 14, no. 3, pp. 351–356, 1984.

[19] J. F. Jackson, D. C. Adams, and U. B. Jackson, "Allometry of constitutive defense: a model and a comparative test with tree bark and fire regime," *The American Naturalist*, vol. 153, no. 6, pp. 614–632, 1999.

[20] P. van Mantgem and M. Schwartz, "Bark heat resistance of small trees in Californian mixed conifer forests: testing some model assumptions," *Forest Ecology and Management*, vol. 178, no. 3, pp. 341–352, 2003.

[21] D. W. Schwilk, M. S. Gaetani, and H. M. Poulos, "Oak bark allometry and fire survival strategies in the Chihuahuan Desert Sky Islands, Texas, USA," *PLoS ONE*, vol. 8, no. 11, Article ID e79285, 2013.

[22] J. G. Pausas, "Bark thickness and fire regime," *Functional Ecology*, vol. 29, no. 3, pp. 315–327, 2015.

[23] J. Laasasenaho, T. Melkas, and S. Aldén, "Modelling bark thickness of *Picea abies* with taper curves," *Forest Ecology and Management*, vol. 206, no. 1–3, pp. 35–47, 2005.

[24] S. G. Pallardy, *Physiology of Woody Plants*, Elsevier, Boston, Mass, USA, 2008.

[25] W. C. Dickison, *Integrative Plant Anatomy*, Academic Press, San Diego, Calif, USA, 2000.

[26] V. R. Franceschi, P. Krokene, E. Christiansen, and T. Krekling, "Anatomical and chemical defenses of conifer bark against bark beetles and other pests," *New Phytologist*, vol. 167, no. 2, pp. 353–376, 2005.

[27] A. Wesolowski, M. A. Adams, and S. Pfautsch, "Insulation capacity of three bark types of temperate Eucalyptus species," *Forest Ecology and Management*, vol. 313, pp. 224–232, 2014.

[28] J. Gignoux, J. Clobert, and J.-C. Menaut, "Alternative fire resistance strategies in savanna trees," *Oecologia*, vol. 110, no. 4, pp. 576–583, 1997.

[29] P. J. van Mantgem, N. L. Stephenson, L. S. Mutch, V. G. Johnson, A. M. Esperanza, and D. J. Parsons, "Growth rate predicts mortality of *Abies concolor* in both burned and unburned stands," *Canadian Journal of Forest Research*, vol. 33, no. 6, pp. 1029–1038, 2003.

[30] W. A. Hoffmann, R. Adasme, M. Haridasan et al., "Tree topkill, not mortality, governs the dynamics of savanna-forest boundaries under frequent fire in central Brazil," *Ecology*, vol. 90, no. 5, pp. 1326–1337, 2009.

[31] J. J. Midgley, M. J. Lawes, and S. Chamaillé-Jammes, "Savanna woody plant dynamics: the role of fire and herbivory, separately and synergistically," *Australian Journal of Botany*, vol. 58, no. 1, pp. 1–11, 2010.

[32] C. E. T. Paine, C. Stahl, E. A. Courtois, S. Patiño, C. Sarmiento, and C. Baraloto, "Functional explanations for variation in bark thickness in tropical rain forest trees," *Functional Ecology*, vol. 24, no. 6, pp. 1202–1210, 2010.

[33] K. L. Dolph, "Nonlinear equations for predicting diameter inside bark at breast height for young-growth red fir in California and southern Oregon," Res. Note. PSW-RN-409, USDA Forest Service, Berkeley, Calif, USA, 1989.

[34] G. S. Biging, "Taper equations for second-growth mixed conifers of Northern California," *Forest Science*, vol. 30, no. 4, pp. 1103–1117, 1984.

[35] E. L. Amidon, "A general taper functional form to predict bole volume for five mixed-conifer species in California," *Forest Science*, vol. 30, no. 1, pp. 166–171, 1984.

[36] K. C. Ryan and E. D. Reinhardt, "Predicting postfire mortality of seven western conifers," *Canadian Journal of Forest Research*, vol. 18, no. 10, pp. 1291–1297, 1988.

[37] G. E. Hengst and J. O. Dawson, "Bark properties and fire resistance of selected tree species from the central hardwood region of North America," *Canadian Journal of Forest Research*, vol. 24, no. 4, pp. 688–696, 1994.

[38] D. A. DellaSala, S. B. Reid, T. J. Frest, J. R. Strittholt, and D. M. Olson, "A global perspective on the biodiversity of the Klamath-Siskiyou ecoregion," *Natural Areas Journal*, vol. 19, no. 1, pp. 300–319, 1999.

[39] S. T. E. Cheng, "Forest service research natural areas in California," Tech. Rep. PSW-GTR-188, USDA Forest Service, Albany, Calif, USA, 2004.

[40] C. G. Parks, S. R. Radosevich, B. A. Endress et al., "Natural and land-use history of the northwest mountain ecoregions (USA) in relation to patterns of plant invasions," *Perspectives in Plant Ecology, Evolution and Systematics*, vol. 7, no. 3, pp. 137–158, 2005.

[41] C. Mesavage, "Measuring bark thickness," *Journal of Forestry*, vol. 67, no. 10, pp. 753–754, 1969.

[42] K. P. Burnham and D. R. Anderson, *Model Selection and Multimodel Inference: A Practical Information-Theoretic Approach*, Springer, New York, NY, USA, 2nd edition, 2002.

[43] R Core Team, *R: A Language and Environment for Statistical Computing*, R Foundation for Statistical Computing, Vienna, Australia, 2012.

[44] IBM Corporation, *SPSS Statistics for Windows, Version 22.0*, IBM, Armonk, NY, USA, 2013.

7

Synchrony in Leafing, Flowering, and Fruiting Phenology of *Senegalia senegal* within Lake Baringo Woodland, Kenya: Implication for Conservation and Tree Improvement

Stephen F. Omondi,[1,2] **David W. Odee,**[1] **George O. Ongamo,**[2] **James I. Kanya,**[2] **and Damase P. Khasa**[3]

[1]*Kenya Forestry Research Institute, P.O. Box 20412, Nairobi 00200, Kenya*
[2]*School of Biological Sciences, University of Nairobi, P.O. Box 30197, Nairobi 00100, Kenya*
[3]*Centre for Forest Research and Institute for Systems and Integrative Biology, Laval University, Sainte-Foy, QC, Canada G1V 0A6*

Correspondence should be addressed to Stephen F. Omondi; stephenf.omondi@gmail.com

Academic Editor: Kihachiro Kikuzawa

Leafing, flowering, and fruiting patterns of *Senegalia senegal* were studied over a period of 24 months from January 2014 to December 2015. The phenological events of the species are bimodal and follow the rainfall patterns. The leafing phase starts during the onset of rains and lasts for 18 weeks. New leaves continued to appear on the new shoots while old leaves persisted to the leaf fall period. Flowering event takes 12 weeks and is concentrated in the months of high relative humidity (April and October) with one-month peak flowering period. Fruiting phase starts at the peak of the rainy seasons (May and November) and peaks in June and December. This phase lasted for 14 weeks. The fruits mature towards the end of the rainy season (January/February and July/August). The fruits open for dispersal mainly in February/March and September during the peak dry season. High synchrony index (SI) was found in leafing (SI: 0.87), flowering (SI: 0.75), and fruiting (SI: 0.85) events among the populations. Temperature, precipitation, and soil moisture content were significantly correlated with the phenological events. Significant variations in floral morphology and fruits traits were also evident. Seed collections should be undertaken in the months of January/February and July/August.

1. Introduction

Phenology is often an overlooked aspect of plant ecology, from the scale of individual species to whole ecosystems [1]. However, phenological studies provide knowledge about the patterns of plant growth and development as well as the effects of the environment and selective pressures on flowering and fruiting behavior [2]. Additionally, flowering of certain plants signals agronomic time and changing phenological patterns may also indicate climate change [3]. Detailed investigations of these events can improve understanding of the strong effects of anthropogenic and environmental factors on life-forms in nature and hence be able to facilitate conservation efforts. Temperature and photoperiods, which are reliable signals of seasons, are probably among the best studied environmental factors [4]. Accurate detection of such environmental cues by plants and the resulting plastic response would enable reproduction to occur when climatic conditions are most suitable. Thus, resources and conditions impose bottom-up selective forces on phenology [3]. In simple terms, for plant reproduction, timing is everything [5]. An individual plant that flowers too early, before it has had time to accumulate sufficient material resources, will have a limited capacity for seed production [2]. Conversely, one that delays flowering might gain higher capacity but might also run out of time to use it before the end of the season.

As emphasized by Okullo et al. [6], biologists have begun examining how phenological patterns are influencing reproductive successes of tree species. The flowering phenology of individual trees (duration of flowering as well as the pattern of flowering intensity) varies continuously between extremes [5]. At one extreme are species with individual trees producing large numbers of new flowers each day over a short period (a week or less), while species with flowering-individuals in the population that produce small numbers of new flowers almost daily for many weeks are at the opposite extreme [7]. Understanding such occurrences is useful in planning improvement programmes and conservation strategies. The need for recognizing and accounting for phenological development in plants in relation to ecological studies has been reported by Abu-Asab et al. [8]. These studies provide information on functional rhythms of plants and plant communities, where the timing of various phenological events may reflect biotic and/or abiotic environmental conditions. These studies are also important from the point of view of the conservation of tree genetic resources and forestry management as well as for a better understanding of plant species and community level interactions. While few studies have focused on African dryland tree species, studies undertaken so far indicate that, in the tropical savanna, some species produce leaves and flowers before the onset of the wet season while others do so after the onset of the season [9–11]. Although such information is limited, knowledge of how these events happen is very important in tree production management strategies and more so for keystone species such as S. senegal.

Senegalia senegal begins its reproductive phenology at relatively young age, mostly at the age of three years at ideal environment [12]. The event, however, varies between populations but these have been reported to occur soon after or just before the rains [11]. Although the phenological data are scarce, the flowering pattern in East Africa is quit variable due to the bimodal rainfall patterns [12]. In some parts of Kenya, the flowering has been observed to occur shortly after the rains and leaf flush but no detailed study has been undertaken to document these events [13]. This includes Lake Baringo ecosystem where commercial exploitation and farmland adoption of the species is viable. Such information is important in conservation and sustainable management of the species [6].

Throughout its distribution in Kenya, S. senegal population has undergone various levels of anthropogenic disturbances and habitat fragmentations more so within Lake Baringo ecosystem [14]. These disturbances have modified the plant communities therein and put pressure on natural regeneration and evolutionary potential of the species [15]. These include the biological processes such as mating systems (pollination and fruit setting) and gene flow. Such disturbed populations will therefore require conservation and management practices that promote sustainable utilization [16]. Execution of viable local conservation strategies based on reproductive processes for S. senegal within the ecosystem is therefore required before the species ecological integrity is lost. This is basically because the success of the species will largely depend on its ability to achieve both its vegetative and its reproductive growths. The aim of the present study was to investigate phenological events in S. senegal within Lake Baringo woodland ecosystem in relation to environmental cues at the individual and population levels. This involved

(1) investigation, interpretation, and documentation of the phenoevents; (2) determining the timing of the phenoevents; (3) establishing the relationship between the phenoevents of the species with climatic variables.

2. Materials and Methods

2.1. Study Site. Lake Baringo woodland ecosystem is found between 035°35′ E, 00°16′ N and 036°00′ E, 00°42′ N. Within the woodland, four S. senegal populations, namely, Kimalel, Kampi Ya Samaki, Lake Bogoria, and Tangulbei, were selected for this study (Figure 1). The ecosystem is characterized by many small hills but majorly step faulted rift valley floor with significant soil type variations [17]. The Tangulbei site exhibits recent volcanic soils while the areas around Kampi Ya Samaki, Kimalel, and Bogoria have deep red soils with high fertility. Some areas show poorly drained with moderately deep to deep clay soils. The temperatures are fairly hot to warm with the mean annual temperatures ranging between 22 and 24°C. The mean minimum and maximum temperature range from 16 to 18°C and 28 to 30°C, respectively. The ecosystem falls under semiarid ecological zone with mean annual rainfall ranging between 450 and 900 mm and mean annual potential evapotranspiration ranging between 1650 and 2300 mm. The vegetation within the ecosystem is majorly bushland and *Acacia* woodland [17].

2.2. Study Species. Senegalia senegal (L.) Willd. (syn. *Acacia senegal*) is an indigenous African *Acacia* species that plays an important role in the dryland economy [18]. The species grows up to 15 m tall and is valued mainly for gum arabic production. Senegalia senegal also play significant role in agricultural production through enhancement of soil fertility and environmental amelioration [12]. The species is also important to the local communities during the dry season as a source of fodder for livestock [19]. Gum arabic is produced by the species through response to injuries caused by animals or incisions by gum collectors. The gum is traded locally and internationally for use in the pharmaceutical, beverage, ink, and lithographic industries as stabilizer and encapsulation agent [18, 20]. Senegalia senegal is widely spread in tropical and subtropical Africa, from South Africa northwards to Sudan [18]. In Kenya, the species grows in the coastal region to the northern parts through rift valley, in dry *Acacia-Commiphora* bushlands [21]. However, commercial exploitation is majorly in the northern and rift valley populations [20]. High densities and sometimes pure stands of this species have been observed in some parts of Turkana and Baringo counties [20, 22].

2.3. Study Design and Recording of Phenological Events. To document the phenological diversity and synchrony/asynchrony within the woodland, 100 m × 100 m temporary plots were established at Kimalel, Kampi Ya Samaki, Lake Bogoria, and Tangulbei populations. These populations represent the wider species distribution range within the woodland. Documentation was conducted in two consecutive years (24 months, January 2014 to end of January 2016). Thirty reproductively mature individual trees (>5 cm girth and having

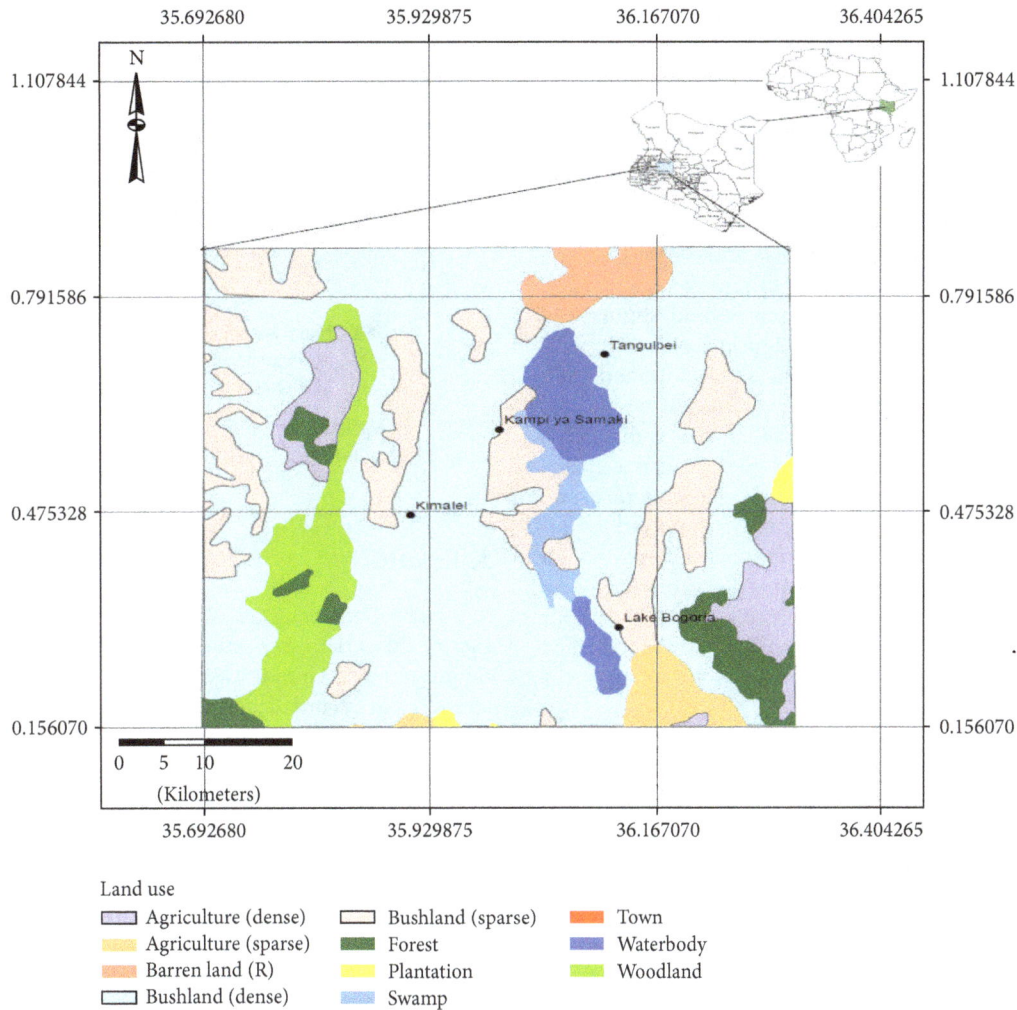

FIGURE 1: Map of Lake Baringo forest ecosystem showing study sites and land use patterns.

a sign of previous-year seed production) per population were selected for the study. Four branches (the branches were distributed in the northern, western, southern, and eastern sides of the tree) per tree were marked and assessed at fortnight intervals. The branches were assessed on one-metre length from the tip. During the assessment, leafing, flowering, and fruiting processes were scored visually. Leaf flush initiation, leaf flush completion, leaf fall initiation, leaf fall completion, leafless period, initiation of flowering, completion of flowering, time lag between start of vegetative (first-leaf flush) and reproductive phases (first-visible flower), initiation of fruiting, completion of fruiting, fruit-fall initiation, and completion of fruit fall were determined. For each of the 120 individual trees monitored, a separate phenological record was maintained. The starting date of a phenophase was assigned to the monitoring date when structures on one or more branches were observed to have entered that phase. The end for the phenophase was assigned to the monitoring date when no branch was observed carrying structure in the phase. At the population level, for each monitoring year, peak flowering and fruiting were used to

refer to the months in which the number of individuals observed in that phenophase reached a maximum. During the same period rainfall, soil moisture content, temperature, and relative humidity data were collected. The data were then summarized into monthly values.

2.4. Floral Morphology. Study of the floral morphology followed the protocol described by Nghiem et al. [23] with some modifications. The flowering peak season was chosen and reproductively mature trees used during the phenology study were sampled for this study. During this season, 30 flower inflorescences per tree were collected at anthesis. The flowers were fixed in methanol: acetic acid solution (3 : 1) for 4 hours and then the solution was replaced by 70% ethanol for transportation to the laboratory. In the laboratory, length of each flower inflorescence was measured (mm). The number of flowers per inflorescence was counted and each flower was observed under binocular dissecting microscope to score for presence or absence of a fully developed pistil. The flower and style lengths were measured for all the flowers.

Another 30 flowers per tree were softened and cleared in a sodium hydroxide solution (0.8 N NaOH) for 10 minutes in an oven held at 60°C and stained in aniline blue for 30 minutes in readiness for dissection. The style and ovary were separated and the ovary was divided into two halves. Following the procedures described by Martin [24], each ovary was placed in a drop of glycerol and viewed by fluorescence microscope under UV light and the number of ovules was counted. The diameters of 30 stigmas and 30 polyads per inflorescence were also determined using light microscopy. The images were then digitally captured and stigma and polyad dimensions were measured using Axiovision 3.1 software. Pollen from 30 inflorescences was collected from the same 30 trees per population and dried in desiccators containing silica gel for 3 hours and then sieved through a stainless steel sieve of 63 mm aperture mesh. Pollen was then placed onto a growth medium of 1% agar, 20% sucrose, and 0.01% boric acid at 26°C for determination of polyad germination and number of pollen tubes per polyad. Germination percentages were recorded by examining three replicates of ~300 polyads per tree by light microscope after 4-hour incubation. A polyad was scored as having germinated when the length of at least one pollen tube was longer than the polyad diameter.

Thirty pods were collected from the 30 trees per population when brown and beginning to dehisce. Each pod was put in a separate plastic sampling bag and the seeds were extracted inside the bag. The pods were measured for length and width. All seeds within the pod were examined and classed either as undeveloped with an empty or wrinkled appearance or as fully developed with a normal filled appearance. The number of undeveloped and developed seeds per pod was recorded. Length and width of the developed seeds were measured using electronic caliper. The seed samples were also weighed to calculate mean seed weight for each combination.

2.5. Data Analyses. Phenology events of the populations were summarized by recording occasions separately and for the four populations combined. Various intra- and interpopulation synchrony indices including leaf development, flower formation, and fruit developments were determined as described by Devineau [25]. Synchrony index was determined as the ratio between the mean individual duration of a phenological phase and the overall duration of the phase. The totals of the different individuals of each population for leafing, flowering, and fruiting were calculated for each month. The ratio of the number of phenological observations to the total number of observed trees provided percentages in each stage. Spearman's rank correlation was used to establish any correlation between phenological events with total monthly rainfall, mean maximum and minimum air temperature, soil water content, and mean relative humidity. Univariate analysis of the inflorescence, flower, pods, and seed parameters was performed. One-way analyses of variance were used to test differences among the populations and their significance was tested through Fisher's least significance difference (LSD). All data were analyzed using GenStat 16th edition software.

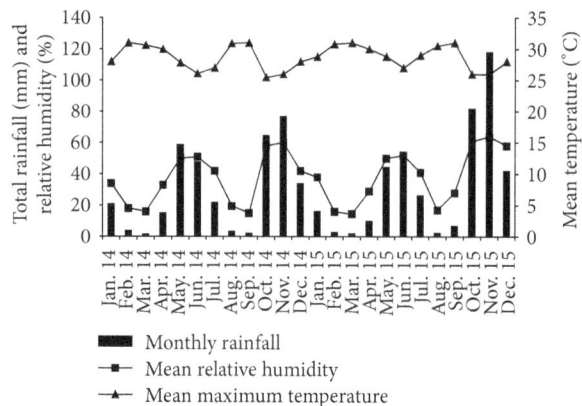

FIGURE 2: Variation in total monthly rainfall, mean maximum temperature, and mean soil moisture content.

3. Results

3.1. Environmental Cues. During the two years of study, significant variation in monthly precipitation was reported ranging from 2.3 mm to 118 mm, although no significant difference was recorded among the populations. The rainfall distribution mainly followed the typical bimodal pattern with the months with higher precipitation being May/June and October/November (Figure 2). Overall, the annual total rainfall differed between the two years with the year 2015 recording more rainfall amounts (650 mm) than the year 2014 (582 mm). The mean temperatures varied significantly and followed the rainfall patterns with the rainy months recording lower temperatures than months with no rains. Mean daily maximum temperatures ranged between 26°C and 31°C and the monthly distributions are as shown in Figure 2. The mean monthly minimum temperatures are relatively constant and falling between 15°C and 20°C. Mean daily relative humidity for each month ranged from 14.5 to 67%. It was observed that the higher the monthly rainfall, the higher the mean relative humidity and the lower the mean maximum atmospheric temperatures (Figure 2). The soil moisture content was high during the rainy months and low during the dry months. These trends were similar during the two years of the study.

Records of leafing, flowering, and fruiting for the two-year study were made for 120 trees (30 each for Tangulbei, Kampi Ya Samaki, Kimalel, and Lake Bogoria). In general, all the phenological phases were periodic and followed the weather patterns. Due to the two rainy seasons in each year, occurrences of two growth seasons per year are reported (Figure 3 and Table 2).

3.2. Leafing Phenology. Within Lake Baringo woodland, leaf initiation in *S. senegal* started with the emergence of leaf buds during the onset of precipitation. This occurred between the last week of September and the first week of October and again in the first week of May for both 2014 and 2015 in all the populations. These months correspond to the beginning of the short rainy season of September/October and the long rainy season of April/May (Figure 3 and Table 1).

TABLE 1: Phenological events of *S. senegal* within Lake Baringo woodland ecosystem for years 2014 and 2015.

Population	2014				2015			
	LI	LFI	PFL	PFR	LI	LFI	PFL	PFR
Tangulbei	Apr. (Sep.)	Aug. (Feb.)	May (Oct.)	Jun. (Dec.)	May (Sep.)	Oct. (Feb.)	Jun. (Oct.)	Jul. (Dec.)
Kampi Ya Samaki	Apr. (Sep.)	Aug. (Feb.)	May (Oct.)	Jun. (Dec.)	May (Sep.)	Oct. (Feb.)	Jun. (Oct.)	Jul. (Dec.)
Kimalel	May (Sep.)	Oct. (Feb.)	Jun. (Oct.)	Jul. (Dec.)	Jun. (Sep.)	Nov. (Feb.)	Jul. (Oct.)	Aug. (Dec.)
Lake Bogoria	May (Sep.)	Oct. (Feb.)	Jun. (Oct.)	Jul. (Dec.)	Jun. (Sep.)	Nov. (Feb.)	Jul. (Oct.)	Aug. (Dec.)

LI: leaf initiation; LFI: leaf fall initiation; PFL: peak flowering month; PFR: peak fruiting month.

TABLE 2: Correlation of *S. senegal* phenological events with climatic factors.

Environmental variables	Leaf initiation		Peak leaf fall		Peak flowering		Peak fruiting	
	Coeff.	P value	Coeff.	P value	Coeff.	P value	Coeff.	P value
Maximum daily temperatures (°C)	0.226	0.106	0.414	0.001**	0.140	0.076	0.212	0.044*
Monthly total precipitation (mm)	0.358	0.031*	−0.618	0.001**	0.347	0.021*	0.492	0.001**
Mean relative humidity (%)	0.116	0.091	−0.172	0.082	0.121	0.218	0.018	0.912
Mean soil moisture content (m³/m³)	0.488	0.001**	−0.322	0.001**	0.278	0.037*	0.398	0.001**

*Significant at $P < 0.05$; **significant at $P < 0.01$.

The leaf initiation started one week earlier in Tangulbei and Kampi Ya Samaki than in Kimalel and Lake Bogoria populations in both seasons of the year 2014. In the year 2015, the event occurred at the same time for all the populations. The leafing duration (leaf initiation to complete leaf fall) lasted for about 18 weeks in both the years and seasons; however, Lake Bogoria population had a shorter leafing duration (17 weeks) than the other populations. The peak leaf fall (complete leaf loss) was observed to occur during the month of March/April and September in both 2014 and 2015 coinciding with the hot and dry seasons (Figure 3 and Table 1). The leaf fall was followed by fresh leaf emergence at the beginning of the subsequent rainy season of May (long rains) and October (short rains). Individual trees did not show any significant difference in leaf shedding patterns between the years.

3.3. Flowering Phenology. The species had two peak seasons in each year for flowering and fruit production which occurred during the short and long rainy seasons. The flowering period in all the populations begun at the onset of the rainy season just immediately after the leaf flush. This phase lasted for about 12 weeks with a one-month peak during the seasons. The flowering occurred between October and November during the short rains and May and June during the long rains. Flowering intensity was similar in both the years although the intensity was lower during the short than the long rainy season. In the year 2014, the first floral buds were observed at the beginning of May and mid-October for Lake Bogoria and Kimalel populations, while, for Kampi Ya Samaki and Tangulbei populations, the first flower buds were observed in the mid of May and mid-October. In the year 2015, the flower initiation occurred during mid of May and beginning of October for all the populations. The peak flowering times were observed during the last week of June to the first week of July in 2014 and mid of July to end of July in 2015. This was also observed from the end of October to the beginning of November in both years.

3.4. Fruiting Phenology. Fruit development proceeded during the rainy season with pod initiation starting in the last week of June and continuing till July while the maturation of pods started in mid-July and end of August for the years 2014 and 2015, respectively. These periods were not significantly different among the populations despite few day differences. The peak fruiting month, when majority of the individual tree had many fruits, was in June/July and December for the year 2014 and in July/August and December for the year 2015 (Table 1). The variation between the years was majorly due to variations in the onset of the seasons. For the two years combined, the peak fruiting month occurred in the months of July and December (Figure 3). Fruiting phenophase generally lasted for about three months (12 weeks) although it lasted for about 14 weeks during the long rains in both years.

3.5. Correlation with Climatic Variable. Spearman's rank correlation between the number of individual trees in different phenophases and the climatic variables is as shown in Table 2. There was positive correlation between leaf initiation and total monthly rainfall ($r = 0.358$, $P < 0.05$) and mean soil moisture content ($r = 0.488$, $P = 0.001$). Peak leaf fall was positively correlated to mean maximum temperature ($r = 0.414$, $P = 0.001$) and negatively related to mean monthly total precipitation ($r = −0.618$, $P = 0.001$). The correlation between numbers of individual trees flowering each month (all plots combined) and monthly rainfall showed that flowering occurred more often in wet than dry months and was positively correlated to mean total monthly rainfall ($r = 0.347$, $P < 0.05$). There was also a positive relationship between flowering and mean soil moisture content ($r = 0.278$, $P < 0.05$). The peak fruiting season was strongly positively correlated to mean total monthly precipitation ($r = 0.492$, $P = 0.001$) and mean monthly soil moisture content ($r = 398$, $P = 0.001$) but weakly positively correlated to mean daily maximum temperature ($r = 0.212$, $P = 0.044$).

Population	J	F	M	A	M	J	J	A	S	O	N	D
Tangulbei	♠●	●	○	●♣	●♣♣♣	●♣♣♣	●♣	○	●	●♣	●♣♣♣	●♣♣♣
Kampi Ya Samaki	♠●	●	○	●♣	●♣♣♣	●♣♣♣	●♣	○	●	●♣	●♣♣♣	●♣♣♣
Kimalel	♠●	●	○	●♣	●♣♣♣	●♣♣♣	●♣	○	●	●♣	●♣♣♣	●♣♣♣
Lake Bogoria	♠●	●	○	●♣	●♣♣♣	●♣♣♣	●♣	○	●	●♣	●♣♣♣	●♣♣♣

● Leafing ♣♣ Peak flowering
○ Complete leaf fall ♣ Fruiting
♣ Flowering ♠♠ Peak fruiting

(a) Monthly phenological events

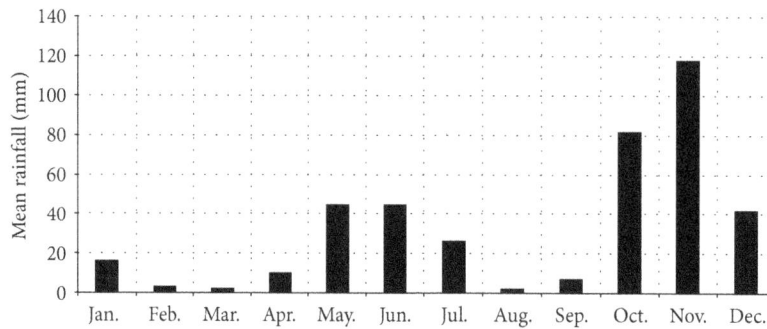

(b) Monthly rainfall data

FIGURE 3: (a) Monthly leafing, flowering, and fruiting events of *S. senegal* and (b) mean monthly rainfall amounts in Lake Baringo woodland ecosystem.

TABLE 3: Synchrony indices for phenological events of *S. senegal* within Lake Baringo woodland ecosystem.

Population	Synchrony index		
	Leafing	Flowering	Fruiting
Tangulbei	0.78	0.85	0.74
Kampi Ya Samaki	0.81	0.74	0.88
Kimalel	0.91	0.80	0.79
Lake Bogoria	0.85	0.78	0.82
Overall	*0.87*	*0.75*	*0.85*

The values are means of the two annual cycles.

3.6. Synchrony of Phenological Events. In determining the synchrony of phenological events of the individuals within and among populations, the results are as shown in Table 3. The leafing stage was synchronous within and among the populations with many individuals initiating leafing at the end of the dry season and the beginning of rainy season and initiating leaf fall at the beginning of the dry season for both of the years. The overall interpopulation synchrony ratio for leaf development was 0.87 (Table 3). The overall interpopulation synchrony ratio for flowering and fruiting phenology was 0.75 and 0.85, respectively (Table 3). Higher synchrony ratio indicates greater coincidence of the phase among individuals or sites.

3.7. Floral Morphology. The floral characteristics were assessed during the peak flowering seasons and the results are as shown in Table 4. The number of flowers per inflorescence ranged between 89 and 134 flowers with a grand mean of

92.7. Significant difference was recorded among populations ($F_{3,1247} = 23.53$; $P < 0.05$). The inflorescence length ranged from 5.9 to 6.2 with a grand mean of 6.1; however, no significant difference was observed among the populations. There were also significant differences in flower length ($F_{3,1827} = 20.66$; $P < 0.05$), stigma diameter ($F_{3,1827} = 9.19$; $P < 0.05$), style length ($F_{3,1827} = 3.96$; $P < 0.05$), and number of ovules per ovary ($F_{3,1208} = 21.28$; $P < 0.05$) among the populations. No significant difference was observed among the populations on inflorescence length (Table 4). High mean number of flowers per inflorescence was found in Kimalel population with the least number observed in Kampi Ya Samaki population. Generally, considering both years and all the seasons, Kimalel population registered the largest values in flower length, stigma diameter, and the mean number of ovules per ovary. The longest style length was observed in Kampi Ya Samaki population (Table 4).

3.8. Pollen Quality. There was no significant difference in polyad diameter among the populations. Tangulbei population showed larger polyad diameter compared to the other populations (Table 5). Pollen germination was as shown in Figure 4. Significant differences in pollen germination percentage were found among the populations ($F_{3,472} = 4.73$; $P < 0.05$) with lower germination rates reported for Lake Bogoria and Kimalel populations. Significant difference was also observed for stigma diameter among the populations ($F_{3,1827} = 9.19$; $P < 0.05$) with Tangulbei population showing large stigma diameter size compared to the other populations. In all the populations, stigma diameter was larger than the polyad diameter.

TABLE 4: Floral characteristics of *S. senegal* within Lake Baringo woodland ecosystem.

Population	FS	SL	FL	SD	STL	OPV
Kampi Ya Samaki	88.99[a]	6.12[a]	7.16[a]	0.24[a]	6.61[b]	4.49[a]
Lake Bogoria	87.68[a]	5.95[a]	7.21[ab]	0.25[a]	6.63[a]	4.43[a]
Tangulbei	89.28[a]	6.21[a]	7.33[c]	0.24[a]	6.63[ab]	4.76[b]
Kimalel	103.21[b]	6.25[a]	7.52[c]	0.26[b]	6.84[ab]	5.05[c]
Difference between years	ns	ns	ns	ns	ns	ns

FS: number of flowers per inflorescence; SL: inflorescence length (cm); FL: flower length (mm); SD: stigma diameter (mm); STL: style length (mm); OPV: number of ovules per ovary; ns: not significant; data followed by the same letter are not significantly different at 95% using Fisher's LSD test.

TABLE 5: Mean pollen germination percentage for the four populations of *S. senegal* within Lake Baringo woodland ecosystem.

Population	PD (mm)	SD (mm)	PG (%)
Kimalel	0.1651[a]	0.24[a]	55.80[ab]
Kampi Ya Samaki	0.1725[a]	0.25[a]	63.32[b]
Lake Bogoria	0.1748[a]	0.24[a]	52.33[a]
Tangulbei	0.1750[a]	0.26[b]	60.60[ab]

PD: polyad diameter; SD: stigma diameter; PG: pollen germination; data followed by the same letter are not significantly different at 95% using LSD test.

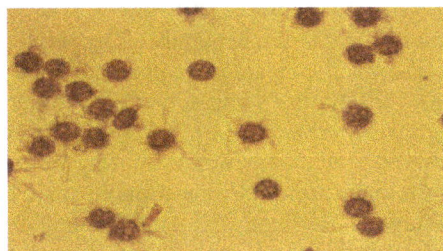

FIGURE 4: Examples of germinated and ungerminated pollen grains of *S. senegal*.

3.9. Pods and Seed Yield. In both years, the pods measured between 3.4 and 16.9 cm long and 1.1 and 3.2 cm wide with the means per population as shown in Table 6. Both the pod length ($F_{3,116}$ = 17.53; P < 0.05) and pod width ($F_{3,116}$ = 39.39; P < 0.05) were significantly different among the populations. Over all the period, Kampi Ya Samaki and Kimalel populations showed larger pod measurements than Tangulbei and Lake Bogoria populations. The number of seeds per pod varied between 2 and 6 per population but did not differ significantly among populations. However, most of the trees from Kimalel population recorded more number of seeds per pod compared to the other populations. The average percentage of developed seeds per pod per tree was similar in all the populations ranging from 65 to 95%. There were no significant differences between years for either trait. Significant difference on seed length ($F_{3,116}$ = 18.97; P < 0.05), width ($F_{3,116}$ = 28.95; P < 0.05), and weight ($F_{3,116}$ = 7.10; P < 0.05) was found among the populations with Lake Bogoria population recording lower values for both the traits

TABLE 6: Pod and seed characteristics of *S. senegal* within Lake Baringo woodland ecosystem.

Population	PL	PW	SPP	SW	FF (%)
Tangulbei	4.6[ab]	2.2[b]	4[a]	66[b]	72[a]
Kampi Ya Samaki	5.2[bc]	1.8[a]	5[a]	71[b]	81[a]
Kimalel	5.8[c]	2.6[c]	5[a]	64[ab]	88[a]
Lake Bogoria	3.9[a]	1.8[a]	3[a]	58[a]	93[a]
Difference among populations	*	*	ns	**	ns

**Significant at P < 0.01; *significant at P < 0.05; ns: nonsignificant; PL: pod length; PW: pod width; SPP: number of seeds per pod; SW: weight of 1000 seeds; FF: fully formed seeds; data followed by the same letter are not significantly different at 95% using Fisher's LSD test.

than the other populations. The average seed weight ranged between 42 and 76 g/1000 seeds and varied significantly among seasons with the long rainy seasons reporting heavier seeds per 1000 seeds than the short rainy seasons ($F_{1,119}$ = 12.9; P < 0.05). The seed weights did not differ significantly between the years.

4. Discussion

4.1. Leafing Phenology. Phenological investigations show that, in most tropical forests, rainfall is one of the most likely environmental changes controlling the periodicity of tree growth and flowering [2]. It is also generally believed that occurrence of rainfall after a period of drought or long dry spell usually initiates plant growth mainly in the dry forest ecosystems such as Lake Baringo woodland [26]. The present study has shown that *S. senegal* usually sheds most of its leaves during the dry season when the soil moisture content is very low and the atmospheric temperatures are high. As a defense mechanism to tolerate drought or the dry conditions, the species drop leaves and regain them during the rainy seasons [12]. With the start of the rains, the tree produces leaf buds that initiate the leafing phenophase. This phenomenon has been reported for many dry forest species [25]. Generally, the onset of rains improves soil moisture content that triggers the tree to begin growth. Once the first new leaves of the season have expanded, the production of leaf buds and young leaves continues constantly until the whole crown is covered with leaves. During this period, there is no distinguishable transition from old to new leaves. This may take between 16 and 18 weeks depending on the length of the rainy season. Complete leaf cover is achieved in the mid of the rainy season and correlated with high water availability both in terms of both rainfall amounts and soil water content (Table 2). This type of development of leaves is more closely connected to changing conditions in water availability than was observed for flowering or fruit production [6]. A high percentage of mature leaves are retained almost throughout the rainy season. The leaf formation and duration was observed to be synchronous within and among the four populations (Table 3). This may be due to the similarity of the dynamics of environmental conditions of the populations. It was noted during the present study that most of the environmental variables occur, generally, at the same time hence triggering the phenophases of the species in the four populations almost simultaneously. The leafing event was not different between the years. However, some small variations were found in terms of duration, which could be explained by variation in the durations of the environmental condition. For example the leafless period during the year 2014 was one week longer than the year 2015; however, the sequences of events were similar. Similar results were also reported for *Lagerstroemia speciosa* by Khanduri [2].

4.2. Flowering Phenology.

Senegalia senegal flowered during the rainy season which was similar to many other *Acacia* species within the woodland and other tropical species so far studied [26]. During the present study, *S. senegal* is reported to initiate flowering few weeks after the beginning of the rains when over 65% of the crown has been covered by new leaves. The peak flowering was actually realized at the peak rainy months which signifies the importance of precipitation to *S. senegal* during the flowering season. Most studies have reported that, for species that flowers during the rainy season, the onset of heavy rains usually act as a cue that triggers flowering ([26] and the references therein). Actually, during the present study, the peak flowering month was positively correlated with the peak rainy months and soil moisture content. Similar results were reported for *S. senegal* by Tandon et al. [27] in India indicating that the species prefers flowering during the rainy seasons. It has also been reported that *S. senegal* in some places may respond by flowering even with unseasonal rains [11]. Principally, most tree species found in dry forest ecosystems normally utilizes the short favorable rainy seasons for leaf development and to accumulate sufficient photosynthate and initiate reproduction before the soil moisture starts to fall in the subsequent drier season [26].

Few studies have examined the possible functional significance of an interrelationship between leafing and flowering/fruiting phenophases in tropical trees; however this occurrence may be attributed to the need for substantial amount of resources to sustain reproduction [28]. *Senegalia senegal*, therefore, just like the other species with similar phenological behavior, requires to undertake photosynthesis to sustain it during the reproductive phase. As described by Sing and Kushwaha [26], flower production and maintenance require considerable expense of energy to form nonphotosynthetic tissues and nectar. This phenomenon therefore requires the availability of foliage for photosynthesis to sustain the physiological activities during flowering. Some amounts of soil moisture will be required during this process; hence the rainy season is the best time for the species to flower. The peak flowering month was not significantly correlated with the relative humidity; however, the phenophase occurred during high relative humidity ranging between 48 and 62%. This finding corroborates the results reported by Stone et al. [29] that relative humidity of between 50 and 60% is correlated with peak pollen availability. Such high relative humidity may be necessary for *S. senegal* to enhance pollen transfer and fertilization.

The present study revealed significant flowering synchrony among populations and individuals within the populations. The synchrony illustrates the plasticity of the individual trees

that may contribute, to a large extent, to population maintenance and connectivity in the woodland. The synchrony may benefit the species by providing an opportunity for pollen transfer within and among the populations hence ensuring high genetic diversity and preventing differentiation. Such genetic impact has been reported for the species by harboring higher genetic diversity with limited population differentiation [16]. The flowering event was similar in both the years in terms of timing and proportion of individual trees with flowers within the months. The two rainy seasons were also not significantly different in flowering intensity. This may mean that the reproductively mature trees flower similarly when triggered by the environmental cues.

4.3. Fruiting Phenology.

The fruiting phase of the species lasted for about three months in both seasons and years. This occurred during the peak rainy season until the seeds were mature and ready for dispersals and probably germinations. This timing of fruiting during the rainy season is to allow for fruit growth and maturation since this stage requires a lot of photosynthates [9]. As the rains subside and the dry season creeps in, almost all the fruits were mature and ripe in readiness for dispersal and even germination. Fruit maturation and presence of suitable conditions for dispersal are closely synchronized in tropical dry forest species because of the pronounced differences of biotic and abiotic conditions between dry and rainy seasons [30].

Senegalia senegal seeds are mainly dispersed by wind and ungulates whose activities are more predominant during the dry seasons. The timing of the season is therefore very important to the species evolution [13]. In most of the dry forest ecosystems, strong winds are common during the dry seasons providing an opportunity for dispersals to wind dispersed seeds like those of *S. senegal* [25]. Furthermore, it is during this same period that the *S. senegal* pods are an important source of fodder for livestock and other herbivores who are also potential seed dispersers of the species [12]. In this study, the greater percentage of individual trees with mature pods (brown pods) was observed towards the end of the rainy season with large number of trees with dry pods occurring during the dry season. The fruiting phenophase was also found to be synchronous within and among populations just as leafing and flowering events. However, the fruits stayed longer in both Kimalel and Kampi ya Samaki populations than in Tangulbei and Lake Bogoria populations. These variations may have been brought about by variations in soil characteristics. The soils found in Kimalel and Kampi Ya Samaki populations are loamy and therefore able to retain moisture for a longer period than the soils in Tangulbei and Lake Bogoria which are majorly sandy and rocky [17]. The soil moisture may have sustained the fruit in green form for a longer period.

4.4. Floral Morphology and Pollen Quality.

In order to understand *S. senegal* reproductive potential, after every flowering season, the floral morphology and pollen viability were studied. The flowers of *S. senegal* are generally creamy white and typically arranged along the inflorescence opening along the axis starting from the base. This observation was similar to the characteristics reported by Fagg and Allison [12] and Chiveu et al. [31]. The floral morphology differed significantly among the populations in all the variables except inflorescence length. The variables however did not differ significantly between the years and seasons. In addition, analysis of the traits within the populations showed no significant differences. In most of the traits, Kimalel population showed larger values compared to the other populations. For instance, the length of flower inflorescence in Kimalel was the longest and it is from the same population that the largest number of flowers per inflorescence was observed. This indicates that the length of the inflorescence may be influencing the number of flowers per inflorescence. This is consistent with research results reported for some Australian *Acacias* with similar floral architecture [23]. Despite Kimalel having the longest inflorescence length, there was no significant difference in the trait among the populations. The longer inflorescence length and large number of flowers per inflorescence may also be linked to the climatic variability within the ecosystem. This is basically because Kimalel population within the two years recorded the highest amounts of monthly rainfall and soil moisture content during the flowering period compared to other populations. Kimalel population also recorded the largest stigma diameter and style length compared to the other populations. The development and growth of these organs may be influenced by the flower size whereby the longer the flower length, the longer the style length and the larger the stigma diameter. Favorable climatic condition may also promote larger sizes of these organs [23].

Generally, pollen quality is one of the very important factors in successful plant reproduction and more so to *Acacia* species whose seed production usually occurs after only a single pollination activity [23]. Over 50% pollen germination was reported in all the populations although there was significant difference among the populations. Lake Bogoria population showed lower pollen germination percentage during the year 2014 short rains compared to the other population. This difference may have been brought about by the fluctuation in environmental conditions required for pollen maturity [32]. During this period of time, Lake Bogoria population experienced sporadic rainfall pattern compared to the other populations. This occurrence may have affected the floral development and maturity by delaying its formation. The flowers might have been caught up by harsh environmental conditions which in the long run affected the pollen quality. Such incidences have been reported by Tandon et al. [27]. However, poor pollen germination may also have been contributed by ageing of the flowers. Flowers collected late after anthesis normally result in poor germination. Probably some of the flowers collected from Lake Bogoria population were old and this might have significantly contributed to the poor germination. Similar results have also been reported for *A. mangium* and *A. auriculiformis* with low germination attributed to both environmental factors and ageing of the flowers [23].

The pollen quality did not differ significantly between the seasons or years. This is contrary to most studies that have shown significant variation between seasons and years ([2] and references therein). However, lack of significant differences within the woodland may be explained by the relatively similar weather patterns experienced during the study period. In the long run, the pollen quality did not have effect on the number of pods set per flower inflorescence pollinated and did not affect either the quality or the number of seeds realized per pod. Furthermore, the pollen viability reported in all the populations was sufficient to produce open pollinated seeds.

4.5. Pods and Seed Production. *Senegalia senegal* produced pods with variable dimensions. The pod length ranged from 3.4 to 8.2 cm and the width varied from 1.1 to 3.2 cm. These values were significantly different among populations, although no differences were found between the years or seasons (long and short rainy seasons). The differences may be attributed to soil factors and environmental variables. Kampi Ya Samaki and Kimalel populations that showed larger pod dimension than the other populations also recorded higher amount of rainfall than the other populations during this study. In this case, rainfall amount could be a factor in pod sizes. Nghiem et al. [23] in their study of fruit morphology of *A. mangium* and *A. auriculiformis* also reported significant variation in these traits and attributed it to varying environmental conditions. However, genetic variability of the individual trees concerned may also play a significant role. Similar variations in pod characteristics were also reported on *S. senegal* by Chiveu et al. [31] and these were attributed to both genetic differences and heterogeneity of environmental conditions among the sites.

Significant variation among populations was also realized on the weight of 1000 seeds. Heavy seeds were found in Kampi Ya Samaki population but no significant difference was realized within populations. The seed weights also did not differ between the seasons and years similar to percentages of fully formed seed. In many studies, seed weight has been viewed to mainly represent genetic differences which may be brought about by adaptation strategy of species. In most cases, some species tend to develop smaller and lighter seeds in drier and harsh environmental conditions compared to those in favorable environments. Similar results were also reported for *S. senegal* from different populations with variable environmental conditions [31]. The smaller and lighter seeds in drier areas reported in this study are contrary to the findings reported by Chaisurisri et al. [33] who correlated seed size to environmental dryness and found out that the seed size increases with dryness. They believed that the drier condition forces the trees to store more food in the seed for use during germination and regeneration. However such adaptation may vary accordingly with species.

5. Conclusions

Despite the fragmentation of *S. senegal* population within Lake Baringo woodland, the species has reported higher degree of phenological synchrony within and among the populations. It is also noted that all the phenophases were environmentally triggered and therefore the synchrony reported here may confidently be attributed to almost similar environmental condition within the woodland. The synchrony may also be an evolutionary strategy of the species to sustain reproduction. Furthermore, the seasonal leafing, flowering time, and fruiting duration, with linkages to leafing and leafless durations, observed in the species suggest the reproductive and survival strategies evolved by the species to adapt to the harsh environment. Although there were significant differences in some aspects of the morphology of flower length, stigma diameter, style length, and ovules per ovary among populations, these differences were only small and did not appear to affect crossing among individuals and populations. It is therefore worth concluding that there were no barriers in phenophase or flower structure to prevent interpopulation reproduction which would enhance genetic diversity and connectivity among populations.

Competing Interests

The authors declare that there are no competing interests regarding the publication of this paper.

Acknowledgments

The study was funded by Kenya Forestry Research Institute (KEFRI) and International Foundation of Science (IFS) Research Grant no. D5452-1 to Stephen F. Omondi as part of his Ph.D. thesis. The authors are grateful to KEFRI Biotechnology Laboratory and Baringo Sub-Regional Centre of the Rift Valley Eco-Regional Research Programme for helping during field sampling and data collection.

References

[1] E. E. Cleland, I. Chuine, A. Menzel, H. A. Mooney, and M. D. Schwartz, "Shifting plant phenology in response to global change," *Trends in Ecology and Evolution*, vol. 22, no. 7, pp. 357–365, 2007.

[2] V. P. Khanduri, "Annual variation in floral phenology and pollen production in *Lagerstroemia speciosa*: an entomophilous tropical tree," *Songklanakarin Journal of Science and Technology*, vol. 36, no. 4, pp. 389–396, 2014.

[3] X. Zhang, M. A. Friedl, and C. B. Schaaf, "Global vegetation phenology from Moderate Resolution Imaging Spectroradiometer (MODIS): evaluation of global patterns and comparison with in situ measurements," *Journal of Geophysical Research*, vol. 111, no. 4, p. 4017, 2006.

[4] N. C. Duke, "Phenological trends with latitude in the mangrove tree *Avicennia marina*," *Journal of Ecology*, vol. 78, no. 1, pp. 113–133, 1990.

[5] R. Milla, P. Castro-Díez, M. Maestro-Martínez, and G. Montserrat-Martí, "Costs of reproduction as related to the timing of phenological phases in the dioecious shrub *Pistacia lentiscus* L.," *Plant Biology*, vol. 8, no. 1, pp. 103–111, 2006.

[6] J. B. L. Okullo, J. B. Hall, and J. Obua, "Leafing, flowering and fruiting of *Vitellaria paradoxa* subsp. nilotica in savanna parklands in Uganda," *Agroforestry Systems*, vol. 60, no. 1, pp. 77–91, 2004.

[7] C. K. Augspurger, "Phenology, flowering synchrony, and fruit set of six neotropical shrubs," *Biotropica*, vol. 15, no. 4, pp. 257–267, 1983.

[8] M. S. Abu-Asab, P. M. Peterson, S. G. Shetler, and S. S. Orli, "Earlier plant flowering in spring as a response to global warming in the Washington, DC, area," *Biodiversity and Conservation*, vol. 10, no. 4, pp. 597–612, 2001.

[9] D. Lieberman, "Seasonality and phenology in a dry tropical forest in Ghana," *Journal of Ecology*, vol. 70, no. 3, pp. 791–806, 1982.

[10] P. A. Huxley, "Phenology of tropical woody perennials and seasonal crop plants with reference to their management in agroforestry systems," in *Plant Research and Agroforestry*, P. A. Huxley, Ed., pp. 503–525, ICRAF, Nairobi, Kenya, 1983.

[11] K. Tybirk, "Pollination, breeding system and seed abortion in some African *Acacias*," *Botanical Journal of the Linnean Society*, vol. 112, no. 2, pp. 107–137, 1993.

[12] C. W. Fagg and G. E. Allison, *Acacia Senegal and Gum Arabic Trade*, Tropical Forestry Papers no. 42, Oxford Forestry Institute, 2004.

[13] E. O. Obunga, *A Study of Genetic Systems of Four African Species of Acacia*, School of Biological Science, University of Sussex, 1995.

[14] L. M. Kiage, K. B. Liu, N. D. Walker, N. Lam, and O. K. Huh, "Recent land-cover/use change associated with land degradation in the Lake Baringo catchment, Kenya, East Africa: evidence from Landsat TM and ETM+," *International Journal of Remote Sensing*, vol. 28, no. 19, pp. 4285–4309, 2007.

[15] J. K. Lelon, *Uptake of Micronutrients by Acacia Senegal Varieties and Its Possible Effects on Gum Arabic Quality*, The University of Nairobi, Nairobi, Kenya, 2008.

[16] S. F. Omondi, E. Kireger, O. G. Dangasuk et al., "Genetic diversity and population structure of *Acacia senegal* (L) Willd. in Kenya," *Tropical Plant Biology*, vol. 3, no. 1, pp. 59–70, 2010.

[17] W. G. Sombroek, H. M. H. Braun, and B. J. A. Van der Pouw, *Exploratory Soil Map and Agro-limatic Zone Map of Kenya. Scale 1:1 000 000*, vol. 1:1 000 of *Exploratory Soil Survey Report no. E1*, Kenya Soil Survey, 1982.

[18] E. A. Y. Raddad, A. A. Salih, M. A. E. Fadl, V. Kaarakka, and O. Luukkanen, "Symbiotic nitrogen fixation in eight *Acacia senegal* provenances in dryland clays of the Blue Nile Sudan estimated by the 15N natural abundance method," *Plant and Soil*, vol. 275, no. 1-2, pp. 261–269, 2005.

[19] L. R. Arce and H. Banks, "A preliminary survey of pollen and other morphological characters in neotropical *Acacia* subgenus *Aculeiferum* (Leguminosae: Mimosoideae)," *Botanical Journal of the Linnean Society*, vol. 135, no. 3, pp. 263–270, 2001.

[20] B. N. Chikamai and J. A. Odera, *Commercial Plant Gums and Resins in Kenya*, Executive Printers, Nairobi, Kenya, 2002.

[21] P. M. Maundu, G. W. Ngugi, and H. C. Kasuye, *Traditional Food Plants of Kenya*, Nairobi, Kenya, 1999.

[22] C. J. Chiveu, O. G. Dangasuk, M. E. Omunyin, and F. N. Wachira, "Genetic diversity in Kenyan populations of *Acacia senegal* (L) willd revealed by combined RAPD and ISSR markers," *African Journal of Biotechnology*, vol. 7, no. 14, pp. 2333–2340, 2008.

[23] C. Q. Nghiem, C. E. Harwood, J. L. Harbard, A. R. Griffin, T. H. Ha, and A. Koutoulis, "Floral phenology and morphology of colchicine-induced tetraploid *Acacia mangium* compared with diploid *A. Mangium* and *A. Auriculiformis*: implications for interploidy pollination," *Australian Journal of Botany*, vol. 59, no. 6, pp. 582–592, 2011.

[24] F. W. Martin, "Staining and observing pollen tubes in the style by means of fluorescence," *Stain Technology*, vol. 34, no. 3, pp. 125–128, 1959.

[25] J.-L. Devineau, "Seasonal rhythms and phenological plasticity of savanna woody species in a fallow farming system (southwest Burkina Faso)," *Journal of Tropical Ecology*, vol. 15, no. 4, pp. 497–513, 1999.

[26] K. P. Singh and C. P. Kushwaha, "Diversity of flowering and fruiting phenology of trees in a tropical deciduous forest in India," *Annals of Botany*, vol. 97, no. 2, pp. 265–276, 2006.

[27] R. Tandon, K. R. Shivanna, and H. Y. Mohan Ram, "Pollination biology and breeding system of *Acacia senegal*," *Botanical Journal of the Linnean Society*, vol. 135, no. 3, pp. 251–262, 2001.

[28] C. P. Van Schaik, J. W. Terborgh, and S. J. Wright, "The phenology of tropical forests: adaptive significance and consequences for primary consumers," *Annual Review of Ecology and Systematics*, vol. 24, no. 1, pp. 353–377, 1993.

[29] G. N. Stone, P. Willmer, and J. A. Rowe, "Partitioning of pollinators during flowering in an African *Acacia* community," *Ecology*, vol. 79, no. 8, pp. 2808–2827, 1998.

[30] L. M. S. Griz and I. C. S. Machado, "Fruiting phenology and seed dispersal syndromes in caatinga, a tropical dry forest in the northeast of Brazil," *Journal of Tropical Ecology*, vol. 17, no. 2, pp. 303–321, 2001.

[31] C. J. Chiveu, O. G. Dangasuk, M. E. Omunyin, and F. N. Wachira, "Quantitative variation among Kenyan populations of *Acacia senegal* (L.) Willd. for gum production, seed and growth traits," *New Forests*, vol. 38, no. 1, pp. 1–14, 2009.

[32] J. Kenrick, *Some Aspects of the Reproductive Biology of Acacia*, University of Melbourne, Parkville, Australia, 1994.

[33] K. Chaisurisri, D. G. W. Edwards, and Y. A. El-Kassaby, "Genetic control of seed size and germination in Stika spruce," *Silvae Genetica*, vol. 41, no. 6, pp. 348–355, 1992.

8

Modelling Analysis of Forestry Input-Output Elasticity in China

Guofeng Wang,[1] Jiancheng Chen,[1] and Xiangzheng Deng[2,3]

[1]*School of Economics and Management, Beijing Forestry University, Beijing 100083, China*
[2]*Institute of Geographic Sciences and Natural Resources Research, Chinese Academy of Sciences, Beijing 100101, China*
[3]*Center for Chinese Agricultural Policy, Chinese Academy of Sciences, Beijing 100101, China*

Correspondence should be addressed to Jiancheng Chen; chenjc_bjfu@hotmail.com

Academic Editor: Piermaria Corona

Based on an extended economic model and space econometrics, this essay analyzed the spatial distributions and interdependent relationships of the production of forestry in China; also the input-output elasticity of forestry production were calculated. Results figure out there exists significant spatial correlation in forestry production in China. Spatial distribution is mainly manifested as spatial agglomeration. The output elasticity of labor force is equal to 0.6649, and that of capital is equal to 0.8412. The contribution of land is significantly negative. Labor and capital are the main determinants for the province-level forestry production in China. Thus, research on the province-level forestry production should not ignore the spatial effect. The policy-making process should take into consideration the effects between provinces on the production of forestry. This study provides some scientific technical support for forestry production.

1. Introduction

The reform of collective forest rights is another major revolution of the rural management system after land reform in China [1, 2]. This reform has endowed farmers with partial forestry rights, so that these farmers are able to use their own forest resources and thereby gain revenue [3–5]. Thereby, in order to calculate the elasticity of labor, capital, and land inputs during the forestry growth, this study used a spatial econometric model to compute the contributions of all elements. The production flexibility and efficiency of capital are always a research hotspot [6–8]. Solow proposed an economy growth accounting model and applied new classical growth theory to economic accounting [9]. The existing research mainly focused on the input-output elasticity of agricultural production [10–12]. However, there is little research about forestry, a special agriculture department, and even the existing findings are controversial [13, 14]. The researchers argued the output elasticity coefficients from the first, second, and third industries of forestry are 1.44, 0.72, and 0.89, respectively, during 1998–2005; 1.66, 0.88, and 0.81, respectively, during 2006–2021; 1.82, 0.96, and 0.91, respectively, during 2022–2030 [15]. These findings should

be further validated from new perspectives and with new methods immediately.

This paper consists of four parts, the first part gives a brief description of relative research, the second part specifies the materials and methods used in this paper, the third part gives out results and discussion, and the last part is the conclusion part. Through this paper, we try to prove that the forestry production in one province influences that in another province.

2. Materials and Methods

2.1. Spatial Autocorrelation Test of Forestry Production. The forestry production in China is found with severe spatial differences and is largely correlated with the differences and fluidity of regional forestry resource [16, 17]. Forestry production is modestly different among regions, but there may be spatial correlations among provinces [18–20]. In order to figure out the correlations and heterogeneity of province-level forestry production, we used global Moran's index:

$$I = \frac{\sum_{i=1}^{n} \sum_{j=1}^{n} W_{ij} \left(Y_i - \overline{Y} \right) \left(Y_j - \overline{Y} \right)}{\sum_{i=1}^{n} \left(Y_i - \overline{Y} \right)^2}, \tag{1}$$

where Y_i and Y_j describe the observations of forestry output in regions i and j, respectively, n represents the number of regions, $\overline{Y}$ is the average observation of forestry outputs, and W_{ij} is the spatial weight.

Under the circumstance of zero correlation, Moran'I was used to construct a standard normal index as follows:

$$Z = \frac{(I - E(I))}{\sqrt{\text{Var}(I)}}, \qquad (2)$$

where $E(I)$ and $\text{Var}(I)$ are decided by the spatial distribution of data and the arrangement of spatial lag matrix. When the Z-value is significant and positive, there is positive space correlation, indicating the presence of regional agglomeration among similar production regions. When the Z-value is significant but negative, there is negative significant correlation, indicating the presence of regional dispersity among similar production regions. When the Z-value is equal to zero, there exists random spatial distribution.

The global Moran's index can partially represent the space autocorrelation. However, owing to the repeated computation or mutual cancellation during computations, we used a local Moran'I index reflecting spatial autocorrelation, the local spatial correlation index, and Moran scatter diagram to further reveal whether or not there exists local spatial agglomeration. Local Moran's index is computed as follows:

$$I_i = \frac{(Y_i - \overline{Y})}{\sum_{i=1}^{n} (Y_i - \overline{Y})^2} \cdot \sum_{j=1}^{n} W'_{ij} (Y_j - \overline{Y}), \qquad (3)$$

where W'_{ij} is the standardized space weight matrix (the sum of each row is one). The expected value of local Moran's index I_i is as follows:

$$E_i(I_i) = -\frac{\sum_{j=1}^{n} W_{ij}}{(n-1)}. \qquad (4)$$

When I_i is larger than the expected value of $E_i(I_i)$, there exists spatial agglomeration of similar forestry outputs around region i or local space positive correlation. When I_i is smaller than $E_i(I_i)$, there exists large differences among similar forestry outputs around region i or local space negative correlation.

Moran scatter diagram shows the 2D scatter plot that visualizes Z (a vector composed of the deviation between the observed value and the mean) and W_z (space weighted average, or space lag vector). The vector-form global Moran'I index is computed as follows:

$$I = \frac{n}{S} \cdot \frac{Z'W'_z}{Z'Z}, \qquad (5)$$

where $S_0 = \sum_{i=1}^{n} \sum_{j=1}^{n} W'_{ij}$; when W'_{ij} is the standardized space weight matrix, then $S_0 = n$; at this moment, the global Moran's index is the linear regression slope of W_z relative to Z. The first and third quadrants on Moran'I scatter plot represent the positive space correlations, while the second and fourth quadrants indicate the negative space

correlations. Specifically, the first quadrant indicates the regions with large observed values are surrounded by large-value regions; the second quadrant indicates the regions with small observed values are surrounded by large-value regions; the third quadrant indicates the regions with small observed values are surrounded by small-value regions; the fourth quadrant indicates the regions with large observed values are surrounded by small-value regions. The first and third quadrants represent typical positive space correlations, while the second and fourth quadrants indicate the local negative space correlations.

LISA (Local Indictors of Spatial Association) analysis is used to figure out the spatial differences in production. When LISA passes the significance test, there is local positive spatial autocorrelation, or this region is surrounded by regions with similar performance, called spatial agglomeration. When this region and its nearby regions are all found with large observed data, it is called a high-high region, and otherwise, it is called a low-low region.

2.2. Selection of Weights for Forestry Space Autocorrelations. The selection of spatial weight W is associated with the results of spatial autocorrelation and spatial regression. W_{ij} is defined as the contiguity or distance of any element from other elements. Currently, there are many types of weight matrices, including contiguity, K-nearest neighbors, and distance threshold. Specifically, contiguity matrices are divided into Rook (contiguity estimated from four directions of east, south, north, and west) and Queen (besides these four directions, it also involves other corners). As for K-nearest neighbors, several points closest to a test point are called its neighbors and each is assigned a weight 1, and other points are given a weight 0. Many researches figured out that different matrix may lead to different results, including spatial coefficient and the signs of the coefficient [21].

2.3. Space Econometric Model in Forestry Economy Growth. According to traditional economics, the economic growth mainly depends on two endogenous factors: labor and capital, but it is affected by technological progress, an exogenous factor. In this model, the land element is considered as an internal factor of economic growth. In other words, the output level Y from each forestry region is decided by the labor input L, land input D, and capital input K. Then, this model is expressed as

$$Y_i = A_i L_i^{\alpha} D_i^{\beta} K_i^{Y} e^{\varepsilon_i}, \qquad (6)$$

where Y_i is the forestry economic development level in region i; A_i is the technical level, L_i is the labor input into forestry; D_i is the land area; K_i is the capital input; α, β, and γ are the corresponding output elasticity, respectively. If $\alpha + \beta + \gamma = 1$, $\alpha + \beta + \gamma > 1$, and $\alpha + \beta + \gamma < 1$, then the return to scale is unchanged, increases, and gradually drops, respectively. Logarithm of both sides of (6) yields

$$\ln Y_i = \ln A_i + \alpha \ln L_i + \beta \ln D_i + \gamma \ln K_i + \varepsilon_i. \qquad (7)$$

2.4. Space Lag Model (SLM) for Forestry Production Function.
The basic model of forestry production does not involve space correlations. Taking spatial effects into account means the regional forestry production is affected not only by the local investment level, but also by the spillover effect from other nearby forestry regions. In this way, SLM is determined:

$$\ln Y_i = \ln A_i + \rho W \ln Y_i + \alpha \ln L_i + \beta \ln D_i + \gamma \ln K_i + \varepsilon_i, \tag{8}$$

where W is the space weight matrix and $W \ln Y$ is the weighted variable from a nearby forestry region. This model reflects the effects of regional forestry production from the input-output levels in nearby regions through the space effect.

2.5. Space Error Model (SEM) for Forestry Production.
SEM takes into account the variables that may be ignored in the decision model, such as human capital, research level, and climate change. The space error model is used to measure the roles that may be played by the spatially interacting errors. SEM is expressed as

$$\ln Y_i = \ln A_i + \alpha \ln L_i + \beta \ln D_i + \gamma \ln K_i + \varepsilon_i,$$
$$\varepsilon_i = \lambda W \varepsilon_i + \mu_i, \tag{9}$$

where W is the space weight matrix, and λ measures the space error effect on regional forestry production due to observational errors.

2.6. Space Units and Data Sources.
This study was targeted at 31 provinces or autonomous regions or municipality cities of Mainland China in 2013. The data were cited from *China Forestry Statistical Yearbook 2013*. The output variable was the total regional forestry production value. Regarding the release time of forestry statistical yearbooks, we used the forest areas in the statistics as the forestry area input. The number of labor forces by the end of 2013 was used as the regional labor force input. The fixed assets investment was used as capital input.

3. Results and Discussion

3.1. Global Moran's Index for Space Correlation of Forestry Production.
To study the interferences of weight indices on the space effect, we used three space weight matrices, and through stepwise distance increment, we tested the attenuation effect of distance (Table 1). Clearly, global Moran'I index gradually decreases and shows the attenuation effect of distance. Moreover, Moran's index estimated from Queen1 weight matrix is 0.3685, indicating the most significant space autocorrelation ($p < 0.003$) and the strong spatial dependence and evident space effect of forestry production.

3.2. Local Moran's Index and LISA Analysis for Space Correlation of Forestry Production.
Table 1 uncovers the overall space autocorrelation in forestry production of each studied region, but local Moran's I cannot be identified by the global

TABLE 1: Global Moran's index for space autocorrelation for forestry productions in different regions.

	Moran's I	Mean	SD	p
W_{ROOK1}	0.3685	−0.0249	0.1112	0.0040
W_{ROOK2}	0.0880	−0.0247	0.0780	0.0700
W_{ROOK3}	−0.0309	−0.0318	0.0696	0.4660
W_{Queen1}	0.3685	−0.0297	0.1041	0.0030
W_{Queen2}	0.0880	−0.0324	0.0791	0.0780
W_{Queen3}	−0.0309	−0.0288	0.0712	0.4930
W_{K1}	0.1031	−0.0233	0.1989	0.2410
W_{K2}	0.0950	−0.0266	0.1454	0.1880
W_{K3}	0.2133	−0.0284	0.1166	0.0290

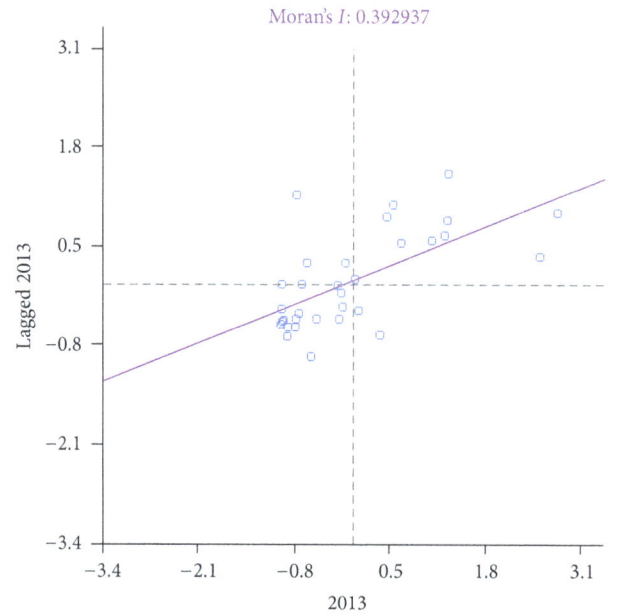

FIGURE 1: Local Moran's index based on W_{Queen1}.

Moran'I index. To further portray the space correlation of forestry production, we used Moran scatter plot and LISA analysis to explore the local space characteristics of province-level forestry outputs (Figure 1).

The first quadrant involves Shandong, Anhui, Hubei, Zhejiang, Jiangxi, Hunan, Fujian, Guangxi, and Guangdong, which are all high-output provinces surrounded by high-output provinces. The second quadrant involves Shanghai, Guizhou, Henan, Yunnan, and Chongqing, which are all low-output provinces surrounded by high-output provinces. The third quadrant involves Chongqing, Hebei, Jilin, Heilongjiang, Xinjiang, Shaanxi, Shanxi, Inner Mongolia, and Tibet, which are all low-output provinces surrounded by low-output provinces. The fourth quadrant involves Liaoning and Sichuan, which are both high-output provinces surrounded by low-output provinces. Clearly, the spatial differences of forestry outputs are very significant among all provinces in China, and the typical characteristics of positive local correlations and accumulation are very significant.

TABLE 2: Local Moran's index based on space weight matrix W_{Queen1} as well as test results.

Province	I_Y	p_y
Beijing	0.3633	0.27
Tianjing	0.3745	0.23
Hebei	0.0130	0.48
Shanxi	0.2943	0.29
Neimenggu	0.3060	0.14
Liaoning	−0.0611	0.39
Jilin	0.0189	0.31
Heilongjiang	0.0534	0.41
Shanghai	−0.0002	0.03
Jiangsu	−0.0382	0.04
Zhejiang	0.0251	0.06
Anhui	1.2997	0.01
Fujian	−0.1454	0.01
Jiangxi	0.7301	0.02
Shandong	0.6203	0.11
Henan	0.3067	0.24
Hubei	−0.2512	0.49
Hunan	0.3949	0.1
Guangdong	2.3471	0.04
Guangxi	0.6460	0.07
Hainan	0.0000	0.01
Chongqing	−0.9785	0.46
Sichuan	0.3683	0.01
Guizhou	0.0096	0.31
Yunnan	0.2826	0.45
Xizang	1.0362	0.22
Shaanxi	0.4459	0.15
Gansu	0.4282	0.09
Qinghai	0.1428	0.04
Ningxia	0.2303	0.09
Xinjiang	0.4537	0.01

Note: I_Y is local Moran's index; p_Y is the adjoint probability of I_Y.

The distribution of space autocorrelation patterns among the 31 provinces is as follows (Table 2): the significant high-high regions include Shandong, Anhui, Jiangxi, Fujian, and Guangdong; the significant low-low regions are Shaanxi, Gansu, and Xinjiang. All other provinces are high-low areas or low-high areas or insignificant areas. It is indicated that the spatial distribution of forestry production in China is featured by a very evident center-periphery mode and by very significant local spatial agglomeration and space correlations.

3.3. Space Econometric Analysis of Forestry Production Elasticity. Then, a space econometric model was used to estimate the elasticity coefficients of labor, land, and capital inputs to province-level forestry production (Table 3). The residual errors of ordinary least squares (OLS) contain significant space correlations ($p < 0.001$). Then, to find out whether the residual errors originated from the internal SLM or SEM, we further validated the models. Results show SLM is very significant, but it is not significant versus SEM. Thus, SEM is more significant.

To identify the effects and contributions of different input elements to forestry production, we built an SEM using

TABLE 3: Space correlation OLS test based on space weight matrix W_{Queen1}.

Test	Trivariate function		Bivariate function	
	Index	p	Index	p
Moran'I (error)	4.8671	0.0000	4.7938	0
Lagrange multiplier (lag)	0.335	0.5628	0.4905	0.4837
Robust LM (lag)	0.2234	0.6365	0.1971	0.6571
Lagrange multiplier (error)	14.9747	0.0001	16.3942	0
Robust LM (error)	14.8631	0.0001	16.1007	0
Lagrange multiplier (SARMA)	15.1981	0.0005	16.5912	0.0002

three elements that decide agricultural production (Table 4), estimating the elasticity coefficients of labor force, land, and capital to the forestry output of each province. First, the forestry production involving the three elements was analyzed via OLS. Results show the fitting degree of SEM is 0.9215, with smallest Akaike information criterion (AIC, 93.93) and Schwarz criterion (SC, 99.79), indicating SEM is most suitable.

Results show neither the estimations nor the significance levels of output elasticity are always the same for any of the three elements, in both OLS and space economic models. The output elasticity of forestry labor from the bivariate SEM is 0.6492 ($p = 0.0058$), while the output elasticity from forestry capital is 0.6603 ($p = 0.003$).

As shown in Table 4, the output elasticity estimated from OLS is the highest from capital (0.8412), indicating a tendency to overestimation. The output elasticity from capital is 0.3968, with a tendency to underestimation. Thus, the elasticities from capital and labor outputs make the SEM more suitable.

The trivariate model also involves the land input, but the forestry production measured from this variable is not very significant ($p > 0.1$); thus, it plays a negative role.

It should be noted that the model used here is based on space economic theory. This method indicates that the space effect should not be ignored. However, the estimation results only reveal the state-level standard but do not take into account the regional differences. In the future, more accurate and precise economic models are needed. Another thing that may be a limit for this paper is that, as to limitation of the data and observations, the result may vary according to different situations, which needs future study.

4. Conclusions

Here we used the sectional data of province-level forestry input-output in 2013 in China. Then, the Moran'I with space autocorrelation and the local Moran'I index with space correlation were used to portray a forestry production spatial distribution mode involving 31 provinces in China. Then,

TABLE 4: Different regression models based on space weight matrix W_{Queen1}.

	Trivariate function			Bivariate function		
	OLS	SLM	SEM	OLS	SLM	SEM
Constant	0.3768	0.3559	0.4777	0.5702	0.5369	0.6657
	(0.7390)	(0.7328)	(0.5558)	(0.6242)	(0.6219)	(0.4199)
$\ln L$	0.8443**	0.8350**	0.9091***	0.3968	0.4006	0.6492***
	(0.0297)	(0.0151)	(0.0016)	(0.1531)	(0.1174)	(0.0058)
$\ln D$	−0.4543*	−0.4418*	−0.3293			
	(0.0955)	(0.0718)	(0.1372)			
$\ln K$	0.7377***	0.6947***	0.6332***	0.8412***	0.7830***	0.6603***
	(0.0025)	(0.0014)	(0.0004)	(0.0007)	(0.0004)	(0.0003)
λ, ρ		0.0395	0.6540***		0.0503	0.6672***
		(0.5711)	(0.0000)		(0.4881)	0.0000
$\log L$	−49.1486	−48.9847	−42.96	−50.7651	−50.5221	−44.0254
R^2	0.8701	0.8714	0.9215	0.8562	0.8585	0.9167
AIC	106.30	107.97	93.93	107.53	109.04	94.0509
SC	112.16	115.30	99.79	111.93	114.91	98.4481

Note: *($p < 0.1$), **($p < 0.05$), and ***($p < 0.01$).

SLM and SEM were used to estimate the elasticity coefficients of province-level forestry input-output. We find the forestry outputs from the 31 provinces are significantly autocorrelated both globally and locally, and the province-level space correlations and heterogeneity are all very significant. The labor force and capital are the detrimental factors on the regional forestry production, but the contributions from lands are not very obvious. The spillover effect of spatial errors plays a significant role in the forestry production of adjacent provinces. In other words, the forestry production of a province, through the spillover of error items, influences the growth of forestry output in nearby provinces. Similarly, it is indicated that SEM is a very suitable space econometric model.

The global and local Moran'I space autocorrelation tests and space econometric estimations all indicate that the spatial effect cannot be ignored in the study on province-level forestry production. With the input-output model, we find the test results of SEM in the provincial production function are all better than OLS, and the precision of regression coefficients is higher. Thus, space autocorrelation method is an efficient method for analysis of forestry production. The space econometric model is more objective in exploration of influence factors.

From the perspective of policy implications, due to the existence of space error spillover effect in forest products and the remaining forestry output, the forestry production behaviors in nearby provinces will affect the agricultural production behaviors of the tested province. The time "incentive" effect in decision-making for regional agricultural production will influence the production game competition among nearby provinces and finally impact the input scales and allocation efficiency of province-level forestry production elements. Thus, the formulation of relevant forestry policies should not ignore the transverse cross-effect among provincial forestry production, and should take into account the interactions of forestry production among nearby provinces.

A regional coordination mechanism should be established to coordinate the rational flow of forestry production elements, to improve the space complementarity and space allocation efficiency of elements, and to promote the regional forestry production ability.

Competing Interests

The authors declare that there is no conflict of interests regarding the publication of this article.

Acknowledgments

This research was supported by the National Natural Science Foundation of China for Distinguished Young Scholars (Grant no. 71225005), the Key Project in the National Science & Technology Pillar Program of China (Grant no. 2013BACO3B00), and the funding support from the Institute of Geographic Sciences and Natural Resources Research, Chinese Academy of Sciences (Grant no. 2012ZD008).

References

[1] A. E. Duchelle, M. Cromberg, M. F. Gebara et al., "Linking forest tenure reform, environmental compliance, and incentives: lessons from REDD+ initiatives in the Brazilian Amazon," *World Development*, vol. 55, pp. 53–67, 2014.

[2] A. M. Larson, D. Barry, and G. Ram Dahal, "New rights for forest-based communities? Understanding processes of forest tenure reform," *International Forestry Review*, vol. 12, no. 1, pp. 78–96, 2010.

[3] P. Jagger, *Forest Incomes after Uganda's Forest Sector Reform: Are the Rural Poor Gaining?* International Food Policy Research Institute (IFPRI), Washington, DC, USA, 2008.

[4] C. Wang, Y. Wen, and J. Wu, "The Socio-economic effect of the reform of the collective forest rights system in Southern China:

a case of Tonggu County, Jiangxi Province," *Small-Scale Forestry*, vol. 13, no. 4, pp. 425–444, 2014.

[5] Y. Zhu, H. Lan, D. A. Ness et al., "Carbon trade, forestry land rights, and farmers' livelihood in rural communities in China," in *Transforming Rural Communities in China and Beyond: Community Entrepreneurship and Enterprises, Infrastructure Development and Investment Modes*, pp. 61–91, Springer International Publishing, Berlin, Germany, 2015.

[6] P. C. Beukes, P. Gregorini, A. J. Romera, G. Levy, and G. C. Waghorn, "Improving production efficiency as a strategy to mitigate greenhouse gas emissions on pastoral dairy farms in New Zealand," *Agriculture, Ecosystems & Environment*, vol. 136, no. 3-4, pp. 358–365, 2010.

[7] J.-P. Chavas, R. Petrie, and M. Roth, "Farm household production efficiency: evidence from the Gambia," *American Journal of Agricultural Economics*, vol. 87, no. 1, pp. 160–179, 2005.

[8] U. Moallem, "Future consequences of decreasing marginal production efficiency in the high-yielding dairy cow," *Journal of Dairy Science*, vol. 99, no. 4, pp. 2986–2995, 2016.

[9] R. M. Solow, "Technical change and the aggregate production function," *The Review of Economics and Statistics*, vol. 39, no. 3, pp. 312–320, 1957.

[10] G. Holloway, D. Lacombe, and J. P. LeSage, "Spatial econometric issues for bio–economic and land–use modelling," *Journal of Agricultural Economics*, vol. 58, no. 3, pp. 549–588, 2007.

[11] B. Hu and M. McAleer, "Estimation of Chinese agricultural production efficiencies with panel data," *Mathematics and Computers in Simulation*, vol. 68, no. 5-6, pp. 474–483, 2005.

[12] M. Lio and M.-C. Liu, "Governance and agricultural productivity: a cross-national analysis," *Food Policy*, vol. 33, no. 6, pp. 504–512, 2008.

[13] D. P. Faith, P. A. Walker, J. R. Ive, and L. Belbin, "Integrating conservation and forestry production: exploring trade-offs between biodiversity and production in regional land-use assessment," *Forest Ecology and Management*, vol. 85, no. 1–3, pp. 251–260, 1996.

[14] J. J. Troncoso and R. A. Garrido, "Forestry production and logistics planning: an analysis using mixed-integer programming," *Forest Policy and Economics*, vol. 7, no. 4, pp. 625–633, 2005.

[15] M. Ma, T. Haapanen, R. B. Singh, and R. Hietala, "Integrating ecological restoration into CDM forestry projects," *Environmental Science & Policy*, vol. 38, pp. 143–153, 2014.

[16] Z. Li, X. Deng, Q. Shi, X. Ke, and Y. Liu, "Modeling the impacts of boreal deforestation on the near-surface temperature in European Russia," *Advances in Meteorology*, vol. 2013, Article ID 486962, 9 pages, 2013.

[17] X. W. Yu, H. Y. Liu, Y. C. Yang, X. Zhang, and Y. W. Li, "Geoserver based forestry spatial data sharing and integration," *Applied Mechanics and Materials*, vol. 295–298, pp. 2394–2398, 2013.

[18] X. Deng, J. Huang, E. Uchida, S. Rozelle, and J. Gibson, "Pressure cookers or pressure valves: do roads lead to deforestation in China?" *Journal of Environmental Economics and Management*, vol. 61, no. 1, pp. 79–94, 2011.

[19] G. Fu, M. Shah, E. Uchida et al., "Impact of the Grain for Green program on forest cover in China," in *Proceedings of the Annual Meeting*, Agricultural and Applied Economics Association, Washington, DC, USA, August 2013.

[20] F. Y. T. S. L. Zengyuan, "Application of spatial statistic analysis in forestry," *Scientia Silvae Sinicae*, vol. 3, article 25, 2004.

[21] L. P. Sun, "Spatial analysis of the evolvement of urban and rural economic disparity in yunnan province, China," *Asian Agricultural Research*, vol. 2, no. 6, 2010.

9

Agricultural and Forest Land Use Potential for REDD+ among Smallholder Land Users in Rural Ghana

Divine O. Appiah,[1] **John T. Bugri,**[2] **Eric K. Forkuo,**[3] **and Sampson Yamba**[1]

[1]*Department of Geography and Rural Development, Kwame Nkrumah University of Science and Technology (KNUST), Kumasi, Ghana*
[2]*Department of Land Economy, Kwame Nkrumah University of Science and Technology (KNUST), Kumasi, Ghana*
[3]*Department of Geomatic Engineering, Kwame Nkrumah University of Science and Technology (KNUST), Kumasi, Ghana*

Correspondence should be addressed to Divine O. Appiah; dodameappiah@gmail.com

Academic Editor: Piermaria Corona

Reducing emissions from deforestation and forest degradation with other benefits (REDD+) mechanism is supposed to address the reversal of forest-based land degradation, conservation of existing carbon stocks, and enhancement of carbon sequestration. The Bosomtwe District is predominantly agrarian with potentials for climate change mitigation through REDD+ mechanism among smallholder farmers. The limited knowledge and practices of this strategy among farmers are limiting potentials of mitigating climate change. This paper assesses the REDD+ potentials among smallholder farmers in the district. Using a triangulation of quantitative and qualitative design, 152 farmer-respondents were purposively sampled and interviewed, using snowballing method from 12 communities. Quantitative data gathered were subjected to the tools of contingency and frequencies analysis, embedded in the Statistical Package for Social Sciences (SPSS) v.16. The qualitative data were analyzed thematically. Results indicate that respondents have knowledge of REDD+ but not the intended benefit sharing regimes that can accrue to the smallholder farmers. Farmers' willingness to practice REDD+ will be based on the motivation and incentive potentials of the strategies. The Forestry Services Division should promote the practice of REDD+ among smallholder farmers through education, to whip and sustain interest in the strategy.

1. Introduction

Deforestation and forest degradation account for about 17% of global greenhouse gas emissions [1], making it third to energy (26%) and industrial (19%) sectors globally and also higher than the transportation sector [2]. The accelerated loss of tropical forests is recognized globally as a major contributor to global warming [3]. Of the land use and climate interaction, the relationship between agriculture and forests and reducing emissions from deforestation and forest degradation (REDD+) programs is of prime concern [3].

The clearing of tropical forest for agriculture contributes significantly to greenhouse gas emission which hastens climate change. Expanding agricultural lands into forest frontiers for most farmers is a cheaper and preferred way of increasing crop production to meet ever increasing food demands [4]. Beyond reducing the degradation of forests, REDD+ incorporates conservation and sustainable management practices [5]. Reducing emissions from deforestation and forest degradation with forest conservation and management (REDD+) provides the opportunity for host countries to gain financially by costing the value of standing forests, curbing deforestation, and encouraging the conservation and sustainable management of forests. Research by the International Institute for Environment and Development (IIED) has shown that REDD+ strategies will be handicapped if they are not in tandem with national agricultural development objectives focusing on adaptation and mitigation [4].

Communities in forested regions in developing countries are some of the most important stakeholders because REDD+ policies will affect their livelihoods, and these stakeholders will likely be directly involved in the implementation and maintenance of REDD+ activities [6]. Through traditional agroforestry practices communities have sustainably managed forests in the past with benefits of increased productivity, sustained soil fertility, erosion control, biodiversity conservation, and income diversification through the harvest and sale of nontimber forest products accruing to them [7]. Reducing emissions from deforestation and forest degradation with its benefits (REDD+) may thus spearhead the restoration of traditional agroecological practices on the continent.

In some African countries, agricultural and environmental sustainability is contextualized in REDD+ policies. In Ethiopia, for example, agricultural systems pose serious threat to the sustainability of the environment and highly contribute to the country's greenhouse gas emissions. Hence REDD+ is anticipated to help reverse this trend [8].

In Tanzania, adopted agroforestry practices include home gardens, alley intercropping, improved fallows, and boundaries [9]. Agroforestry practices in rural communities in Southern Africa include improved fallows, rotational woodlots, and indigenous fruit trees in the parklands system [10]. Bryan et al. [11] and Gledhill et al. [12] emphasized that agroforestry comes with a "triple win" of climate change mitigation, agricultural adaptation, and increased productivity as does REDD+.

REDD+ has been accepted among most African countries because of possible financial benefits through carbon financing to support the forestry sector on the continent. Nevertheless, tenure conflicts based on cultural inheritance pose a threat that hampers development efforts in most African countries [13]. The Agriculture, Forest, and Land Use (AFOLU) and reducing emissions from deforestation and forest degradation (REDD+) interconnection is increasingly gaining attention on the African continent but the realization of actual deforestation and reduction in forest degradation has yet to completely unfold [14].

In [15] Djagbletey and Adu-Bredu found in Nkoranza in Ghana that ownership of teak farms was dominated by natives because tree planting on a parcel of land by an individual customarily implied his or her ownership of the land. Settlers and migrants were therefore less actively involved in tree planting initiatives [15]. According to Adaba [16], in Northern Ghana, families establish woodlots on family lands as alternative sources of income and fuel wood. Communal woodlots are however not popular because individual and family access and utilization of these communal woodlots are usually restricted.

The Cancun agreement states clearly that respect for the rights of local people and the conservation of biodiversity and natural forests must be upheld in the implementation of REDD+ initiatives [17]. REDD+ through avoided deforestation has the potential to reduce GHG emissions. It could conversely result in leakages and increased degradation in adjoining marginal lands [18]. Herein lies the need for agroforestry to absorb such leakages by augmenting the benefits of forests to forest communities and agriculture.

Carbon sequestration in trees initially increases as trees grow but eventually declines as the trees age [19]. Agricultural and forested lands present major carbon sequestration opportunities if the appropriate land use and management practices are adopted [7]. Since agricultural extensification could threaten REDD+, interventions should focus not only on forest but also on forest-farm frontier [17].

Carbon emission reduction through REDD+ can contribute significantly to land-based mitigation in two ways: firstly reducing land-based greenhouse gas emissions and secondly sequestering carbon dioxide through reforestation and agroforestry [20]. Decision on land use at the grassroots involving stakeholders as smallholder must be a key target of REDD+ interventions [17]. The Energy and Resources Institute [21], therefore, recommends that, in order for REDD+ to be effective, there is the need for stratification considering prevailing land use options and patterns.

In this regard, some countries have integrated REDD+ with prevailing land uses including reforestation, afforestation, agroforestry, and assisted natural regeneration using these as key drivers of REDD+ interventions [22]. Smallholders are an important contributor to deforestation. Issues related to land rights are perhaps the most complex and have far-reaching impacts on forest governance, communities, and REDD+ outcomes [23]. To this Kotru [24] asserts that clarity of tenure and hence right to benefits present challenges at the community level. Private land ownership has facilitated the adoption of agroforestry systems in Masaka District since the majority of farmers hold private land. In contrast, other types of land ownership may hinder the adoption of agroforestry systems.

Sebukyu and Mosango [25] put forward that agroforestry practices through REDD+ have benefits of soil fertility restoration, among others, greatly reducing the need for inputs such as fertilizers. Many developing countries are however not adequately prepared to utilize their forest and forest-frontier potentials to benefit from the REDD+ market [26]. More emphasis on educating farmers on the benefits of sustainable resource management and specifically agroforestry and conservation practices would reduce hindrances to the adoption [27]. Critical areas assessed for the purposes of this paper therefore included awareness of REDD+ and its benefits among smallholder farmers in the Bosomtwe District, willingness of smallholder farmers to engage in forest management practices, possible motivation for adoption of REDD+ activities, and land tenure and its potential implication on access to REDD+ benefits.

Problematizing the Concept of REDD+ in the Bosomtwe District. Factors constraining the development of agroforestry among smallholder farmers in the Bosomtwe District are variegated depending on ecological and socioeconomic factors confronting the smallholder farmers. The adoption of agroforestry practices is influenced by many factors and one category of these factors is the characteristics and conditions of the farmer, as has been espoused by Oino and Mugure [28]. Mbwambo et al. [9] explain, for example, that land size, tenure, access to extension services, capital, crop yield, and household income are key in determining farmers adoption of agroforestry.

Rapid population growth within the Bosomtwe District and the Kumasi Metropolis has necessitated increased food supply from nearby agrarian districts as Bosomtwe.

Periurban developments within the district are increasingly expanding into adjoining arable agricultural lands. This requires that agriculture be sustained by expansion into forest frontiers. Other activities as small-scale mining and wood demands for firewood and charcoal production are on the rise with a current rate of 56.9% within the district.

The Bosomtwe District being predominantly agrarian with forest cover has potentials for climate change mitigation through REDD+ mechanism among smallholder farmers. Thus, knowledge and practice of smallholder farmers of this strategy as an alternative livelihood potential are critical and practices of this strategy as alternative livelihood potential to their farming activities. Consequently, a substantial proportion of the forest cover is being depleted. Considering the foregone argument, the focus of this paper is to assess the REDD+ potentials among smallholder land users in selected rural communities in the Bosomtwe District of the Ashanti Region of Ghana. In earlier study, the rate of periurbanization was identified as one of the various factors, having depleting impacts on forest land cover in the district [29].

Smallholder land users have access to original forest land covers as well as their potential to engage in other forest regrowth systems in their food production activities. However these potentials appear largely untapped. Furthermore, the enormous potentials of mitigating climate change by reducing emission of agriculture-based carbon dioxide with favorable effects on the local warming and climate variability and change are not adequately studied in the Bosomtwe District. This is the point of departure in making original contribution to the literature in general and in Ghana in particular. The main focus of this paper is therefore to assess the agricultural and forest land use potential for REDD+ among smallholder farmers in the Bosomtwe District of the Ashanti Region of Ghana.

2. Materials and Methods

2.1. Profile of the Study Area. The Bosomtwe District is located in the central part of the Ashanti Region. It lies within Latitude 6° 28′N–Latitude 6° 40′N and Longitude 1° 20′W–Longitude 1° 37′W. *Kuntanase* is the District Capital. It spreads over a land area of 330 km² (Figure 1). The district is bounded to the north by *Atwima Nwabiagya* and Kumasi Metropolis and to the east by *Ejisu-Juaben* Municipal. The southern section is bounded by *Amansie* West and East Districts, all in the Ashanti Region of Ghana.

Lake Bosomtwe, the largest natural (crater) lake in West Africa, is located in the district [30]. The lake is also one of the main sources of livelihood for 24 communities living around it [29]. With the exception of the lake which has an outer ridge that maintains a constant distance of 10 km from the center of the lake and stands at an elevation of 50 to 80 m, the rest of the district has no other varying unique topographical features.

The drainage pattern of rivers and streams draining the Bosomtwe District is dendritic and centripetal in outlook. Around Lake Bosomtwe, there is internal drainage where the streams flow from surrounding highlands into the lake in a centripetal fashion. The streams form a dense network due

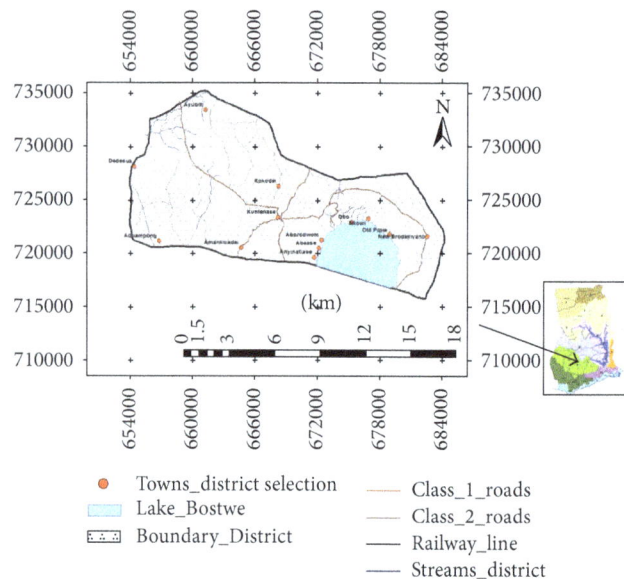

Figure 1: The map of the Bosomtwe District in Ghana.

to the double maxima rainfall regime. Notable rivers in the district are Rivers *Oda, Butu, Siso, Supan,* and *Adanbanwe.*

The district is within the moist semideciduous forest ecological zone with a major and a minor rainfall regime. The major rainfall regime is from March to July while the minor one is from September to November. The zone has mean annual rainfall and mean monthly temperature of about 1,900 mm and 36°C, respectively. Relative humidity ranges between 60% and 85% [31].

The district falls within the moist semideciduous forest zone where different species of tropical hard woods with high economic value can be found. The trees species found in the district include *wawa* (*Triplochiton scleroxylon*), *denya* (*Cylicodiscus gabunensis*), *mahogany* (*Khaya ivorensis*), *asanfena* (*Aningeria* spp.), and *onyina* (*Ceiba pentandra*). However, due to extensive farming activities in the area, the original vegetation has been degraded to a mosaic of secondary forest, thicket, and regrowth with abandoned farms of food crops and vegetables.

In certain parts of the district, however, the original forest cover has been turned into secondary forest and grassland through indiscriminate exploitation of timber and inappropriate farming practices such as the slash and burn system and illegal gold mining activities.

The population of the district according to the Ghana Statistical Service Census is 93,910 with an urban to rural population ratio of approximately 1 : 2 [32]. Proximity of the district to the Kumasi Metropolis is greatly encouraging the growth of settlement in the district. Moreover, the district's tourism potential has drawn a lot of investments in infrastructure development and other socioeconomic activities into the district [29].

2.2. Sampling Design, Instruments, and Data Analysis. The study analyzes agricultural and forest land use potential for

REDD+ among smallholder farmers in the Bosomtwe District. The purposive cluster sampling technique was used to select 12 communities from which 152 smallholder farmers were sampled and data solicited from them. The respondents for the study were sampled using snowball sampling technique because of the difficulty of locating the smallholder farmers as the target population. This was to ensure that the required target respondents (smallholder farmers) were accessed [33]. Using a semistructured partially precoded questionnaire, we administered proportionately to the communities based on their respective population sizes according to the 2011 District Assembly *Scalogram*. The *Scalogram* is the table that details the socioeconomic profile of the district in terms of the percentage availability of social and economic infrastructure. Field observation was also done to facilitate the understanding of possible physical features of the communities that have bearing on the study.

The quantitative data gathered were subjected to frequencies analysis embedded in the Statistical Package for Social Sciences (SPSS) *v.16*, for Windows application. The results are displayed in tables, charts, and graphs. The diagrams generated in the SPSS were exported to Excel for editing for better visual presentation. Open-ended qualitative responses were integrated in the discussions under the various thematic treatments of the sections of the paper.

3. Results and Discussions

3.1. Smallholder Farmers' Knowledge of REDD+. Elbehri et al. [34] espoused that some arguments remain that agriculture is a driver of deforestation in REDD+. The knowledge of REDD+ among smallholder farmers was ascertained. This revealed that awareness of reducing emissions from deforestation and forest degradation (REDD+) and its benefits was diminutive among smallholder farmers in the Bosomtwe District. From Figure 2, only one percent of respondents indicated knowing what REDD+ meant and stood for and the benefits thereof for developing countries (and other stakeholders). The remaining 99% have never heard of REDD+. According to Madeira [6] although the focus of international discourse on REDD+ is on credit design and policy, its success depends on the practicality of implementation in host countries and among local stakeholders.

This is equally premised on their awareness and understanding of what REDD+ is and the appreciation of their role in the implementation of these policies at the local level. Streed et al. [35] posit that ill-informed stakeholders and beneficiaries could unknowingly sell their carbon rights to others even with a policy of equitable distribution of benefits. It is therefore very necessary that these stakeholders be appropriately educated on what REDD+ is, associated responsibilities, and its accruing benefits accordingly.

Although the potentials to achieve climate change mitigation using REDD+ begin with its design, actualization of its purposes lies with the commitment of grassroot stakeholders to this cause [6]. In soliciting for the willingness of farmers to trade off their immediate gains of cutting down trees for future REDD+ benefits, it was found that the majority of the farmers are willing to utilize immediate benefits of cutting

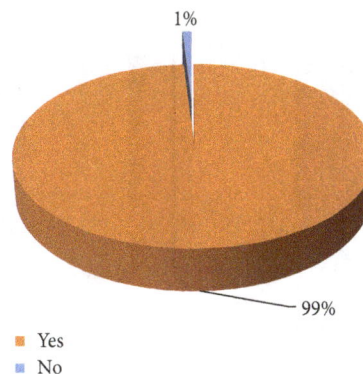

FIGURE 2: Knowledge of REDD+ in the Bosomtwe District.

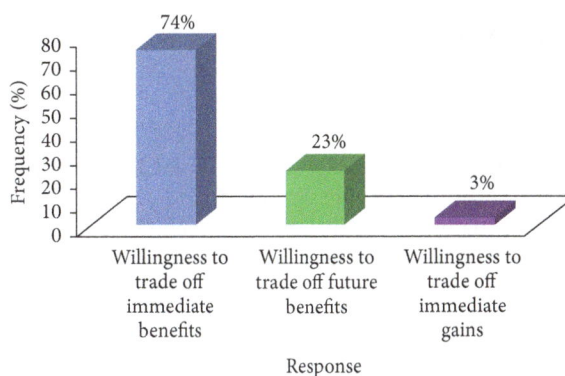

FIGURE 3: Decision on trading off future gains from tress for present needs.

down trees and planting new ones instead. A few were willing to trade off future benefits of REDD+ for present gains (without replanting) and much fewer respondents willing to trade immediate gains of cutting down trees for future benefits from the trees (Figure 3). Farmers did not expect to benefit immediately from preservation of trees. Consequently, trading off their immediate and primary source of livelihood for benefits that may not be forthcoming in the short term did not seem motivating enough to engage in the REDD+ mechanism. Some farmers were also much more willing to cut down the tress owing to the perceived notion of some shade intolerant crops which may not thrive well under trees.

In the same vein, it was found that their willingness to be involved in REDD+ activities was based on training and other supportive measures. This is because most of the smallholder farmers (being 66%) indicated that they were very willing to be involved in REDD+ initiatives if the necessary training and support are provided while 29% were quite willing to engage in REDD+ activities. Only seven percent were not willing to be involved in these activities. This agrees with findings by Banerjee-Woien [36] who asserts that, in Indonesia, the willingness of indigenous people was essential in determining the success of REDD+ initiatives.

Using three benefits of REDD+ as benchmarks, the motivation for smallholder farmers in the Bosomtwe District to be involved in REDD+ initiatives was ascertained. These were

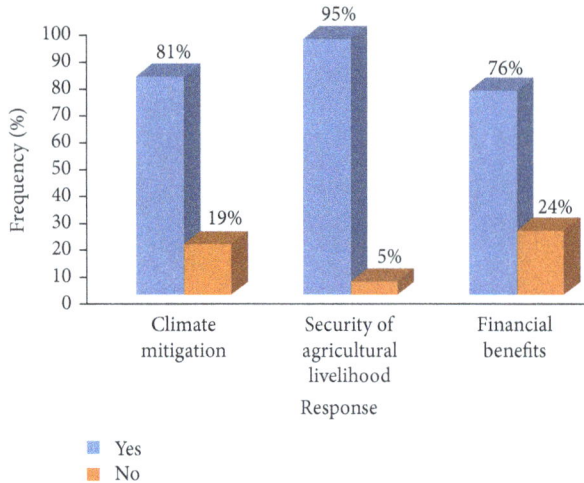

FIGURE 4: Motivation for willingness to be engaged in REDD+ initiatives.

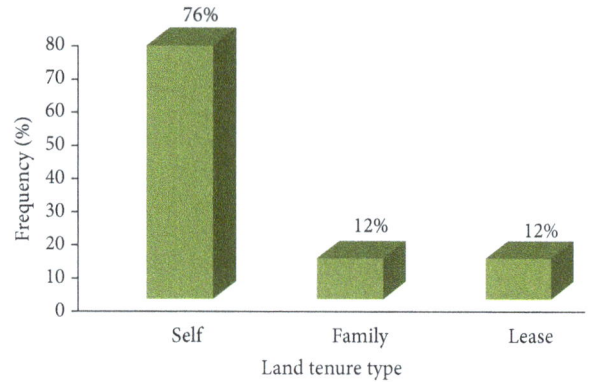

FIGURE 5: Land tenure arrangements among smallholder farmers.

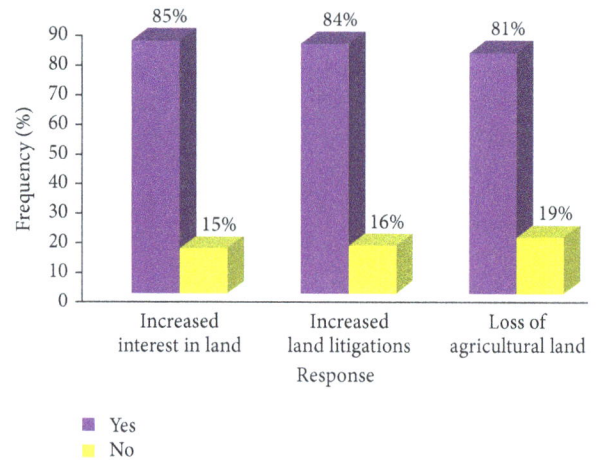

FIGURE 6: Implications for access to REDD+ benefits.

climate mitigation, security of agricultural livelihood, and possible financial benefits. It was found that interest in REDD+ initiatives is underpinned by benefits of climate mitigation, security of agricultural livelihood, and possible financial benefits that could accrue to smallholder farmers and the local communities as a whole. This is shown in Figure 4.

It is quite obvious that smallholder farmers in the district keenly have their livelihood at heart, as the need to secure agricultural livelihood recorded the highest responses, as a motivating factor for the adoption of REDD+ initiatives. Therefore, whether REDD+ does, in fact, deliver on its promised benefits and avoid adverse impacts strongly depends on, among others, fair and equitable benefit sharing, land and carbon tenure in favor of communities, and full and effective participation [37].

3.2. Relationship between Land Tenure Arrangement and REDD+ Benefits.

Creating effective carbon benefit sharing must not lose sight of prevailing land tenure systems, forest and natural resource related livelihoods, and territories [38]. Respondents were required to indicate the land tenure systems they subscribe to and their potential implications for REDD+ in relation to land use. It was found that land tenure system in Bosomtwe District among smallholder farmers is predominantly self-owned as can be seen in Figure 5. Hence 76% of smallholder farmers own their farmlands, 12% of farmlands belong to the families of the farmers, and 12% are leased.

This presupposes that it is quite clear who should be considered in the formulation and implementation of agroforestry projects as well as their respective roles and benefits.

Reference [38] notes that the challenge of potential tenure conflicts could be aggravated by the increased value of land due to carbon benefits accruing from the implementation of REDD+. There is little anticipation among smallholder farmers of increased interest in farmland by land owners because farmland is predominantly self-owned and benefits would accrue to them. Hence, 85% do not anticipate an increased

interest in farmland by land owners while 15% think otherwise. This trend is equally made manifest in their response to the possibility of increase in land litigations due to access to REDD+ benefits. Whiles the majority were not anticipating an increase in land litigation, a few of the respondents do anticipate that an increase in land litigation may likely be an outcome. Figure 3 shows smallholder farmers response to possible implications of access to REDD+ benefits.

Incidentally, while only a small proportion of the respondents do not anticipate a reduction in agricultural land anticipating a competing land use change with the introduction of REDD+ activities, a substantial proportion do anticipate otherwise (Figure 6). Those who anticipated a reduction in agricultural land attributed it to the shade that trees provide and the root systems of trees being less favorable for the tilling of land. The majority who indicated no reduction in agricultural land explained that trees were more helpful and improved soil moisture and nutrients which would buffer reduction in land area due to trees planted.

The Forest Investment Program (FIP), one of the climate funds that target forests and REDD+ activities, had a target ranging from US $1 billion to US $2 billion to support, among other activities, REDD+, afforestation, and sustainable forest management [39]. Grieg-Gran [4] suggests that forest conservation initiative should necessarily support alternative income generation activities among forest communities, sustainable agricultural practices, and agroforestry. Engaging in afforestation and other forest regeneration projects is likely to have implication for agricultural activities at the household scale. Although 64% of smallholder farmers are willing to practice agroforestry as compared to 27% willing to use part of their farms for afforestation, only one percent are willing to convert their entire current cropping system into farm for afforestation, with seven percent not willing to engage in agroforestry or afforestation. The latter's justification for opting out was based on concerns of the possible loss of agricultural land through afforestation and agroforestry.

This is because land is scarce and crops do not do well in the shade. They therefore anticipate that, with the adoption of either practice, agriculture on such lands will invariably not be sustainable. Devoting the entire land to afforestation thereby becomes an inevitable option. It however remains that sustainable agricultural and forest land use and conservation are favorable for the majority of smallholder farmers in the district.

4. Conclusion

Conservation agriculture constitutes an important component for successful REDD+ programs. Awareness of REDD+ and its benefits is low among smallholder farmers in the Bosomtwe District. However, farmers are willing to engage in activities that are geared towards sustainable environmental resource use and conservation. Herein lies the potential for the implementation of REDD+ in the district for local, national, and global benefits. The majority of smallholder farmers in the Bosomtwe District are willing to utilize immediate benefits of cutting down trees and plant new ones instead because they do not expect to benefit immediately from preservation of trees.

Consequently trading off their immediate and primary source of livelihood for benefits that may not be forthcoming in the short term does not seem appropriate. Benefits from REDD+ reaching the grassroots will accrue to farmers as land is generally self-owned. This is one motivating factor for their willingness to engage in REDD+ and related projects. By contributing to the literature, this paper proffers the need for policy to be directed towards the protection of marginal agricultural lands. This would ensure continuous local and regional monitoring of agriculture and forest land use practices in the Bosomtwe District and Ghana.

Conflict of Interests

There is no conflict of interests regarding the publication of this paper.

Acknowledgments

The authors acknowledge the financial support from the West African Science Service Centre for Adapted Land Use (WAS-CAL) as well as the German Ministry of Higher Education and Research. They also thank the Department of Geography and Rural Development for their provision of office space and logistics. They thank Miss Lois Antwi-Boadi for her careful proofreading of this paper.

References

[1] IPCC, "Climate change 2007: impacts, adaptation, and vulnerability," Contribution of Working Group II to the Fourth Assessment Report, Cambridge University Press, Cambridge, UK, 2007.

[2] UN-REDD, "Frequently Asked Questions and Answers—The UN-REDD Programme and REDD+," November 2010, http://www.unep.org/forests/Portals/142/docs/UN-REDD%20FAQs%20%5B11.10%5D.pdf.

[3] Japan International Cooperation Agency (JICA) and International Tropical Timber Organization (ITTO), *Reducing Emissions from Deforestation and Forest Degradation in Developing Countries; The Role of Conservation, Sustainable Management of Forests and Enhancement of Forest Carbon Stocks in Developing Countries*, 2nd edition, 2012, http://www.itto.int/files/user/pdf/publications/Other%20Publications/op-20%20e%20j.pdf.

[4] M. Grieg-Gran, Beyond forestry: why agriculture is key to the success of REDD+, A Briefing, November 2010, http://www.iied.org/pubs/display.php?o=17086IIED.

[5] H. Reid, M. Chambwera, and L. Murray, "Tried and tested: learning from farmers on adaptation to climate change," Gatekeeper, 153, 2013, http://pubs.iied.org/pdfs/14622IIED.pdf.

[6] E. C. M. Madeira, *Policies to Reduce Emissions from Deforestation and Degradation (REDD+) in Developing Countries: An Examination of the Issues Facing the Incorporation of REDD+ into Market-Based Climate Policies*, Resources for the Future, Washington, DC, USA, 2008.

[7] B. Bishaw, H. Neufeldt, J. Mowo et al., *Farmers' Strategies for Adapting to and Mitigating Climate Variability and Change through Agroforestry in Ethiopia and Kenya*, edited by: C. M. Davis, B. Bernart, A. Dmitriev, Forestry Communications Group, Oregon State University, Corvallis, Ore, USA, 2013.

[8] A. Wilkes, T. Tennigkeit, and K. Solymosi, *National Integrated Mitigation Planning in Agriculture: A Review Paper*, Mitigation of Climate Change in Agriculture (MICCA), Food and Agriculture Organization of the United Nations (FAO), Rome, Italy, 2013, http://mahider.ilri.org/handle/10568/27782.

[9] J. S. Mbwambo, P. L. Saruni, and G. S. Massawe, "Agroforestry as a solution to poverty in rural Tanzania. Lessons from Musoma Rural District, Mara Region, Tanzania," *Kivukoni Journal*, vol. 1, no. 2, pp. 15–30, 2013.

[10] K. F. Kalaba, P. Chirwa, S. Syampungani, and O. C. Ajayi, "Contribution of agroforestry to biodiversity and livelihoods improvement in rural communities of Southern African regions," in *Tropical Rainforests and Agroforestry under Global Change, Ecological and Socio-Economic Valuations*, T. Tscharntke, C. Leuschner, E. Veldkamp, H. Faust, E. Guhardja, and A. Bidin, Eds., pp. 461–476, Springer, New York, NY, USA, 2010.

[11] E. Bryan, C. Ringle, B. Okoba, I. Koo, M. Herrero, and S. Sivestri, *Agricultural Land Management: Capturing Synergies among Climate Change Adaptation, Greenhouse Gas Mitigation, and Agricultural Productivity*, International Food Policy Research Institute IFPRI, Washington, DC, USA, 2011.

[12] R. Gledhill, C. Herweijer, D. Hamza-Goodacre, J. Grant, C. Webb, and J. Steege, "Agricultural carbon markets: opportunities and challenges for Sub-Saharan Africa," Rockefeller Foundation, 2011, https://www.pwc.co.uk/assets/pdf/agricultural-carbon-markets.pdf.

[13] Sahara and Sahel Observatory, *Comprehensive Framework of African Climate Change Programmes*, Sahara and Sahel Observatory, 2010, http://www.unep.org/roa/amcen/Amcen_Events/4th_ss/Docs/AMCEN-SS4-INF-3.pdf.

[14] Common Market for Eastern and Southern Africa (COMESA), "Programme on Climate Change Adaptation and Mitigation in the Eastern and Southern Africa (COMESA-EAC-SADC) Region," 2011, http://www.sadc.int/files/9613/5293/3510/COMESA-EAC-SADC_Climate_Change_Programme_2011.pdf.

[15] G. D. Djagbletey and S. Adu-Bredu, "Adoption of agroforestry by small scale teak farmers in Ghana—the case of Nkoranza district," *Ghana Journal of Forestry*, vol. 20, no. 21, pp. 1–13, 2007.

[16] G. B. Adaba, *Natural resource management, governance and globalisation [M.S. thesis]*, Centre For Transdisciplinary Environmental Research, Stockholm University, Stockholm, Sweden, 2005.

[17] J. N. H. Scriven and Y. Malhi, "Smallholder REDD+ strategies at the forest-farm frontier: a comparative analysis of options from the Peruvian Amazon," *Carbon Management*, vol. 3, no. 3, pp. 265–281, 2012.

[18] R. W. Gorte and J. L. Ramseur, "Forest carbon markets: potentials and drawbacks," CRS Report for Congress RL 34560, Congressional Research Service, Washington, DC, USA, 2008.

[19] FAO, "Managing forests for climate change: FAO, working with countries to tackle climate change through sustainable forest management," 2010, http://www.fao.org/docrep/013/i1960e/i1960e00.pdf.

[20] Tanzania National REDD+ Task Force, *Preparing for the REDD+ Initiative in Tanzania: A Synthesized Consultative Report Compiled by Institute of Resources Assessement*, University of Dar es Salaam for National REDD+ Task Force, Dar es Salaam, Tanzania, 2009.

[21] The Energy and Resources Institute (TERI), "Adaptation to climate change in the context of sustainable development, United Nations Department of Economic and Social Affairs, Division for Sustainable Development, Climate Change and Sustainable Development: A Workshop to Strengthen Research and Understanding, New Delhi, 7-8 April," 2013, http://www.teriin.org/events/docs/adapt.pdf.

[22] G. Kissinger, M. Herold, and V. De Sy, *Drivers of Deforestation and Forest Degradation: A Synthesis Report for Redd+ Policymakers*, Lexeme Consulting, Vancouver, Canada, 2012.

[23] E. Jurgens, W. Kornexl, C. Oliver, T. Gumartini, and T. Brown, "Integrating communities into REDD+ in Indonesia," Working Paper, PROFOR, Washington, DC, USA, 2013.

[24] R. Kotru, "Nepal's national REDD+ framework: how to start?" *Journal of Forest and Livelihood*, vol. 8, no. 1, pp. 1–6, 2009.

[25] V. B. Sebukyu and D. M. Mosango, "Adoption of agroforestry systems by farmers in Masaka District of Uganda," *Ethnobotany Research and Applications*, vol. 10, pp. 59–68, 2012.

[26] E. Streed, S. Hajost, and M. Sommervile, "USAID program brief forest carbon, markets and communities," A Program Managed by the Natural Resources Management Office Of USAID, 2012, http://www.fcmcglobal.org/documents/FinanceandCarbonMar-ketsLexiconFinal40clean.pdf.

[27] World Agroforestry Centre, *Socio-Economic Analysis of Farmers' Potential for Adoption of Evergreen Agriculture in Bugesera District, Rwanda*, World Agroforestry Centre, Nairobi, Kenya, 2008.

[28] P. Oino and A. Mugure, "Farmer oriented factors that influence adoption of agroforestry practices in Kenya: experiences from Nambale District, Busia County," *International Journal of Science and Research*, vol. 2, no. 4, pp. 450–456, 2013.

[29] D. O. Appiah, J. T. Bugri, E. K. Forkuor, and P. K. Boateng, "Determinants of peri-urbanization and land use change patterns in Peri-Urban Ghana," *Journal of Sustainable Development*, vol. 7, no. 6, pp. 95–109, 2014.

[30] O. D. Anim, Y. Li, A. K. Agadzi, and P. N. Nkrumah, "Environmental issues of Lake Bosomtwe impact crater in Ghana (West Africa) and its impact on ecotourism potential," *International Journal of Scientific & Engineering Research*, vol. 4, no. 1, pp. 1–9, 2013.

[31] S. Prakash, P. Wieringa, B. Ros et al., "Socio-economics of forest use in the tropics and subtropics: potential of ecotourism development in the Lake Bosumtwi Basin a case study of ankaase in the Amansie East District, Ghana," SEFUT Working Paper 15, Albert-Ludwigs-Universität Freiburg, Freiburg im Breisgau, Germany, 2005.

[32] Ghana Statistical Service, *2010 Populaiton and Housing Census, Summary Report of Final Results*, Ghana Statistical Service, Accra, Ghana, 2010, http://www.statsghana.gov.gh/docfiles/2010-phc/Census2010_Summary_report_of_final_results.pdf.

[33] H. Katz, "Global surveys or multi-national surveys? On sampling for global surveys," in *Proceedings of the Thoughts for the Globalization and Social Science Data Workshop*, p. 6, UCSB, November 2006, http://www.global.ucsb.edu/orfaleacenter/conferences/ngoconference/Katz_for-UCSB-data-workshop.pdf.

[34] A. Elbehri, A. Genest, and M. Burfisher, *Global Action on Climate Change in Agriculture: Linkages to Food Security, Markets and Trade Policies in Developing Countries*, Trade and Markets Division, FAO, Rome, Italy, 2011.

[35] E. Streed, S. Hajost, and M. Sommervile, "USAID Program Brief Forest Carbon, Markets and Communities: A Program Managed by the Natural Resources Management Office of USAID," 2012, http://www.fcmcglobal.org/documents/Finance-andCarbonMarketsLexiconFinal40clean.pdf.

[36] T. Banerjee-Woien, *Trust: a precondition for successful implementation of REDD+ initiatives? [M.S. thesis]*, University of Agder, Kristiansand, Norway, 2010.

[37] Tanzania Natural Resource Forum, "REDD+ Realities: Learning from REDD+ pilot projects to make REDD+ work," 2011, http://www.tnrf.org/files/e-REDD%20Realities.pdf.

[38] UN-REDD+, "The UN-REDD+ Programme Strategy 2011–2015," 2013, http://www.tnrf.org/files/e-REDD%20Realities.pdf.

[39] T. Griffiths, *Seeing REDD+? Avoided Deforestation and the Rights of Indigenous Peoples and Local Communities*, Forest Peoples Programme, Gloucestershire, UK, 2007.

10

Tree Species Richness, Diversity, and Vegetation Index for Federal Capital Territory, Abuja, Nigeria

Aladesanmi D Agbelade,[1] **Jonathan C. Onyekwelu,**[2] **and Matthew B. Oyun**[2]

[1]*Department of Forest Resources and Wildlife Management, Faculty of Agricultural Sciences, Ekiti State University, PMB 5363, Ado Ekiti, Ekiti State, Nigeria*
[2]*Department of Forestry and Wood Technology, School of Agriculture and Agricultural Technology, Federal University of Technology, PMB 704, Akure, Ondo State, Nigeria*

Correspondence should be addressed to Aladesanmi D Agbelade; aladesanmi2008@gmail.com

Academic Editor: Piermaria Corona

This study was conducted to investigate the tree species richness and diversity of urban and periurban areas of the Federal Capital Territory (FCT), Abuja, Nigeria, and produce Normalized Difference Vegetation Index (NDVI) for the territory. Data were collected from urban (Abuja city) and periurban (Lugbe) areas of the FCT using both semistructured questionnaire and inventory of tree species within green areas. In the study location, all trees with diameter at breast height (dbh) ≥ 10 cm were identified; their dbh was measured and frequency was taken. The NDVI was calculated in ArcGIS 10.3 environment using standard formula. A cumulative total of twenty-nine (29) families were encountered within the FCT, with 27 occurring in Abuja city (urban centre) and 12 in Lugbe (periurban centre) of the FCT. The results of Shannon-Wiener diversity index (H') for the two centres are 3.56 and 2.24 while Shannon's maximum diversity index (H_{max}) is 6.54 (Abuja city) and 5.36 (Lugbe) for the urban (Abuja city) and periurban (Lugbe) areas of the Federal Capital Territory (FCT). The result of tree species evenness (Shannon's equitability (E_H) index) in urban and periurban centres was 0.54 and 0.42, respectively. The study provided baseline information on urban and periurban forests in the FCT of Nigeria, which can be used for the development of tree species database of the territory.

1. Introduction

Urban forests are made up of the trees, shrubs, and other vegetative covers that play important role in human life. Urban forests serve important roles such as tree species diversity conservation and protection of fragile ecosystem; development of parks and event centres for relaxation and social engagements; provision of vegetable and fruits/seeds for foods and medicines; and purification of air, wind break, and beautification of the environment [1]. Research in developing countries has revealed that trees are planted around houses for fruits, nuts, leaves, fuelwood, fodder, vegetables, shade, and windbreaks [2]. In traditional African settlements, it is usually common to plant trees in village squares to provide shade for social meetings, ceremonies, recreation, and religious worships. The growing urban population in Nigeria is redefining urban forestry practices

and has presented new challenges and opportunities for researchers.

Urbanization in developing countries is on the increase and this has resulted in social burden in the urban cities due to their limited capacity to adapt to socioeconomic challenges of the new environments [3]. Global projections indicate that future trends in urbanization could triple between 2000 and 2030 [4–7]. According to 2011 revision of World Urbanization Prospects, the urban areas of the world is expected to gain 1.4 billion people between 2011 and 2030, 37% of which will come from China (276 million) and India (218 million). The report predicts that between 2030 and 2050 another 1.3 billion people will be added to the global urban population. With a total addition of 121 million people, Nigeria will be the second major contributor next to India (270 million) [8]. The negative effect of urban expansion includes threat to ecosystem biodiversity and carbon emissions due to tropical deforestation and expansion of social infrastructure [9].

The increasing population of Abuja city, Nigeria's Federal Capital Territory (FCT), has led to a wide range of challenges which has now put lots of pressure on land, forest and forest resources, and green spaces and exposed the city to different environmental hazards. Environmental hazards such as air pollution, wind, and water erosion are on the increase as a result of deforestation and population increase within the FCT and this is evident in different parts of Nigeria cities [2, 10]. In our growing urbanized environments, the maintenance and development of urban vegetated areas are among the challenges of sustainable urban planning [2, 11].

Research has suggested that urban forest and urban green areas could be another effective means of biodiversity conservation and ecosystems potentials in terms of physiological, sociological, economic, and aesthetic benefits [12]. Urban forests could also be used to reduce the challenges posed by urbanization such as food insecurity, energy shortage, deteriorating air quality, high temperatures, health hazards, and increased noise levels [2]. Other benefits derived from urban forests include healthy environment which translates to healthy citizens, beautification of the environment and scenery, cooler air temperature, reductions in ultraviolet radiation, and social and ecological benefits [9, 11, 13]. This study investigates tree species richness, diversity, and vegetation index for Federal Capital Territory, Abuja, Nigeria.

2. Methodology

2.1. Location of Study Area. The study was conducted in Abuja, the Federal Capital Territory (FCT) of Nigeria. The city lies between latitude and $9°03'$ and $9°07'$N and longitude $7°26'$ and $7°39'$E in the North central region of Nigeria. Abuja experiences tropical wet and dry climate. The FCT experiences a warm, humid rainy season and a blistering dry season (harmattan), occasioned by the northeast trade wind, with the main features of dust haze, intensified coldness, and dryness. Rainfall in the FCT reflects the territory's location on the windward side of the Jos Plateau and the zone of rising air masses. Balogun [14] opined that, due to the hilly and mountainous nature of Abuja city, orographic activities bring heavy and frequent rainfall during the rainy season. The rainy season begins in March and ends in November, with peak in September, during which abundant rainfall is received. Mean annual rainfall in Abuja ranges from 1000 mm to 1600 mm. Mean monthly temperature ranges between 25.8°C and 30.2°C [14, 15]. The soils of the study area are basically Alluvial and Luvisols, which supports growth of tree species such as *Khaya* spp., *Parkia biglobosa*, *Delonix regia*, *Eucalyptus* spp., *Azadirachta indica*, and *Gmelina arborea* [16].

Lugbe is one of the popular suburban settlements in Abuja. It is in the Abuja Municipal Area Council (AMAC). It is largely residential and densely populated. Lugbe is about 17 minutes' drive from the Central Business District of Abuja and 13 minutes' drive to the Abuja Airport. Though Lugbe is not in the Federal Capital City (FCC), its proximity to the city centre and also to the Abuja Airport has brought it into

lime light and attracted significant development to the area. The area is developing very fast and it houses the National Space Development and Research Agency, Federal Housing Authority (FHA) Estate, and Voice of Nigeria Transmission Station.

3. Data Collection

The city of Abuja was purposefully selected as the urban sector for this research based on its high infrastructural development, population density, and its economic importance in Nigeria while the closest satellite town (Lugbe) was selected as the periurban settlement. Both semistructured questionnaire and biodiversity assessment were used for data collection. The questionnaire was used to obtain information on the socioeconomic and environmental benefits of urban forest. Twenty respondents were purposefully selected in each urban and periurban centre of the FCT, which translated to a total of forty (40) questionnaires for this study. Administration of questionnaires was done using snowball sampling; respondents were those who own tree(s) or who have association(s) with tree(s). This was supported by earlier study by [11] which used snowball sampling methods due to the peculiarity of urban forest development in Ibadan metropolis. These questionnaires were administered in form of interview guide, such that respondents were requested to complete and return them immediately, thus resulting in 100% retrieval. The study covers 20% of Abuja Federal Capital Territory (FCT) built-up centres. The biodiversity assessment entails complete enumeration of tree species in public parks/garden, private gardens/home gardens, avenue/roadside trees, school grounds, public and private institutions, and any space with conglomerates of trees. Within the selected urban and periurban centre, all trees with diameter at breast height (dbh) ≥ 10 cm were identified; their diameters at breast height, diameters at the base, middle, and top, and total height were measured while their frequencies were taken. All tree species in each city were assigned to families using Keay [17] as guide.

4. Data Processing and Analysis

4.1. Computation of Normalized Difference Vegetation Index (NDVI). Normalized Difference Vegetation Index (NDVI) is one of the most widely used vegetation indices, which is applicable in satellite analysis and monitoring of vegetation cover [18, 19]. In this study, the NDVI was calculated in ArcGIS 10.3 environment using

$$\text{NDVI} = \frac{\text{NIR} - \text{VIS}}{\text{NIR} + \text{VIS}}, \quad (1)$$

where NDVI is Normalized Difference Vegetation Index, NIR is near infrared, and VIS is visible red reflectance.

The value of the pixels varies between −1 and +1; the higher the value of NDVI, the richer and healthier the vegetation cover of such environment.

4.2. Computation of Growth Parameters and Biodiversity Indices. The computations of the following growth parameter and biodiversity indices were undertaken.

4.2.1. Basal Area.

The basal area of all trees in this study area was calculated using

$$BA = \frac{\pi D^2}{4},\tag{2}$$

where BA is basal area (m^2), D is diameter at breast height (cm), and π is pie (3.142). The total BA for the city was obtained by adding all trees BA in the city.

4.2.2. Volume.

Volume of individual trees was estimated using tree volume equation developed Newton's formula as follows [20]:

$$V = \pi h \frac{Db^2 + 4\left(Dm^2\right) + Dt^2}{24},\tag{3}$$

where V is tree volume (m^3), Db, Dm, and Dt are diameters (m) at the base, middle, and top of each tree, and h is total tree height (m).

4.2.3. Species Relative Density (RD).

Species relative density, which is an index for assessing species relative distribution [21], was computed with

$$RD = \left(\frac{n_i}{N}\right) \times 100,\tag{4}$$

where RD (%) is species relative density; n_i is the number of individuals of species i; and N is the total number of all individual trees of all species in the entire community.

4.2.4. Species Relative Dominance (RD$_o$).

Species Relative Dominance (RD$_o$ (%)), used in assessing relative space occupancy of a tree, was estimated using [22]

$$RD_o = \frac{\left(\sum Ba_i \times 100\right)}{\sum Ba_n},\tag{5}$$

where Ba$_i$ is basal area of all trees belonging to a particular species i and Ba$_n$ is basal area of all trees in a city.

4.2.5. Importance Value Index (IVI).

The Importance Value Index (IVI) of each species was computed with the relationship in the following equation [21]:

$$IVI = \frac{(RD + RD_o)}{2}.\tag{6}$$

4.2.6. Species Diversity Index.

Species diversity index (H') was computed using the Shannon-Wiener diversity index in the following equation [23, 24]:

$$H' = -\sum_{i=1}^{s} P_i \ln\left(P_i\right),\tag{7}$$

where H' is Shannon-Wiener diversity index; s is the total number of species in the community; P_i is the proportion of S made up of the ith species; and ln is natural logarithm.

4.2.7. Shannon's Maximum Diversity Index.

Shannon's maximum diversity index was calculated using [24]

$$H_{max} = \ln(S),\tag{8}$$

where H_{max} is Shannon's maximum diversity index and S is the total number of species in the community.

4.2.8. Species Evenness.

Species evenness in each city was determined using Shannon's equitability (E_H), which was obtained using [23]

$$E_H = \frac{H'}{H_{max}} = \frac{-\sum_{i=1}^{s} P_i \ln\left(P_i\right)}{\ln(S)}.\tag{9}$$

4.2.9. Sorensen's Species Similarity Index.

Sorensen's species similarity index between two cities was calculated using [25, 26]

$$SI = \left(\frac{2C}{a+b}\right) \times 100,\tag{10}$$

where C is the number of species in sites a and b and a and b = number of species at sites 1 and 2, respectively.

4.3. Data Analyses.

After retrieval, the questionnaires were coded to obtain quantitative values for statistical analysis. Descriptive analysis was used to summarise the data while correlation analysis was used to investigate the relationships between some biodiversity indices and growth variables. Student's t-test was used to test for significant difference in the growth variables of individual trees in urban and periurban areas for the study. All statistical analyses were undertaken using Statistical Package for Social Sciences (SPSS 20.0) software package. Normalized Difference Vegetation Index of the entire city was calculated and analysed in ArcGIS 10.3 environment to determine the level of greenness of the city between 2000 and 2015.

5. Results

5.1. Biodiversity Indices and Growth Variables.

A cumulative total of twenty-nine (29) families were encountered within the FCT, with 27 occurring in Abuja city (i.e., urban centre) and 12 in Lugbe (i.e., periurban centre) of the FCT (Table 1). Within the urban and periurban areas, families with high number of tree species include Fabaceae, Moraceae, Euphorbiaceae, Combretaceae, Arecaceae, and Myrtaceae. A pooled total of 69 in Abuja city and 20 in Lugbe were identified within the FCT. Numbers of tree species were higher in the urban centre (Abuja city) than in the periurban area (Lugbe) (Table 1). The results of Student's t-test showed that both number of family and tree species were significantly higher in the urban area than the periurban area of the FCT. Based on the results of tree growth parameters, the trees in the urban area were larger than those in the periurban area. Results in Table 1 show that mean dbh (59.3 cm), basal area (51.03 m^2), volume (752.8 m^3), and maximum dbh (212.3 cm) of tree in Abuja city (urban area) were significantly higher than the values recorded for the corresponding parameters

TABLE 1: Summary result of biodiversity indices and growth parameters for tree species in urban and periurban areas of FCT.

Benefits derived	Federal Capital Territory (FCT)	
	Urban forest	Periurban forest
Number of individual trees	695[a]	213[b]
Number of species	69[a]	20[b]
Number of families	27[a]	12[b]
Mean Dbh (cm)	59.3[a]	16.2[b]
Basal area (m^2)	51.03[a]	13.70[b]
Maximum Dbh (cm)	212.3[a]	190.7[b]
Volume (m^3)	752.8[a]	191[b]
Diversity index (H')	3.56[a]	2.24[b]
Max diversity (H_{max})	6.54[a]	5.36[b]
Species evenness (E_H)	0.54[a]	0.42[a]

Values followed by similar letters are not significantly different ($p > 0.05$).

in Lugbe (periurban area). Also, the frequency of occurrence of individual trees in urban area of Abuja was significantly higher than that of the periurban centre of the city.

The results of Shannon-Wiener diversity index (H') for the two centres are 3.56 and 2.24 while Shannon's maximum diversity index (H_{max}) is 6.54 (Abuja city) and 5.36 (Lugbe) for the urban (Abuja city) and periurban (Lugbe) areas of the Federal Capital Territory (FCT). The result of tree species evenness (Shannon's equitability (E_H) index) in urban and periurban centres was 0.54 and 0.42, respectively.

The result also revealed that the species relative density (RD) for individual trees in the urban and periurban centres of the FCT ranged from 0.14 to 10.41% in Abuja city and from 0.47 to 33.80% Lugbe (Tables 2 and 3). Tree species with high relative density (RD) in Lugbe were *Gmelina arborea*, *Parkia biglobosa*, and *Mangifera indica*, accounting for 33.80%, 12.21%, and 9.39%, respectively, while in Abuja city *Gmelina arborea* (10.40%), *Terminalia ivorensis* (6.07%), and *Delonix regia* (5.35%) had high relative density. Species Relative Dominance (RD$_o$) varied from 0.04 to 6.94% in Abuja city and from 0.15 to 20.85% in Lugbe. Within the urban areas (Abuja city), tree species with high Relative Dominance (RD$_o$) were *Spathodea campanulata* (6.94%), *Nauclea latifolia* (5.68%), and *Azadirachta indica* (5.60%) while, in the periurban areas (Lugbe) of the FCT, *Azadirachta indica* (20.85%), *Ricinodendron heudelotii* (16.77%), and *Nauclea diderrichii* (10.59%) dominated. Tree species with high Importance Value Index were *Gmelina arborea* (5.82%), *Azadirachta indica* (3.96%), and *Khaya senegalensis* (3.74%) in Abuja city while *Azadirachta indica* (14.18%), *Parkia biglobosa* (10.09%), and *Ricinodendron heudelotii* (8.62%) were species with high importance in the floristic composition of Lugbe (Tables 2 and 3).

5.2. Benefits Derived from Urban Forests. Urban forest benefits are numerous, important, and beneficial to human livelihood, reduce health related problems, and contribute to the amelioration of both micro- and macroclimates. Table 4 indicates that people are aware of the various benefits derived from urban forest. The majority of respondents (70%) in Lugbe (periurban area) and Abuja (urban area) opined that they derive fresh air from trees around them. A much higher percentage (85%) of respondents in the periurban area derived fuelwood for their cooking purposes from urban forest when compared to the 10% of respondents in urban area who indicated that they sourced fuelwood from urban forests. About 50% and 65% of the respondents in urban and periurban centres, respectively, derived edible fruits from trees around them. Between 40 and 50% of the respondents in urban and periurban centres of the FCT indicated that they derived vegetables from trees around them, which they used for their diet. High percentage (60 to 75%) of respondents in both urban and periurban centres use tree as windbreak to protect their buildings and other structures. The percentage of respondents who use trees for shade for the purpose of social gathering in both urban (Abuja; 60%) and periurban areas (Lugbe; 75%) is relatively high, which is similar to the percentage of those who use urban forest facilities for relaxation/garden/bar/joint purposes (70 to 60%) (Table 4). About 40% and 50% of the respondents in Abuja city and Lugbe, respectively, make use of different parts of trees around them for medicines for curing different diseases (Table 4).

6. Relationship between Tree Growth Variables

Generally, there was positive and significant linear relationship between tree growth variables in both urban (Abuja city) and periurban (Lugbe) areas of the FCT (Tables 5 and 6). The correlation coefficient ranged from 0.16 to 0.96 in the periurban area (Lugbe) and from 0.26 to 0.97 in the urban area (Abuja city). The highest correlation coefficient was obtained between logarithmic transformed basal area and logarithmic transformed volume for trees in Lugbe (periurban area), while it was between height and basal area as well between logarithmic transformed basal area and logarithmic transformed volume for trees in Abuja city (urban area). Very weak correlation was observed between height and basal area (0.16), height and diameter (0.18, 0.26), and diameter and basal area (0.29) for urban and periurban centres in the FCT.

7. Results of Thematic Map Production and Green Area Index of Federal Capital Territory

The thematic maps of the Federal Capita Territory (FCT) were produced from its satellite imagery using ArcGIS 10.3, which led to the generation of Normalized Difference Vegetation Index of the city for the years 2000 and 2015 (Figures 1(a) and 1(b)). Apart from the boundary of each FCT, the locations visited for data collected and NDVI were shown in the maps. The boundary was digitized to determine the size of the FCT, which was georeferenced with the attribute data collected from the field. The maps produced in this research reflected the level of flexibility of GIS in creating, calculating,

TABLE 2: Biodiversity indices and growth parameters of individual trees in the urban areas (Abuja city) of the Federal Capital Territory, Nigeria.

S/N	Tree species	Family	FQ	MHt	MDbh	B.A	Vol.	RD	RD$_O$	IVI
1	*Adansonia digitata*	Bombacaceae	3	4.2	51.0	1.84	7.72	0.43	3.60	2.02
2	*Albizia falcata*	Fabaceae	1	19.8	16.0	0.02	0.40	0.14	0.04	0.09
3	*Albizia ferruginea*	Fabaceae	1	17.0	138.0	1.50	25.43	0.14	2.93	1.54
4	*Alstonia boonei*	Apocynaceae	1	14.6	86.4	0.59	8.56	0.14	1.15	0.65
5	*Anacardium occidentale*	Anacardiaceae	1	18.8	32.0	0.08	1.51	0.14	0.16	0.15
6	*Anogeissus leiocarpa*	Combretaceae	2	17.1	84.1	2.22	37.96	0.29	4.35	2.32
7	*Azadirachta indica*	Meliaceae	16	19.1	11.9	2.86	54.56	2.31	5.60	3.96
8	*Bambusa vulgaris*	Poaceae	3	16.6	12.8	0.12	1.93	0.43	0.23	0.33
9	*Bauhinia monandra*	Fabaceae	5	12.7	11.7	0.27	3.44	0.72	0.53	0.63
10	*Bauhinia polyantha*	Leguminosae	1	18.5	105.0	0.87	16.02	0.14	1.70	0.92
11	*Blighia sapida*	Sapindaceae	1	13.4	65.0	0.33	4.45	0.14	0.65	0.40
12	*Bosqueia angolensis*	Moraceae	24	18.8	3.2	0.46	8.66	3.47	0.90	2.19
13	*Caesalpinia pulcherrima*	Fabaceae	3	11.7	6.0	0.03	0.30	0.43	0.05	0.24
14	*Calliandra calothyrsus*	Leguminosae	7	8.6	8.3	0.26	2.26	1.01	0.52	0.76
15	*Callitris intratropica*	Cupressaceae	1	16.8	112.0	0.99	16.55	0.14	1.93	1.04
16	*Cassia pleurocarpa*	Caesalpiniaceae	7	8.4	4.6	0.08	0.69	1.01	0.16	0.59
17	*Casuarina equisetifolia*	Casuarinaceae	11	10.8	6.6	0.41	4.41	1.59	0.80	1.19
18	*Ceiba pentandra*	Bombacaceae	6	9.5	21.6	1.31	12.48	0.87	2.57	1.72
19	*Citrus aurantifolia*	Rutaceae	9	14.2	7.9	0.39	5.61	1.30	0.77	1.04
20	*Cleistopholis patens*	Annonaceae	5	15.6	17.4	0.59	9.27	0.72	1.17	0.94
21	*Cocos nucifera*	Arecaceae	11	8.4	5.3	0.26	2.22	1.59	0.52	1.05
22	*Cola gigantean*	Malvaceae	1	15.8	109.0	0.93	14.75	0.14	1.83	0.99
23	*Corymbia citriodora*	Myrtaceae	1	13.4	36.0	0.10	1.36	0.14	0.20	0.17
24	*Dacryodes edulis*	Burseraceae	1	12.6	74.8	0.44	5.54	0.14	0.86	0.50
25	*Daniellia oliveri*	Leguminosae	20	12.8	7.2	1.62	20.70	2.89	3.17	3.03
26	*Delonix regia*	Fabaceae	37	14.2	1.2	0.17	2.38	5.35	0.33	2.84
27	*Drypetis* spp.	Nymphalidae	4	15.0	23.1	0.67	10.06	0.58	1.31	0.95
28	*Elaeis guineensis*	Arecaceae	1	14.2	47.8	0.18	2.55	0.14	0.35	0.25
29	*Eucalyptus camaldulensis*	Myrtaceae	1	11.0	42.0	0.14	1.52	0.14	0.27	0.21
30	*Eucalyptus citriodora*	Myrtaceae	32	17.6	4.3	1.46	25.65	4.62	2.86	3.74
31	*Eucalyptus forrestiana*	Myrtaceae	10	11.0	4.2	0.14	1.50	1.45	0.27	0.86
32	*Euphorbia dendroides*	Euphorbiaceae	1	8.0	52.0	0.21	1.70	0.14	0.42	0.28
33	*Ficus goliath*	Moraceae	14	15.5	5.0	0.38	5.93	2.02	0.75	1.39
34	*Gmelina arborea*	Verbenaceae	72	14.0	1.2	0.63	8.83	10.40	1.24	5.82
35	*Hildegardia baterii*	Malvaceae	1	15.7	89.6	0.63	9.90	0.14	1.24	0.69
36	*Hura crepitans*	Euphorbiaceae	8	11.6	10.8	0.58	6.74	1.16	1.14	1.15
37	*Jatropha curcas*	Euphorbiaceae	1	4.0	16.0	0.02	0.08	0.14	0.04	0.09
38	*Khaya senegalensis*	Meliaceae	32	17.6	4.3	1.46	25.65	4.62	2.86	3.74
39	*Lophira alata*	Ochnaceae	1	15.2	115.0	1.04	15.79	0.14	2.04	1.09
40	*Lophira procera*	Ochnaceae	3	12.3	15.4	0.17	2.07	0.43	0.33	0.38
41	*Mangifera indica*	Anacardiaceae	20	11.1	2.1	0.13	1.48	2.89	0.26	1.58
42	*Milicia excelsa*	Moraceae	2	19.6	57.5	1.04	20.36	0.29	2.04	1.16
43	*Milicia regia*	Moraceae	1	14.4	90.1	0.64	9.18	0.14	1.25	0.70
44	*Millettia thonningii*	Fabaceae	5	13.5	14.5	0.41	5.59	0.72	0.81	0.77
45	*Morinda lucida*	Rubiaceae	1	14.8	68.8	0.37	5.50	0.14	0.73	0.44
46	*Moringa oleifera*	Moringaceae	2	17.0	16.5	0.09	1.45	0.29	0.17	0.23
47	*Nauclea diderrichii*	Rubiaceae	2	15.5	68.0	1.45	22.49	0.29	2.84	1.57
48	*Nauclea latifolia*	Rubiaceae	2	14.3	96.1	2.90	41.45	0.29	5.68	2.99
49	*Parkia biglobosa*	Fabaceae	26	16.3	4.5	1.09	17.80	3.76	2.14	2.95
50	*Pueraria phaseoloides*	Leguminosae	1	13.5	132.0	1.37	18.48	0.14	2.68	1.41

TABLE 2: Continued.

S/N	Tree species	Family	FQ	MHt	MDbh	B.A	Vol.	RD	RD_O	IVI
51	*Pinus caribaea*	Pinaceae	7	11.9	8.0	0.25	2.95	1.01	0.49	0.75
52	*Pinus caribaea*	Pinaceae	6	7.1	14.7	0.61	4.36	0.87	1.20	1.04
53	*Plumeria alba*	Apocynaceae	19	12.1	3.7	0.39	4.78	2.75	0.77	1.76
54	*Polyalthia longifolia*	Annonaceae	31	13.2	2.5	0.48	6.32	4.48	0.94	2.71
55	*Psidium guajava*	Myrtaceae	3	19.4	14.2	0.14	2.77	0.43	0.28	0.36
56	*Ravenala madagascariensis*	Strelitziaceae	1	15.4	36.0	0.10	1.57	0.14	0.20	0.17
57	*Ricinodendron heudelotii*	Euphorbiaceae	1	6.8	171.0	2.30	15.62	0.14	4.50	2.32
58	*Roystonea dunlapiana*	Arecaceae	24	16.6	5.6	1.40	23.31	3.47	2.75	3.11
59	*Roystonea regia*	Arecaceae	21	17.8	3.8	0.49	8.79	3.03	0.97	2.00
60	*Senna siamea*	Fabaceae	6	21.2	14.8	0.62	13.13	0.87	1.21	1.04
61	*Spathodea campanulata*	Bignoniaceae	1	22.2	212.3	3.54	78.60	0.14	6.94	3.54
62	*Tectona grandis*	Lamiaceae	11	16.0	10.7	1.09	17.50	1.59	2.14	1.87
63	*Terminalia catappa*	Combretaceae	37	7.9	0.7	0.05	0.37	5.35	0.09	2.72
64	*Terminalia ivorensis*	Combretaceae	42	9.1	0.7	0.08	0.70	6.07	0.15	3.11
65	*Terminalia macroptera*	Combretaceae	31	11.6	3.5	0.92	10.71	4.48	1.81	3.14
66	*Terminalia superba*	Combretaceae	9	12.7	13.6	1.17	14.85	1.30	2.29	1.80
67	*Treculia africana*	Moraceae	1	6.0	85.8	0.58	3.47	0.14	1.13	0.64
68	*Vachellia nilotica*	Fabaceae	17	9.2	3.6	0.30	2.74	2.46	0.58	1.52
69	*Vitex doniana*	Verbenaceae	2	16.4	42.5	0.57	9.31	0.29	1.11	0.70
			692			51.02	752.75			

FQ: number of tree stems in the city, B.A.: basal area of trees in the city, Vol.: volume of trees in the city.

TABLE 3: Biodiversity indices and growth parameters of individual trees in the periurban areas (Abuja city) of the Federal Capital Territory, Nigeria.

S/N	Tree species	Family	FQ	MHt	MDbh	B.A	Vol.	RD	RD_O	IVI
1	*Albizia falcate*	Fabaceae	1	19.8	16.0	0.02	0.40	0.47	0.15	0.31
2	*Azadirachta indica*	Meliaceae	16	19.1	11.9	2.86	54.56	7.51	20.85	14.18
3	*Cleistopholis patens*	Annonaceae	5	15.6	17.4	0.59	9.27	2.35	4.34	3.34
4	*Cocos nucifera*	Arecaceae	11	8.4	5.3	0.26	2.22	5.16	1.93	3.55
5	*Delonix regia*	Fabaceae	17	14.2	2.7	0.17	2.38	7.98	1.22	4.60
6	*Elaeis guineensis*	Arecaceae	1	14.2	47.8	0.18	2.55	0.47	1.31	0.89
7	*Euphorbia dendroides*	Euphorbiaceae	1	8.0	52.0	0.21	1.70	0.47	1.55	1.01
8	*Ficus goliath*	Moraceae	4	15.5	17.5	0.38	5.93	1.88	2.79	2.34
9	*Gmelina arborea*	Verbenaceae	72	14.0	1.2	0.63	8.83	33.80	4.60	19.20
10	*Hildegardia baterii*	Malvaceae	1	15.7	89.6	0.63	9.90	0.47	4.60	2.54
11	*Jatropha curcas*	Euphorbiaceae	1	4.0	16.0	0.02	0.08	0.47	0.15	0.31
12	*Mangifera indica*	Anacardiaceae	20	11.1	2.1	0.13	1.48	9.39	0.97	5.18
13	*Nauclea diderrichii*	Rubiaceae	2	15.5	68.0	1.45	22.49	0.94	10.59	5.76
14	*Parkia biglobosa*	Fabaceae	26	16.3	4.5	1.09	17.80	12.21	7.97	10.09
15	*Ravenala madagascariensis*	Strelitziaceae	1	15.4	36.0	0.10	1.57	0.47	0.74	0.61
16	*Ricinodendron heudelotii*	Euphorbiaceae	1	6.8	171.0	2.30	15.62	0.47	16.77	8.62
17	*Senna siamea*	Fabaceae	6	21.2	14.8	0.62	13.13	2.82	4.52	3.67
18	*Terminalia superba*	Combretaceae	9	12.7	13.6	1.17	14.85	4.23	8.53	6.38
19	*Treculia Africana*	Moraceae	1	6.0	85.8	0.58	3.47	0.47	4.22	2.35
20	*Vachellia nilotica*	Fabaceae	17	9.2	3.6	0.30	2.74	7.98	2.18	5.08
			213			13.7	191.0			

FQ: number of tree stems in the city, B.A.: basal area of trees in the city, Vol.: volume of trees in the city.

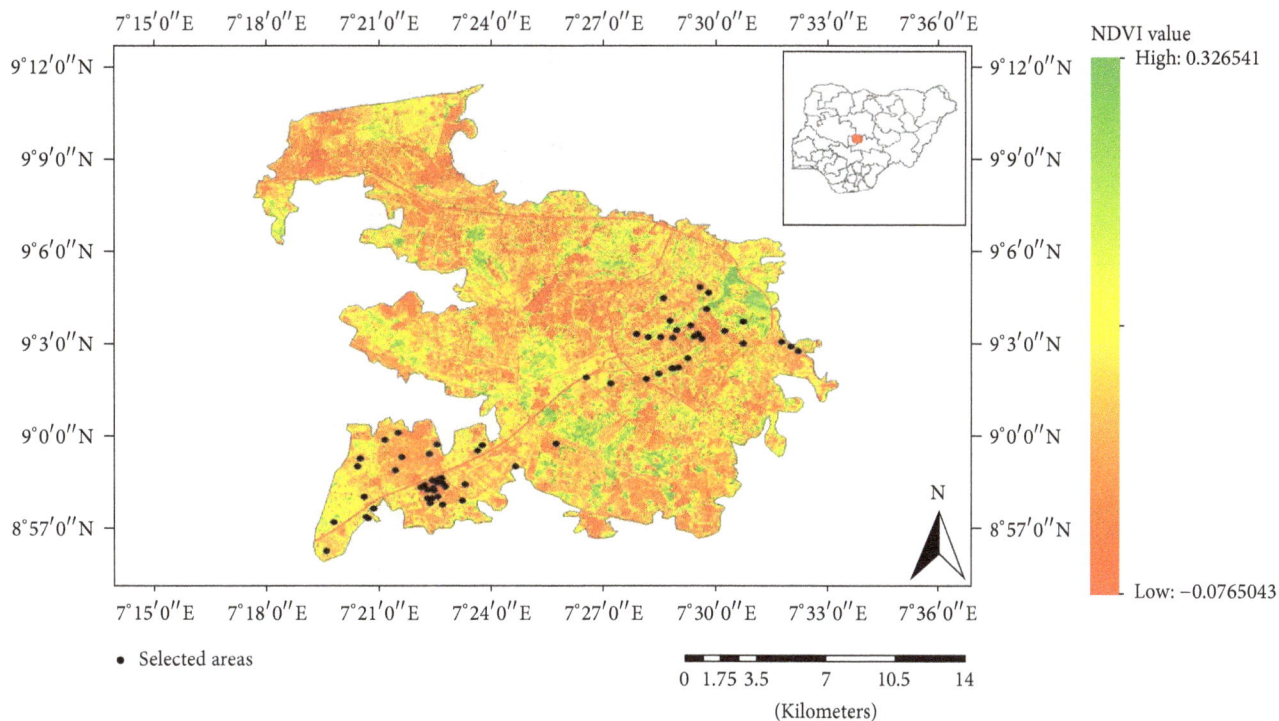

(a)

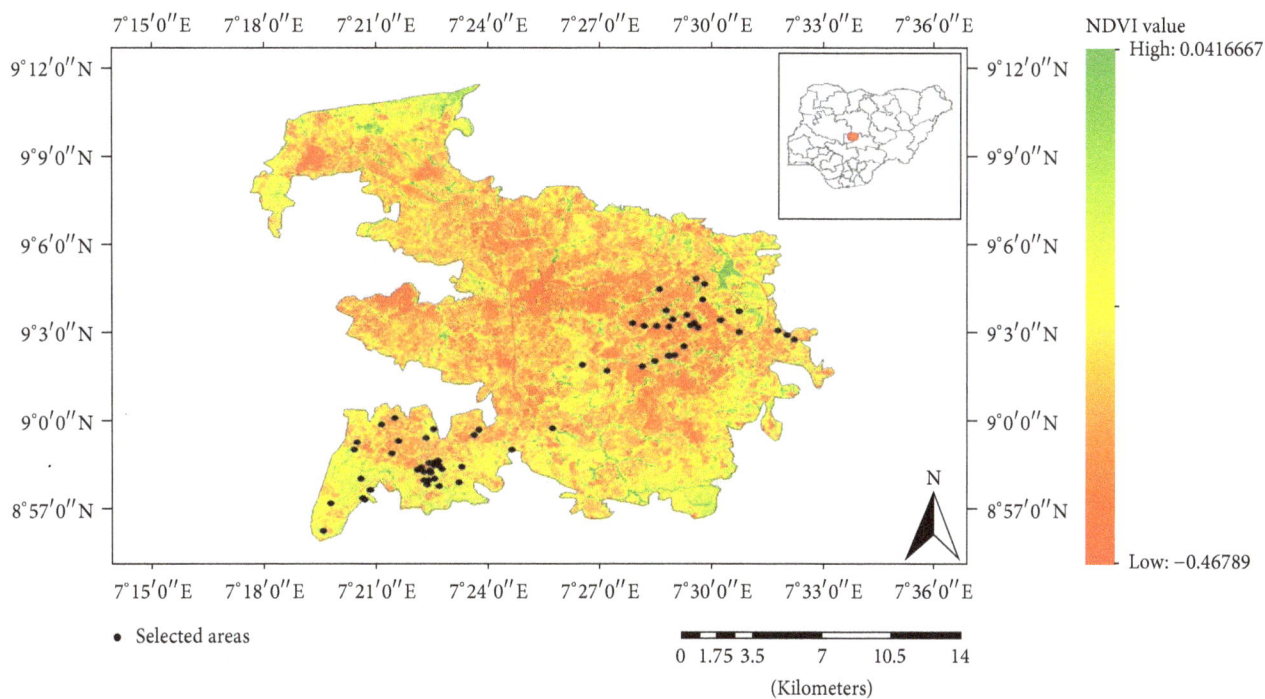

(b)

Figure 1: (a) Normalized Difference Vegetation Index (NDVI) map for Abuja for the year 2000 and sampling locations. (b) Normalized Difference Vegetation Index (NDVI) map for Abuja for the year 2015 and sampling locations.

and manipulating different research data for production of different maps. Normalized Difference Vegetation Index (NDVI) values of the city were calculated in ArcGIS 10.3 environment to determine the level of greenness of the city.

The NDVI value (0.327) for 2000 was relatively low but moderate. There was drastic reduction in the NDVI value (0.042) for 2015 which may be an indication of reduction and scattering in the vegetative cover of the FCT as a result

TABLE 4: Benefits derived by respondents (%) from urban and periurban forests in the Federal Capital Territory, Nigeria.

Benefits derived	Abuja (FCT)	
	Urban area (%)*	Periurban area (%)*
Medicine (herbs)	40	50
Relaxation/garden	70	60
Animal fodder	15	50
Fuel wood (cooking)	10	85
Shade (meetings)	65	70
Vegetable (soup)	50	40
Fresh air	70	70
Beautification	90	50
Edible fruits (food)	50	65
Windbreak	60	75

*Respondents were allowed to choose more than one option (i.e., multiple responses).

of infrastructural development and population increase. The low value of NDVI in year 2015 has greater effects on the Land Surface Temperature (LST) and climate change of the city. The lower the NDVI value, the higher the LST, which tend to expose the city to more higher temperature and the climate change on the populace. Increase in NDVI value and reduction in LST value will create conducive environment for increase in tree diversity and green space coverage.

8. Discussion

8.1. Biodiversity Indices and Tree Growth Yield Variables. Forest ecosystems are pivotal to the functioning and conserving of biosphere, as they are the origin of many plants and animals [27]. The urban forests include trees planted along streets, compounds, school gardens, parks, riverbanks, city cemeteries, city vacant lots, adjacent wood land, and anywhere else where trees grow in urban areas [28, 29]. Urban forests are among the in situ conservation methods that are essential in protecting and conserving tree species for the greatest benefits of the society. The results of this study confirmed that urban forest is a repository of many indigenous tropical hardwood and exotic tree species in different families, judging by the tree species richness of Abuja city, which is similar to or higher than what has been reported in some natural forest ecosystems in Nigeria [30–33]. For example, [30, 31] reported 31 and 51 trees species in tropical rainforest ecosystems of southwestern Nigeria. A total of 56, 55, and 54 tree species were found in Sapoba, Shasha, and Ala forest reserves in Nigeria, respectively [32, 33]. Thus, the large numbers of tree species that are found in tropical forests [30, 31] could also be said to be characteristics of urban forest landscapes. This similarity of tree species richness of urban centres and natural forest ecosystems underscores the importance of urban forests in biodiversity conservation and thus is evidence that urban forests can be both reservoirs and contributor to global biodiversity conservation.

Biodiversity indices of urban and periurban forests are generated in order to appreciate the level of diversity and abundance of species in built-up environment. IIRS [34] noted that biodiversity indices are generated to bring the diversity and abundance of species in different habitats to similar scale for comparison and the higher the value, the greater the species richness.

Konijnenddijk et al. [35] had noted that the levels of biodiversity in urban and periurban areas are often surprisingly high even when compared with forest trees diversity. The Shannon-Wiener diversity and Evenness indices of this study are higher than the values of Parthasarathy [36] and Yang et al. [37]. In another study, Duran et al. [38] obtained Shannon-Wiener diversity index range of between 2.69 and 3.33 which is within the range of the values reported for urban and periurban areas of FCT in this study. The Shannon-Wiener diversity index of Abuja city (i.e., urban area) in the study is within the range of the values reported for some natural tropical forest ecosystems [30, 31, 33] as well as the values obtained by Agbelade et al. [11] for urban forests in Ibadan, southwestern Nigeria. However, the Shannon-Wiener diversity index of Lugbe (i.e., periurban area) is lower than the values reported by the studies mentioned above. The implication of this is that urban centres have higher species diversity than periurban areas, which is further confirmed by the higher species richness in Abuja city (urban area) compared to that of Lugbe (periurban area).

Though Alvey [39] reported that, traditionally, urban forest areas have been regarded as locations of low biodiversity that are dominated by nonnative species; evidence from this study as well as those from published information is mounting that urban and suburban areas can contain relatively high levels of biodiversity [40–42]. Contrary to the conclusion of Alvey [39], there are indications from the results of this study that urban and periurban forests have good store of both native and nonnative species. However, a higher percentage of the very common species are exotic species. Our results showed that *Azadirachta indica*, *Eucalyptus* species, *Acacia* species, and *Gmelina arborea* are the four most common species in Abuja, which is in agreement with the earlier study of Fuwape and Onyekwelu [2]. Three of these four most common species are exotic species. The tree planting and landscaping mechanism in Abuja may be responsible for the high level of tree species diversity encountered in it. The infrastructural development in Abuja city had low impact on the level of tree species diversity, which could be attributed to the careful planning, managing, and planting of trees. Thus it could be concluded that infrastructural development in Abuja city did not negatively affect the extent of biodiversity in it. This could imply that landscaping of the city was considered alongside with the infrastructural development. The spatial arrangement of trees as well as the choice and combination of tree species clearly distinguishes Abuja as a well-landscaped city. Nowak and Walton [43] noted that expanding urbanization increases the importance of urban forests in terms of their extent and the critical ecosystem services they provide to sustain human health and environmental quality in and around urban areas. In Guangzhou city, China, Jim and Liu [44] found over 250 plant species

TABLE 5: Correlation coefficients for tree growth variables in periurban area (Lugbe).

	Height	Diameter	Basal area (BA)	Volume (Vol.)	Ln BA	Ln Vol.
Height	1.00					
Diameter	.18	1.00				
Basal area (BA)	.16	.96	1.00			
Volume (Vol.)	.41	.86	.90	1.00		
Ln BA	.17	.93	.81	.71	1.00	
Ln Vol.	.44	.88	.76	.74	.96	1.00

TABLE 6: Correlation coefficients for tree growth variables in urban area (FCT, Abuja).

	Height	Diameter	Basal area (BA)	Volume (Vol.)	Ln BA	Ln Vol.
Height	1.00					
Diameter	.26	1.00				
Basal area (BA)	.97	.26	1.00			
Volume (Vol.)	.87	.47	.93	1.00		
Ln BA	.94	.25	.83	.73	1.00	
Ln Vol.	.91	.49	.80	.76	.96	1.00

after surveying over 115,000 plants in the parks on university grounds and along streets. This affirmed that there are relatively high numbers of stems found in urban centre, which could be higher than the number in conservation centres. Araújo [45] also asserted that human actions directly or indirectly increase the total number of species through introduced species and increase in landscape heterogeneity.

8.2. Benefits Derived from Urban and Periurban Forests. Tree species encountered in urban and periurban centres of the Federal Capital Territory, Abuja, are used for food (e.g., edible fruits/seeds, vegetables), nutrition supplement, medicinal substances, fuelwood, and animal fodder. Onyekwelu and Olaniyi [10] had observed that urban forestry practices improve food security of poor urban inhabitants through collection of wild edible plants as vegetable, planting of low-care fruit bearing trees, including a gardening component, in multifunctional parks, or creating edible public parks. There are environmental benefits derived from urban forest such as purification of air (fresh air) wind break, provision of shade, beautification, relaxation parks, and gardens. Fuelwood, animal fodder, and edible fruits are among urban forest products sold for income generation. The higher percentage of respondents who collected fuelwood and animal fodder in Lugbe (periurban centres) compared to Abuja city (urban centre) could be compared. This could be an indication that higher percentage of periurban residents is more interested in income generation from urban forests. Income generation from urban and periurban forest products in Akure, Nigeria, ranged from $30 to $150 [10, 46], which could further attest economic viability of urban forests in developing countries. Parks and recreation centres employ people as service men and women; thus the livelihood of these people depends on the income from these parks and recreation centres. Recreational centres differ from one another because of their

relatively small size and type of use with relative amount of income generation for the people and government [10]. Recreation centres are suitable places for causal meetings, lunch outing, association meeting, family relaxation, business meetings, and holiday relaxation. Fuwape and Onyekwelu [2] observed that parks and recreation centres in cities across West Africa serve as small businesses centres, community meeting, religious worship centres, and shades for groves in some urban and periurban centres.

8.3. Green Index Mapping Using Normalized Difference Vegetation Index (NDVI). The low value of Normalized Difference Vegetation Index (NDVI) generated for Abuja proved that the vegetation coverage is scattered around the cities. In urban area, green spaces play important role in the quality of life and healthy nature of the ecosystem, which will improve the healthy nature of communities. The Federal Capital Territory (FCT) NDVI value indicated the presence of shrubs and scattered trees around the city. Xu et al. [47] and Miguel-Anyanz and Biging [48] opined that NDVI values of 0.1 to −0.1 represent degraded land; moderate values of 0.2 to 0.3 represent shrubs and grass land while 0.6 to 0.8 represent tropical forest land with vegetation coverage. Gillies and Carlson [49] reported that vegetation coverage has different impacts on recreation potential and microclimate of the environment, as well as improving the socioeconomic values of green spaces. Normalized Difference Vegetation Index (NDVI) is important parameter for the determinant of urban and periurban climate change and the Land Surface Temperature (LST), which relate the temperature generation and the cooling system of the vegetation coverage [50]. High level of tree species richness will enhance healthy vegetation coverage and reduces temperature of the atmosphere through evapotranspiration processes in green vegetation, which would be achieved if the high tree species richness in Abuja city is both maintained and improved. There is also need to substantially improve the vegetation cover of Lugbe, which is expected to contribute to reducing the usual high seasonal temperature. Miller [51] and Türk and Hastaoğlu [52] reported that areas where Land Surface Temperature (LST) is low, the Normalized Difference Vegetation Index (NDVI) values measurement are usually high which a sign of healthy vegetation coverage of the city. This can as well translate to healthy environment with healthy humans and other animals; it can also increase economic values of households.

9. Conclusion and Recommendation

The result of this research has provided baseline information on urban and periurban forests in the Federal Capital Territory (FCT) of Nigeria (Abuja), which can be used for the development of tree species database of the territory. The potentials of urban and periurban forests in conserving biodiversity and providing essential products and services towards environmental management, economic empowerment, and social services to the society were revealed. Different goods (edible fruits/seeds, vegetables, fuelwood, herbs, animal fodder, etc.) and services (parks, windbreak, pollution reduction, beautification, etc.) were provided by urban forests in the FCT. The sale of tree products provided much needed income, especially for periurban inhabitants.

This similarity of tree species richness of Abuja city (urban centre) and some natural forests confirms and underscores the importance of urban forests in biodiversity conservation; it is evidence that urban forests can be reservoirs and contributor to biodiversity conservation. The urban centre (Abuja city) had higher tree species richness and diversity than the periurban areas (Lugbe). The high species richness and diversity in Abuja city, despite its high infrastructural development, showed that infrastructural development in the city did not negatively affect its biodiversity conservation potential. The spatial arrangement of trees as well as the choice and combination of tree species shows that Abuja is a well-landscaped city. The maintenance and improvement of the high tree species richness and diversity in Abuja city will enhance healthy vegetation coverage and reduce temperature. There is need to substantially improve the vegetation cover of Lugbe in order to reduce its usually high seasonal temperature.

Forestry extension services should do more in educating the people on the benefits, importance, and contributions of urban forest to the environment and the people. It is important for government at all levels to be involved in planting trees in urban areas and create measures for the development of urban forests. Therefore, during the process of construction, expansion and infrastructure development attention should be paid to conserving trees rather than cutting them down.

Competing Interests

The authors declare that there is no conflict of interests regarding the publication of this paper.

Acknowledgments

This research was supported by the International Foundation for Science (IFS), Stockholm, Sweden, through a grant (D/5609-1) to Aladesanmi D. Agbelade. The authors also acknowledge International Institute of Tropical Agriculture (IITA), Ibadan, Nigeria, for scholarship in data processing and Normalized Difference Vegetation Index (NDVI) Map Production.

References

[1] D. J. Nowak and J. F. Dwyer, "Understanding the benefits and costs of urban forest ecosystems," in *Urban and Community Forestry in the Northeast*, J. E. Kuser, Ed., pp. 25–46, Springer, Dordrecht, The Netherlands, 2nd edition, 2007.

[2] J. A. Fuwape and J. C. Onyekwelu, "Urban forest development in West Africa: benefits and challenges," *Journal of Biodiversity and Ecological Sciences*, vol. 1, no. 1, pp. 77–94, 2011.

[3] UN-Habitat, *State of African Cities; Re-Imagining Sustainable Urban Transitions*, UN-Habitat Rapidly Expanding Regional Ststes of Cities Report Series, Design and Layout by Michael Jones Software (MJS), Nairobi, Kenya, 2014.

[4] United Nations, *World Urbanization Prospects: The 2014 Revision, Department for Economic and Social Affairs*, United Nations, New York, NY, USA, 2014.

[5] S. Angel, S. C. Sheppard, and D. L. Civco, *The Dynamics of Global Urban Expansion*, World Bank, Transport, Urban Development Department, Washington, DC, USA, 2011.

[6] M. Rajkumar and N. Parthasarathy, "Changes in forest composition and structure in three sites of tropical evergreen forest around Serigaltheri, Western Ghats," *Current Science*, vol. 53, pp. 389–393, 2008.

[7] K. C. Seto, A. Reenberg, C. G. Boone et al., "Urban land teleconnections and sustainability," *Proceedings of the National Academy of Sciences*, vol. 109, no. 20, pp. 7687–7692, 2012.

[8] P. O. Akunnaya and O. Adedapo, "Trends in urbanisation: implication for planning and low-income housing delivery in Lagos, Nigeria," *Architecture Research*, vol. 4, no. 1, pp. 15–26, 2014.

[9] J. F. Dwyer, D. J. Nowak, M. H. Noble, and S. M. Sisinni, "Connecting people with ecosystems in the 21st century: an assessment of our nation's urban forests," General Technical Report PNW-GTR-490, U.S. Department of Agriculture, Forest Service, PNRS, Portland, Ore, USA, 2000.

[10] J. C. Onyekwelu and D. B. Olaniyi, "Socio-economic importance of Urban and peri-urban forests in Nigeria," in *Proceedings of the 6th Annual Conference of SAAT, FUTA*, Adebayo, Ed., pp. 200–210, Akure, Nigeria, November 2012.

[11] A. D. Agbelade, J. C. Onyekwelu, and O. Apogbona, "Assessment of Urban tree species population and diversity in Ibadan, Nigeria," *Environmental and Ecology Research*, vol. 4, no. 4, pp. 185–192, 2016.

[12] C. C. Konijnendijk, R. M. Ricard, A. Kenney, and T. B. Randrup, "Defining urban forestry—a comparative perspective of North America and Europe," *Urban Forestry & Urban Greening*, vol. 4, no. 3-4, pp. 93–103, 2006.

[13] L. M. Westphal, "Urban greening and social benefits: a study of empowerment outcomes," *Journal of Arboriculture*, vol. 29, no. 3, pp. 137–147, 2003.

[14] O. Balogun, *The Geography of Its Development*, The Federal Capital Territory University Press, Ibadan, Nigeria, 2001.

[15] P. E. Adakayi, "Climate," in *Geography of Abuja, Federal Capital Territory*, P. D. Dawam, Ed., Famous/Asanlu Publishers, Abuja, Nigeria, 2000.

[16] F. Ujoh, I. D. Kwabe, and O. O. Ifatimehin, "Understanding urban sprawl in the Federal Capital City, Abuja: towards sustainable urbanization in Nigeria," *Journal of Geography and Regional Planning*, vol. 3, no. 5, pp. 106–113, 2010.

[17] R. W. J. Keay, *Trees of Nigeria*, Oxford University Press, Oxford, UK, 1989.

[18] H. Liu and A. R. Huete, "A feedback based modification of the NDVI to minimize canopy background and atmospheric noise," *IEEE Transactions on Geoscience and Remote Sensing*, vol. 33, pp. 457–465, 1995.

[19] M. C. Imhoff, W. T. Lawrence, C. D. Elvidge et al., "Using nighttime DMSP/OLS images of city lights to estimate the impact of urban land use on soil resources in the United States," *Remote Sensing of Environment*, vol. 59, pp. 105–117, 1997.

[20] B. Husch, T. W. Beers, and J. A. Keenshaw Jr., *Forest Mensuration*, John Wiley & Sons, Hoboken, NJ, USA, 4th edition, 2003.

[21] M. B. Brashears, M. A. Fajvan, and T. M. Schuler, "An assessment of canopy stratification and tree species diversity following clearcutting in central Appalachian hardwoods," *Forest Science*, vol. 50, no. 1, pp. 54–64, 2004.

[22] M. P. Aidar, J. R. Godoy, J. Bergmann, and C. A. Joly, "Atlantic Forest succession over calcareous soil, Parque Estadual Turístico do Alto Ribeira—PETAR, SP," *Revista Brasileira de Botânica*, vol. 24, no. 4, pp. 455–469, 2001.

[23] M. Kent and P. Coker, *Vegetation Description and Analysis: A Practical Approach*, John Wiley & Sons, Chichester, UK, 1992.

[24] Y. Guo, P. Gong, and R. Amundson, "Pedodiversity in the United States of America," *Geoderma*, vol. 117, no. 1-2, pp. 99–115, 2003.

[25] P. C. Nath, A. Arunachalam, M. L. Khan, K. Arunachalam, and A. R. Barbhuiya, "Vegetation analysis and tree population structure of tropical wet evergreen forests in and around Namdapha National Park, northeast India," *Biodiversity and Conservation*, vol. 14, no. 9, pp. 2109–2135, 2005.

[26] T. Sørenson, "A method of establishing groups of equal amplitude on similarity of species content," *Biologiske Skrifter K. Danske Videnskbernes Selskab*, vol. 5, no. 4, pp. 1–34, 1948.

[27] European Union, "Forest biodiversity as a challenge and opportunity for climate change adaptation and mitigation," in *Informal Meeting of EU Environment Ministers*, 12 pages, Ljubljana, Slovenia, April 2008.

[28] C. M. Shackleton, "Urban forestry—a Cinderella science in South Africa?" *Southern African Forestry Journal*, vol. 208, pp. 1–14.

[29] J. C. Onyekwelu, "Urbanization and challenges of urban forestry. Green economy: balancing environmental sustainability and livelihoods in an emerging economy," in *Proceedings of the 36thannual conference of the Forestry Association of Nigeria, Uyo, Akwa Ibom State*, Popoola et al., Ed., pp. 402–419, November 2013.

[30] J. C. Onyekwelu, R. Mosandl, and B. Stimm, "Tree species diversity and soil status of primary and degraded tropical rainforest ecosystems in South-Western Nigeria," *Journal of Tropical Forest Science*, vol. 20, no. 3, pp. 193–204, 2008.

[31] D. L. Owen, "The glossary of forestry terminology," in *South African Forestry Handbook 2000*, D. L. Owen, Ed., pp. 724–734, SAIF, Pretoria, South Africa, 2000.

[32] R. G. Lowe, "Volume increment of natural moist tropical forest in Nigeria," *Commonwealth Forestry Review*, vol. 76, no. 2, pp. 109–113, 1997.

[33] V. A. J. Adekunle, "Conservation of tree species diversity in tropical rainforest ecosystem of southwest Nigeria," *Journal of Tropical Forest Science*, vol. 18, pp. 91–101, 2006.

[34] IIRS (Indian Institute of Remote Sensing), *Biodiversity Characterization at Landscape Level in Western Ghats India Using Satellite Remote Sensing and GIS*, Department of Space Dehradun, Indian Institute of Remote Sensing, National Remote Sensing Agency, Dehradun, India, 2002.

[35] C. Konijnenddijk, S. Sadio, T. Randrup, and J. Schipperijn, "Urban and peri-urban forestry in a development context-strategy and implementation," *Journal of Arboriculture*, vol. 30, pp. 269–276, 2004.

[36] N. Parthasarathy, "Changes in forest composition and structure in three sites of tropical evergreen forest around Sengaltheri, Western Ghats," *Current Science*, vol. 80, no. 3, pp. 389–393, 2001.

[37] K.-C. Yang, J.-K. Lin, C.-F. Hsieh et al., "Vegetation pattern and woody species composition of a broad-leaved forest at the upstream basin of Nantzuhsienhsi in mid-southern Taiwan," *Taiwania*, vol. 53, no. 4, pp. 325–337, 2008.

[38] E. Duran, J. A. Meave, D. J. Lott, and G. Segura, "Structure and tree diversity patterns at landscape level in a Mexican tropical deciduous forest," *Boletin De Sociedad Botanica De Mexico*, vol. 79, pp. 43–60, 2006.

[39] A. A. Alvey, "Promoting and preserving biodiversity in the urban forest," *Urban Forestry and Urban Greening*, vol. 5, no. 4, pp. 195–201, 2006.

[40] A. Balmford, J. L. Moore, T. Brooks et al., "Conservation conflicts across Africa," *Science*, vol. 291, pp. 2616–2619, 2001.

[41] J. Cornelis and M. Hermy, "Biodiversity relationships in urban and suburban parks in Flanders," *Landscape and Urban Planning*, vol. 69, pp. 385–401, 2004.

[42] I. Kühn, R. Brandl, and S. Klotz, "The flora of German cities is naturally species rich," *Evolutionary Ecology Research*, vol. 6, pp. 749–764, 2004.

[43] D. J. Nowak and J. T. Walton, "Projected urban growth (2000–2050) and its estimated impact on the US forest resource," *Journal of Forestry*, vol. 103, pp. 383–389, 2005.

[44] C. Y. Jim and H. T. Liu, "Species diversity of three major urban forest types in Guangzhou City, China," *Forest Ecology and Management*, vol. 146, pp. 99–114, 2001.

[45] M. B. Araújo, "The coincidence of people and biodiversity in Europe," *Global Ecology and Biogeography*, vol. 12, pp. 5–12, 2003.

[46] J. C. Onyekwelu, "Biodiversity, socio-economic and cultural importance of trees in emerging Nigerian urban centres: case study of Akure city, Nigeria," in *Proceedings of the 15th World Forestry Congress*, Technical Paper, 8 pages, Durban, South Africa, September 2015.

[47] H. Xu, X. Wang, and G. Xiao, "A Remote Sensing and Gis Integrated Study On Urbanization with Its Impact On Arable Lands: Fuqing City, Fujian Province, China," *Land Degradation & Development*, vol. 11, no. 4, pp. 301–314, 2000.

[48] J. S. Miguel-Anyanz and G. S. Biging, "Comparison of single-stage and multi-stage classification approaches for cover type mapping with TM and SPOT data," *Remote Sensing of Environment*, vol. 59, pp. 92–104, 1997.

[49] R. R. Gillies and T. N. Carlson, "Thermal remote sensing of surface soil water content with partial vegetation cover for incorporation into climate models," *Journal of Applied Meteorology*, vol. 34, pp. 745–756, 1995.

[50] R. R. Gillies, T. N. Carlson, J. Cui, W. P. Kustas, and K. S. Humes, "A verification of the triangle method for obtaining surface soil water content and energy fluxes from remote measurements of the Normalized Difference Vegetation Index (NDVI) and surface radiant temperature," *International Journal of Remote Sensing*, vol. 18, pp. 3145–3166, 1997.

[51] R. W. Miller, *Urban Forestry: Planning and Managing Urban Green Spaces*, Prentice-Hall, Upper Saddle River, NJ, USA, 2nd edition, 1997.

[52] T. Türk and K. Hastaoğlu, "Mobile GIS application in urban areas and forest boundaries: a case study," in *Proceeding of the 5th International Symposium on Mobile Mapping Technology*, Padua University, Padua, Italy, May 2007.

11

A Hidden Pitfall for REDD: Analysis of Power Relation in Participatory Forest Management on Whether It Is an Obstacle or a Reliever on REDD Pathway

Angelingis Akwilini Makatta,[1] Faustin Peter Maganga,[2] and Amos Enock Majule[2]

[1]*Ministry of Natural Resources and Tourism, Forestry Training Institute, Arusha, Tanzania*
[2]*Institute of Resource Assessment, University of Dar es Salaam, Dar es Salaam, Tanzania*

Correspondence should be addressed to Angelingis Akwilini Makatta; nginaanyangala@gmail.com

Academic Editor: Piermaria Corona

Power relation among stakeholders is a key concept in collaborative approaches. This study aims to examine the reality of the acclaimed power sharing in Participatory Forest Management (PFM) and implication of existing power relation to the national REDD+ programme in Tanzania. The study involved a review of PFM policy and legal supporting documents; meta-analysis of previous studies done at two sites known to have succeeded in PFM; and empirical study at Kolo-Hills forests. Methods used include the meta-analysis of existing literature; Household Questionnaire Survey; Focused Group Discussion; and key person unstructured interviews. Results revealed that a large part of the PFM processes involved power struggle instead of power sharing. REDD+ pilot was perceived to have succeeded in improving PFM only in villages where the majority of the community about 70% experienced higher levels of inclusiveness and power balance with other PFM stakeholders in PFM processes. Power imbalance and power struggle were also noted in the REDD+ project adoption processes. Thus power relations exercised under PFM fall under potential obstacle rather than a reliever to the REDD+ programme. The study recommends reviewing of PFM legal frameworks to strengthen community empowerment for effectiveness of REDD+ on PFM platform.

1. Introduction

The role of forest communities as the main stakeholder in forest resource management programmes for pursuing biodiversity conservation and socioeconomic objectives has been extensively announced in recent decades [1–4]. Various terms have been used in different countries to express delegation of power to lowest level of community to manage natural resources proximately. In Tanzania Participatory Forest Management (PFM) is an approach involving multiple stakeholders in decision making over resources management, which is characterized by the transfer of power from central (state authority) and structures to lower levels (village authorities). PFM can be conceived as community empowerment to manage forest resources [5, 6]. Central to the approach is sharing of power among the stakeholders towards predetermined common objectives at global, national, and local levels [6–8]. It represents a paradigmatic shift in natural resources governance ideology from "fortress" state-centred protectionist conservation to "inclusive" people-centred conservation approaches. Power relation among stakeholders is a key concept for an understanding of status of collaborative approaches to natural resources management including Participatory Forest Resource Management. Power relation is among basic determinant factors for success or failure of participatory natural resource governance approaches [9–11]. Recent experience from China shows that where collaborative approaches are adopted on papers while decisive power is retained by central government the outcomes have been an accelerated degradation of resources and biodiversity loss contrary to expected outcome of improved management and biodiversity conservation [12, 13]. Study elsewhere in Tanzania by Vihemäki [14] expresses power relationships between state and community actors as a shaping agent to PFM as well as a resultant product shaped by PFM approaches.

Since PFM approaches are given prime consideration in strategies for pursuing goals of various multilateral environmental treaties including the UN Framework Convention on Climate Change (UNFCCC), it is indispensable to scrutinize how power relations can help to explain the processes and outcomes of PFM and REDD+ initiatives.

The term "power" has diverse definition, but all meant to describe the ability of a person or a group of people to influence and control the behaviour of others in the direction that is against their will. According to Giddens [15, 16], power is defined as a capacity of agents to achieve outcomes in social practices. Lukes [17] regards power as the ability to inspire people's minds interests that are contrary to their own good.

Power resource refers to particular attribute(s) that an actor is embedded in or holds, which endow him/her with potential to influence and achieve his/her determination. Communities involved in PFM need power to both exclude other users and also regulate utilization so that resources can be used sustainably.

According to Benjaminsen and Svarstad [18] there are several forms of power resources including economic power resource, legal rights to land and natural resources, political influence power resource, influence on governmental institutions, discursive power resource, knowledge power resource, "weapon of the weak," and identity power resource. Barnett and Duvall [19] and also Nuijten [9] summarize power from all those power resources into three main categories of strategic power, institutional power, and structural power. Power operates in actors and institutions whereby actors use one or multiple forms of power resources [18].

Power relation in this paper refers to ways in which actors and institutions have exercised power from their power resources in PFM processes from introduction, implementation, and distribution of benefits. Actors involved in PFM processes include central government forest agency officials; district council authorities; donor agencies officials; politicians and village communities through their Village Natural Resources Committees (VNRCs) or Village Forest Management Committees (VFCs). Institutions refer to policy and legal tools in support of PFM including National Forest Policy [20]; Forest Act number 14 [21]; Local Authorities Act number 7 [22]; Land Act 1999 number 4 [23]; and Village Land Act number 5 [24].

The objective of this paper is to examine power relation's potential to influence outcomes of PFM and the implication to REDD+ programme in Tanzania. This paper was intended to ascertain main kind of power relations exercised amongst PFM key stakeholders in striving to achieve their interests.

2. Materials and Methods

2.1. Meta-Analysis of Existing Literature

2.1.1. Power Relation Exercised by the Government Actors and Institutions in PFM Processes. In the late 1990s Tanzania like many other countries of developing world adopted decentralization of forest resources governance approaches collectively referred to as Participatory Forest Management (PFM) as a sturdy effort towards achievement of sustainable

forest management at reduced cost. This was executed through radical institutional reforms in the forest sector including formulation Forest Policy of 1998 followed by the enactment of Forest Act of 2002 and forest legislation. Other tools for implementation of the policy were the National Forest Programme (NFP) (2001 to 2010) that provided a strategic framework for implementation of forest sector policy and reinforce the role of stakeholders from the public and private sectors. Also Community Based Forest Management Guidelines are issued by MNRT in 2001 to display step-by-step procedures for CBFM establishment and finally the Forest Regulations of 2004 to operationalize the Forest Act of 2002. All these institutions aimed towards fast and effective transfer of management responsibilities to communities' lowest level of governance unity. This simply means only transferring forest resources management burdens rather than benefits to the local level.

To ensure the attainment of community low cost forest management without compromising government central control over the resources, power transferred to community was limited to enable them to serve interests of the government on conservation and not their livelihood interests [25]. The approaches to involve communities in forest management were set strategically in such a way that they look very attractive to community and also to donors for financial and logistics support. Community empowerment and livelihood benefits for poverty alleviation were put at front in PFM promotion as dual objectives of the approaches but when it came to implementation livelihood benefits were the spin-off of community conservation efforts. Bullock [26] noted that PFM was trademarked by devolving village autonomy in benefit sharing and revenue management. The livelihood benefits reported at PFM sites include access of clean water as a result of increased water flow from water sources and in rare cases increase in access of nontimber products like wild fruits. Due to power imbalance between the forest owner (government) and the managing collaborator (communities), the former has used its power to exploit the latter in terms of energy and time for about 20 years of JFM implementation while keeping silence on unclear benefit sharing terms [27].

The trick of the government to restrict devolution of substantial power to community can be noted in the limited provisions of the Forest Policy and the Forest Act. The National Forest Policy [20] while it is encouraging forest adjacent communities to engage in forest management responsibilities with local and central government forest reserves for reward of appropriate user rights does not state who is provider of working resources to communities. The policy is silent on the role of government forest agency: Tanzania Forest Service Agency (TFS) and other government institutions on the management of central government forest reserves under Joint Forest Management (JFM). Likewise the policy does not recognize the village as a key community institution for JFM but recognizes an organized community (Policy Statement 3 (p.17)), while in actual practice JFM is permitted by signing agreement for comanagement between village leaders and government forest agency officials. The policy strongly encourages devolution of management responsibilities of unreserved forests to village governments as a means to improve management but silent on devolving financial and material resources for the same. According to Agrawal and Ostrom [28] rights and capacities that are transferred to actors at lower levels are the major determinants of the outcomes of devolution approaches to forestry resources management. The forest legislation provides a clear legal procedure for assertion and gazettement of Village Land Forests including Village Forest reserves, Community Forest Reserves, and private forest reserves. However, the Forest Policy, Forest Act, and legislation are silent on degazettement and degazettement procedures of Village Land Forest Reserves when in future the village communities need to utilize the land for alternative social and economic investments.

The Forest Policy and Forest Act legalized JFM as comanagement approach involving the community (invited partner in management) and government (the forest owner). It requires management responsibilities to be handled to communities but it is keeping silence on sharing of economic gains from the forests and also on practical management activities/responsibilities of the government part. Similarly the government has been infringing its own approved Forest Act because the Forest Policy and Forest Act require signing of Joint Management Agreement (JMA) as a prerequisite for comanagement to be implemented but in most cases JFM has been implemented without approval of JMA by government part. The government has been reluctant in signing JMA particularly those relating to JFM in productive National Forest Reserves (NFRs). Only a few number of agreements have been signed by the government out of several hundred villages developed JMAs around a range of forests managed by central or local government for more than two decades. Furthermore, power of communities to hold the government or the government forest agency accountable in case of violation of the JMA is barred.

Although the Forest Act [21] gives power to village communities to reserve and gazette Village Land Forest Reserves (VLFRs) on village land, communities can only gain the power when the village is registered and has a Certificate of Village Land [24]. The registration conditionality limits power to unregistered majority of villages in Tanzania. Conversely those villages may have limited opportunity to benefit from Community Based Forest Management (CBFM) and REDD+ programme carbon credits.

Likewise although Forest Act [21] provisions seem to empower communities with full ownership and management responsibility under CBFM regime, the same Act gives power to the Director of Forestry or local governments to take over the management of forests that are under CBFM or JFM if the forests are deemed to be mismanaged [21]. This creates unsecured community access and user rights to forest resources under CBFM and JFM regimes.

The greediness of the government to continue holding ultimate power over the resources was found to be the pivotal barrier to PFM effectiveness, yet it was deemed to escalate in REDD+ implementation [29, 30].

2.1.2. Power Relation Exercised by the Community under PFM.
In the beginning of PFM community readily accepted the seemingly offered opportunity of being empowered to manage forest resources. The community gladly accepted the approach because it was their desire to gain back control over the resources they enjoyed before but lost under the colonial government with the establishment of centralised resources governance. They were also in need of fulfilling their livelihood demands from the forest resources.

The first PFM evaluation report [31] revealed that communities under PFM gained power over the resources and effectively excluded the outsiders from unsustainable use of the resources. Consequently, within few years of PFM implementation, degraded forests were reported to recover through regeneration, wild animals that disappeared from degraded forests had returned to the forests, and streams originating from PFM forests were reported to have increased water flow while incidences of illegal activities in those forests were seldom. All these were evidences for positive outcomes of PFM on conservation. However the same report disclosed that the outcome of PFM on improved livelihood of participating communities was negligible.

Thereafter, government efforts were directed toward scaling up of PFM throughout the country. The second evaluation report [27] three years after the first report had nothing new from the former other than increased area coverage in terms of number of hectares, villages, and districts involved in PFM. However ever since, nothing has been reported on the performance quality of PFM in early adopted sites. This makes the reality of PFM performance quality remain gloomy. For those who are keen to visualize through the "window of the government practices" they can see the reality. The tendency of government surveillance team to conduct frequent and sudden forest protection patrols in JFM forests has been seen as a way of recentralizing while decentralizing on the management of forest resources owned by the government. This has been necessary attempts of the government to contain accelerated exploitation on protective forest reserves under JFM. Though it is not publically declared, the government has noticed that the communities' tolerance has ended and the community had employed "weapon of the weak" power whereby silently they exploited the forest resources as compensation to their management efforts. The Village Forest Natural Resources/Village Forest Management Committees (VNRCs/VFCs) used to oversee controlled utilization of forest resources for sustainability turned to be promoters of exploitation and informers to the community to alert them to the strategies of forest agency to contain escalated exploitation of resources by communities.

Through personal communication to several forest managers of PFM forests concerning current performance of PFM in controlling deforestation and forest degradation, they all said the approach was very ambitious and unworkable. They said instead of containing illegal activities, it has worsened it. They mentioned the reason for the situation now getting worse than before JFM is that with PFM community has been exposed to resources and knows the stock of the forest at every point. Also they are very aware of the resources and manpower capability of the district forest departments on containing their planned crime over the forests. Therefore communities are sure of their security while doing illegal exploitation of timber resources on protected forests. For instance, on 22 March 2015, I visited Monduli district catchment forests office and I did witness about 5 M^3 of stacked wood outside the office. The district catchment manager informed that the wood was obtained from a single day patrol in the forest conducted without informing the community by staff from her office in collaboration with staff from wildlife department (personal communication). This is similar to the findings from study elsewhere in Tanzania by Brockington [32] suggesting that "village forest management committee are not functioning well and forests are not well protected". Similar experience has happened in China whereby the protected areas built under the Grain for Green Project (GGP) conservation programme were at risk of undergoing greater levels of damage by local residents because of dissatisfaction of the communities with little conservation cost which is compensation from the central government [13].

2.1.3. Power Relation Experienced at Two Renowned PFM Success Sites of Angai and SULEDO Forests.
Analysis of existing literature at two popular sites of Angai and SULEDO known to have succeeded PFM revealed various power relationships that were exercised by different stakeholders in the course of establishing and implementing PFM at these sites.

For Angai, PFM process involved sort of power struggle among three actors/stakeholders: local government of Liwale district authority and community in villages that border the forest and Finnish Funded Rural Integrated Programme Support (RIPS) NGO. Government used its institutional as well as structural power in trying to influence gazettement of Angai forests as a Local Government Forest Reserve (LGFR) instead of Village Land Forest Reserve (VLFR). RIPS with its structural and financial power was able to break through its interest of involving 13 villages adjacent to forests and thus changed the targeted forest ownership from district council to community forest. The district authority silently disagreed with the establishment of Community Forest Reserve; hence it used its institutional power to delay processes for establishing the AVLRF for about six years [33]. Under strong push by institutional and financial power of RIPS, being backed by the Forest Policy the establishment of AVLRF was realized [33]. This was followed by the process to issue a Village Land Certificate to each respective village government. This is because while Land Act [23] recognizes customary rights of occupancy even if the land is not registered and the landholder has no certificate, Village Land Act [24] recognizes village land right of occupancy on registered village, attested by certificate of registration. The certificate renders villages with power to control and excluding others from the resources. Again this took more than seven years. However, even after having the certificate, communities' power to utilize forest resources for livelihoods was still constrained by the institutional power of district authority granted under the local government Act number 7 [22]; Forest Policy [20]; and Forest Act [21] to approve community proposal for extractive utilization of timber resources only after having a detailed forest management plan. Preparation of the management plant was highly technical and financially demanding for village communities to afford, and thus it took again two years to accomplish the work after securing financial support

from Finnish development cooperation through Tanzanian National Forest Programme (NFP).

The accomplishment of AVLRF management plan raised hope for communities to start utilizing timber resources as fruits of their management efforts after 15 years. On the contrary higher government power tiers had identified Angai forests among sites for pilot REDD+ project which is against exploitation of timber products and instead expecting to award community for being more restrictive on harvesting so as to accumulate more carbon stock [29]. Due to complexity in REDD+ implementation process, guarantee of community rewards through REDD+ cobenefits seems to be uncertain [30].

For the case of SULEDO forests, historically the area has been occupied by Maasai tribe and recognised as special area for pastoral activities since back to colonial era. The central government in 1983 with its institutional power decided to change status of the area and exclude the Maasai community by turning the area into a proposed forest reserve [34]. Due to strong strategic power of Maasai society it was not easy for the government to gazette the forests without support from the community. However community in nine villages used their institutional power rendered to them through the Forest Act to maintain control of the resources by gazetting the forest as Community Forest Reserve. Institutional power gained by the communities at SULEDO forests excluded outsiders from the right to directly access and benefit from the resources [35] including the district authority. Having lost direct control over the income from forests at SULEDO, Kiteto district authority using institutional, structural, and financial power silently strives to fail community efforts through delaying processes for timber harvesting proposed by village councils and limiting facilitation to district forest officer who is responsible for assisting communities forest management technical issues [34]. According to [33], potential revenue from sustainable timber harvesting at community managed forests amounts to USD 15,000 and USD 70,000 USD per village per year for SULEDO and Angai VLFRS forests, respectively. Based on this fact, the government recommended that the harvesting of timber is to be dealt by a private sector operator through lease or tender arrangements, for the reason that such amount of revenue is too huge for village governments to handle. Clearly this seems to be an attempt of using knowledge as power to snatch the revenue from control of the community. One could question why the recommendation was not focusing on building capacity of village governments regarding financial management.

The community using "weapon of the weak" power silently decline participation in forest management activities as a result of discouragement from district authority. This created a loophole for corrupt village leaders to use structural power and few elites with strong strategic power to grab most of opportunities from PFM including appropriation of income collected from the forests (elite capture) [34]. Similar to Angai situation, the dream of SULEDO communities to benefit from commercial timber harvesting has been diminishing with the introduction of REDD+ initiatives spearheaded by national and international conservation super powers.

2.2. Empirical Study on Power Relation under PFM at Kolo-Hills at Forests

2.2.1. Location of the Study Area.
This study was conducted as part of big research project under Climate Change Impacts, Adaptation and Mitigation Programme Tanzania (CCIAM). The study site is found in Kondoa district located between latitudes $4°10'-5°44'$ south and longitudes $34°54'-36°28'$ east, in Dodoma region of central Tanzania [36, 37]. The district climate condition is semiarid with minimum and maximum temperature of $16°C$ and $29°C$, respectively, while annual rainfall ranges between 500 and 800 mm [36–38].

The study was based at Kolo-Hills forests commonly known as Isabe and Salanga forest reserves and the surrounding village of Mnenia, Masawi, Humai, Kisese-Disa, and Kikore. Selection of Kolo-Hills site is based on having experience of both PFM and also a REDD+ pilot project. The forest reserves are found 30 km east-south of Kondoa district council township and cover about 346 km^2. The villages and the forest are located at $4°54'983''S$ and $35°47'937''E$ at elevation of range between 1650 and 2000 M above the sea level [39].

2.2.2. Research Design.
Household survey was conducted to 250 household heads randomly selected from five villages with disproportionate sampling whereby 50 households from each village were selected from village registers with an aid of random numbers generated using a computer. Sampling unit was a household, defined as all members related and unrelated who share the same dwelling unit or a group of people sharing resources, expenditures, and responsibilities [40]. Apart from the household survey, key person unstructured interviews were conducted.

Key persons were 2 village leaders (chairperson and village executive officer) of each village, and some elders in the community; two district authority forest officers; the district land, natural resources, and environmental officer; and two staff members from Wildlife Foundation (AWF) which is responsible for implementation of REDD+ pilot named ARKFor at Kolo-Hills forests. Again researchers' observation was used alongside household survey to observe social setting and daily community interaction with forest resources proximately. The use of several methods in data collection is a triangulation technique to reduce error due bias that may occur in data collection and thus increase reliability of the data [41]. Questionnaires and checklists of questions are survey tools used for data collection. Both questionnaires and checklists were prepared in English and Kiswahili and administered in Kiswahili (the national language in Tanzania) by the authors. Kiswahili language is fluently spoken by the study communities apart from their mother tongue.

Response to questions was measured using rank of points which correspond to the respondent's perceived level of performance using a five-point Likert scale. Respondents were asked to indicate values which reflected level of agreement with a statement (0 = strongly disagree, 4 = strongly agree) or a rating of the frequency of occurrence of an event or behaviour (0 = not at all, 4 = always). A high score indicated a greater agreeing or higher frequency of an event occurrence or level of performance.

The data was collected from communities of 5 villages: Mnenia, Humai, Masawi, Kisese-Disa, and Kikore (indicated by yellow dots in Figure 1).

Exploration as well as analysis of the data from household survey was done using SPSS Version 16 and Microsoft Excel Office 2007 computer programmes. Content and Structural-Functional Analysis technique was applied to summarize text information from key person unstructured interview into meaningful statements presented in results.

3. Results and Discussions

3.1. Results

3.1.1. Results from Unstructured Interviews.
Unstructured key persons interviews conducted to two staff members from Kondoa district forest department revealed intense use of coercive power over communities who resisted adoption of REDD+ pilot project at their forest area. Box 1 includes the information which was given under condition that their names will be kept anonymous for their job security.

Local government at district authority strives to maintain control over forest resources which are under comanagement with communities through institutional power. Lack of inclusiveness and clarity of REDD+ pilot project objectives created doubt on security of communities' access rights to forest resources; thus some villages refused to adopt it. Among the villages resisting REDD+ pilot project is Kisese-Disa. Following their resistance, local authority government of Kondoa district council used various kinds of its power to force the acceptance of the project.

Based on the Local Authorities Act number 7 of 1982 village governments are subordinated to district councils. Thus village governments depend on district council's goodwill when it comes to approval of community decisions as well as social and economic welfare support. Consequently the government of Kondoa district restricted supply of social and economic services to those villages resisting REDD+ pilot project as an instrument to threaten them to accept the project.

The existence of resistance towards the REDD+ pilot project at Kolo-Hills by some villages was also mentioned in the Mid-Term Review Report of the project but according to the report, the resistance was mitigated through engaging large numbers of community members in project activities and creating patrols that are made up of members from multiple villages. The study revealed contradicting reasons for the resistance. While the Mid-Term Review Report mentioned personal interests of some village leaders who benefitted from deforestation activities as the reason, information from interview with common people of the community disclosed that resistance to REDD+ pilot project was due to fear of losing access to daily needs from the forest and fear of imposition of severe punishment as opposed to those locally agreed in by-laws concerning violation of forest management and utilization rules. They said community viewed REDD+ project as a strategy to extend Tarangire National Park to include Kolo-Hills forests which would lead to wildlife-community conflicts as it happened in various

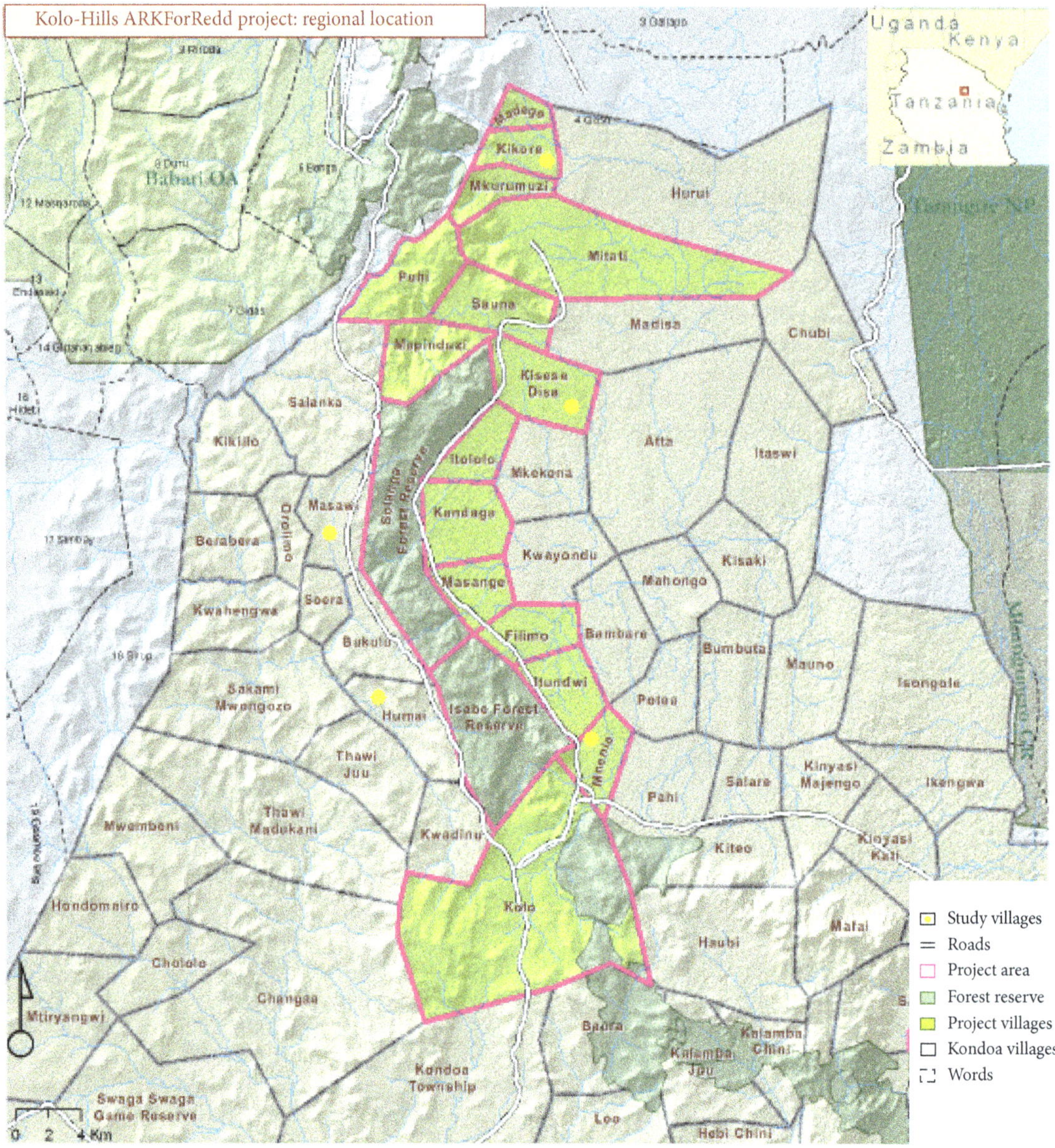

(Source: ARKFor feasibility study report 2010).

FIGURE 1: Map of the study area showing ARKFor REDD+ pilot and CCIAM project area and study villages.

parts of the country. The interviewees insisted that resistance to REDD+ project was based on aforementioned reasons that were enfolded by inadequate information (clarity) about the project but not grounded on political influence as mentioned by project leaders.

Information from unstructured interview with staff from AWF and Kondoa district about number of villages involved in PFM and REDD+ pilot project was contradicting. While AWF staff informing that all 21 villages have agreed to participate and are involved in the project, the district land and natural resources management officer informed that there are villages that resisted the project and are not participating in the project activities, and one of the villages is Kisese-Disa. The conflicting information seems to be an attempt of AWF

to mask their ineffectiveness in REDD+ pilot implementation by overrating the success including area coverage in terms of villages involved.

The explanation from the three villages (Boxes 2–4) discloses prevalence of community dissatisfaction on power imbalance among the government at district council and village communities as copartners of forest management under PFM. Additionally the community complains about the little incentives from PFM due to many restrictions on taking products from the forest.

3.1.2. Results from Household Questionnaire Survey. About 60% of all respondents ($n = 250$) from five villages ranked preparation time for adoption of PFM and REDD+ initiatives;

They said "there are some villages that willingly accepted PFM and REDD+ pilot while others resisted REDD+ project. In Kisese-Disa village the resistance was very strong because of political influence from their Parliamentary representative member. At first instant a team for REDD+ promoters at Kolo Hills called an audience with district administrative staff of different fields including the parliamentary representative member and explained their plan to implement REDD+ pilot project at Kolo-Hills forests. Their proposal was agreed after putting clear the project objectives and the overall framework for implementation that would involve villages bordering the forests. The meeting participants agreed to encourage people to adopt the project for their economic benefit from anticipated carbon credit for conserving the forests. Contrarily that Member of Parliament went back to Kisese-Disa village and start advocating hateful issues attributed to REDD+ pilot project and the promoters. For that reason this village resisted REDD+ at their village forest management area.

However REDD+ project administration in collaboration with Kondoa district authority forced this village to accept and collaborate with other villages in REDD+ project forest management activities including sending representatives to joint forest guard team. This team is directly accountable to REDD+ pilot project and it is responsible to ensure protection of overall forests at Kolo Hills. Though Kisese-Disa village maintained their resistance to the REDD+ project, the forest guard team composed of members from all villages persistently conduct patrol over the whole forest area under REDD+ pilot including forest areas of villages who refused the project.

Kisese-Disa village wanted to show their disagreement to those who conduct patrol activities at their forest area and some youngsters of the village were sent by elders to kidnap one of female forest guard member of the joint team while the team was patrolling at Kisese-Disa forest area. The rest of the team conveyed the kidnapping information to REDD+ project administration (AWF) and Kondoa district. Immediate support from Kondoa district including the two of us and armed professional guards from TARANGIRE National park sent by AWF arrived at the event site and start to search for the guard who was kidnapped. Following that, the Kisese-Disa youngsters immediately decided to release the guard safely to the forest. When the team tried to communicate to her, she told them she was free and safe.

However the team continued to search for the kidnap suspects up to the evening. Around 8:00pm the armed Tanzania National Parks (TANAPA) guards from Tarangire National Park together with us started to ambush the suspects in their home dwellings, forcing whoever found there to hand the suspects. We managed to catch few suspects and TANAPA staff taught them a lesson. They commanded them to do severe physical exercises to the extent that some suspects released bowel materials to their clothes unconsciously. Noticing that after taking them onto their truck, the TANAPA guards ordered them to drop out of the truck and wash it thoroughly and after that they released them without any court case proceeding on the matter".

Explaining more on the set strategies to force the village to accept REDD+, they said "we believe it will reach a point where the troublesome villages will surrender and accept the project since there is a planned persistent silent punishment of isolating and neglecting them from various social and economic development support from the district. For instance currently Kisese-Disa village is obliged to participate in protection of Kolo-Hills forests in collaboration with other villages but it is excluded from beneficiaries of carbon credit and other economic incentives expected to come through REDD+ programme and PFM".

Box 1: Unstructured interview conducted with two staff members from Kondoa district responsible for the management of Kolo-Hills forests.

power balance among stakeholders; and good understanding (clarity) about the initiatives below average scores. This implies that the three factors are areas of weakness which are potential barrier for effectiveness of both PFM and REDD+ initiatives. Of the three factors, power balance was ranked as the most weakly exercised. Through ANOVA statistical test, significant difference of scores between villages was revealed at $p < 0.05$. About 70% of respondents from Mnenia and Kikore rated the three factors as good performance while more than 80% of respondents from Kisese-Disa and Masawi villages ranked the three factors below average scores which implies poor performance (Figure 2).

However, PFM success after REED+ project seems to be more pronounced in villages that experienced higher levels of power balance among stakeholders; good understanding (clarity) about the initiatives; and adequate preparation time for adoption of PFM and REDD+. For instance, Mnenia village which scores higher in all three factors had highest perceived PFM success.

Table 1 displays some REDD+ challenges and opportunities experienced by the community at Kolo-Hills including increased restrictions over forest resources after REDD+ pilot and additional benefits. However for those villages that experienced high level of power imbalance they also experienced high level of additional restrictions with little or no additional benefits from REDD+ pilot project.

3.2. Discussion. Although from the literature Tanzania has been commended for being at the forefront in refined laws for people-centred rather than state-centred in power holding in natural resources management issues, this has been rhetoric while the reverse seems to be reality. This can be verified from

He said "In the beginning, we agreed to work with district council forest staff as they said the government has decided to give back the forests to local people to protect them and collect our needs from the forest like firewood. We were asked to formulate Village forest management Committee which would work with the district people on behalf of the community in protecting the forest. We were asked to plan the way of using the forests without destroying them and we did. The problem is that thereafter there was misunderstanding among village government (village council) leaders and that forest management committee. The whole village leadership collapsed and we elected new village government. Therefore, all forest management arrangements ended with the previous village leadership. However we continued to protect our forests and get our needs there until when district people wanted to give the forest to African Wildlife Foundation (AWF) with the name of REDD+ project. We refused to work with AWF because they are Tarangire National Park people. It is hard to believe TANAPA people you know! They can take our forest as they did to other places we heard in Arusha and Kilimanjaro regions. Although we refused to work with AWF as our partners, they forced us to accept REDD+ project and have imposed many restrictions on the forests. To date if we enter the forest we get caught by a team of forest guards of AWF trained from every village and sometimes with park rangers of Tarangire national Park who join district people in patrolling the forest. When youngsters are caught cutting poles for repairing or constructing a house they are beaten hardly without taking them to court, which is against the Forest Act and our village by-laws".

Box 2: Unstructured interview with one of key persons from Kisese-Disa village (a village leader). How can you comment on the status of PFM and REDD at your village?

"What REDD+ project people are doing is not constructive at all! Before the project we had established effective system for protecting forests under Community Based Forest Management (CBFM). It is well know that village government has authority as a manager and owner of VLFRs on behalf of the community. It is the one which is responsible to ensure forests are well managed based on the village forest management plan document. Surprisingly, REDD+ pilot project came and re-establish new institution which is joint team of forest guards that is over the village government and directly answerable to them. You can imagine how awkward it is; a group of individuals from different villages mealy selected randomly are given power to collect fines from forest defaulters without consultation of village government leaders of the particular forest area. Nobody knows at village what is happening on the forest because at village office nowadays there has been no either a report of patrols performance, offences or that of defaulters and action taken over them. Because forest patrols are done at forest landscape wide not village based; there have been cases where fine for a crime done at a certain village is collected at another village. This is a kind of overthrowing the village government from its legal authority".

Box 3: Unstructured interview with one of key persons from Kikore who is a member of village standing committees.

"In average PFM approach is going well, except for the issue of benefits to community. It is very annoying because we as the partners to the government we forego our productive activities and attend meetings to discuss issues of forests, we do patrolling activities and all about the wellbeing of the forest but we are always restricted of many things. We agreed to not graze in the forest because cows may destroy small trees. We are happy to have good forest around us and wishing to maintain them unspoiled for our grannies. However we maintain living trees and not dead ones! Why government restricts us to take dead trees for our use? They say we are allowed to gather fire wood but we should go into the forest empty handed without cutting tools. How is it possible to take whole dead trees without tools? There are lots of big timber trees which have died under natural causes. Here we have our primary school that is under shortage of desks and even firewood for preparing lunch for pupils. The dead trees could be used to solve some of school demands. The village government asked the district forest officer to come to verify and oversee the process of removing those dead logs from the forest but for months they didn't even turn up".

Box 4: Unstructured interview with one of key persons from Masawi village providing a view about PFM and REDD+ pilot project situation.

TABLE 1: Experienced additional restrictions and benefits on forests after REDD+ pilot project.

					Villages					
	Mnenia	Freq	Masawi	Freq	Humai	Freq	Kisese-Disa	Freq	Kikore	Freq
Additional restrictions from REDD+										
Tree cut for construction	√	10	√	13	√	13	√	15	√	20
Entrance (short cut foot path)	√	9	√	9	√	3	√	22	√	12
Firewood collection	√	6	√	8	√	2	√	1	√	0
Grazing	√	7	√	15	√	9	√	22	√	18
No restriction	√	6	√	3	√	5	x	0	√	7
Agriculture farming	√	2	√	2	x	0	x	0	√	9
Sum		34		47		27		60		59
Additional benefits from REDD+										
Income from IGAs	√	16	√	4	√	2	x	0	√	10
Ecological services	√	17	√	4	√	9	x	0	√	11
Improved farming	√	3	√	2	√	3	x	0	√	5
No additional benefit	√	2	√	24	√	11	√	30	√	10
CO trial payment	√	25	√	18	√	7	x	0	√	15
Extension education	√	16	x	0	√	5	x	0	√	4
Forest products	√	5	x	0	x	0	x	0	√	13
Sum		82		28		26		0		58

Freq: frequency of the item was mentioned.

√: item was mentioned.

x: item was not mentioned.

Note: assumption under the table displays that the more frequent an item was mentioned, the more it benefits or costs the community.

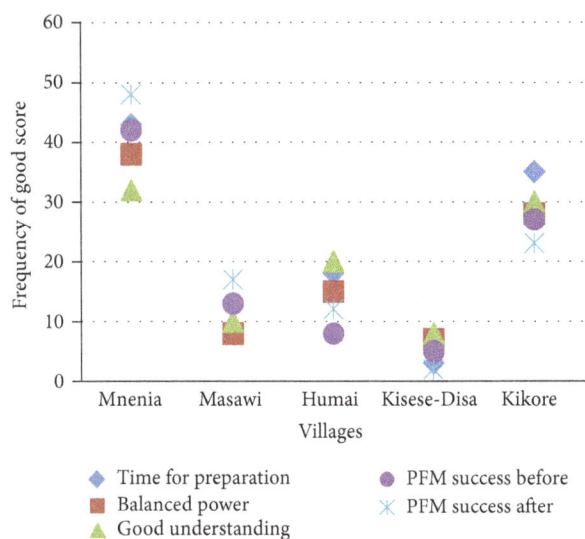

FIGURE 2: Comparison of perceived success of PFM before REDD+ and with REDD+ pilot project under different status of time for preparation; balanced power; good understanding of the programmes.

initial national programme to support PFM implementation with the goal "natural resources contributed on sustainable basis towards reduced income poverty, vulnerability amongst the poorest groups and improved quality of life and social well-being in Tanzania" and objective to "increased benefits to rural communities based on sustainable natural resource management in Tanzania." The programme which was known as the Management of Natural Resources Programme was implemented with financial support of the Government of Norway for about 13 years from 1994 to 2006 when it was phased out with negligible impact on PFM contribution to local communities' livelihood. The reason explained was that livelihood centred objective was theoretical for donor's aid focus but when it came to implementation the efforts and utilization of funds were concentrated towards conservation while ignoring support to community livelihood [42].

The prevailing PFM circumstances whereby the government has delegated responsibility to people while it retains facilities and financial resources for implementation can be viewed as a kind of exploitation which opposes power sharing. Similarly study elsewhere by Mustalahti [33] commented that "participatory and decentralised forest management can hardly guarantee rights of the local people particularly in a situation where devolution and transfer of power to community level governing board (Village for Tanzania) without transferring financial, material and technical resources necessary for local government to fulfil its obligations, such kind of empowerment is meaningless." Also "if central governments grant local governments the rights to make and implement decisions but in practice withhold resources or otherwise check local ability to do so, then discretionary powers have not been effectively transferred" [43]. The oppressive situation in some PFM communities where people are forced to adopt new management arrangement at their PFM managed forests as found at Kisese-Disa village in this study is contrary to power sharing. People have been distressed for the sake of project REDD+ project administration to display illusive success on paper with justification that all 21 villages have accepted and implemented the project. In fact, on these grounds REDD+ pilot project at Kolo-Hills forests has demonstrated blueprint to failure rather than success for the forthcoming REDD+ programme.

The fact that government under JFM and CBFM approaches has room to question and sue the community in case of failure to implement forest management as per the management plan approved by the government while the community do not have such room in case of government failure to meet its roles seems to oppose the idea of power sharing and instead reveals inequitable accountability. This can be witnessed where Forest Act gives mandate to the Director of Forestry or local governments to take over the management of forests that are under JFM or CBFM when the forests under PFM are deemed to be mismanaged (Forest Act 2002, s. 8 and 41–48). In contrast for about 20 years of PFM implementation the government have failed to formalize benefit sharing mechanism especially on productive forests managed under JFM regime [44]. While the JFM regime is aimed at cutting off management costs from government on already gazetted government forest reserves, the underlying policy goal for CBFM is to bring large areas of unprotected woodlands and forests progressively under village management and protection through establishment of VLRFs [20, 27]. Accordingly the silence restriction on harvesting of timber products from the ever known best CBFM success sites including Angai forests and SULEDO as noted in study by Mustalahti and Lund [45] gives no doubt that all what has been done to convince communities to adopt PFM as win-win approach might be elusive. Under such skewed power relationships, power struggle instead of power sharing, and competition instead of collaboration between community and government, one can view PFM as a strategy for the government through its forest agencies to deliberately use power to burden the poor and marginalized rural citizens with conservation costs. This has been through exploitation of their energy, time, and resources to meet conservation interests under PFM with no clear terms on benefit sharing.

Effective PFM implementation necessitates a firm commitment on shifting from competitive psychological orientation which permits inequality and emphasizes a win-lose struggle for superiority and instead orients towards a cooperative psychological orientation which emphasizes equal powers and win-win mutual relations. Among benefits of power sharing according to [30] are an increase in decision acceptance, commitment, and quality and enhanced satisfaction and commitment. Hence by unification of interests and balancing power among stakeholders PFM is likely to achieve its multiple integrated objectives.

4. Conclusion and Recommendation

From the findings we concluded that where PFM approaches are adopted on papers while decisive power is retained by central government the outcomes have been an accelerated degradation of resources instead of improved conservation. The large part of PFM processes involved power relations that are contrary to power sharing. Both the government and local communities have been exercising power struggle in fulfilling their interests over the forest resources. This power struggle between community and government stakeholders of PFM has been transferred to REDD+ pilot project. The findings are empirical evidence that this supports our description of power relations exercised under PFM as an obstacle rather than reliever to the forthcoming REDD+ programme. Unless power relation trend is reversed, practical REDD+ on PFM platform is erroneous.

To improve from the weak situation disclosed by this study, we recommend the policy makers to amend the Forest Policy, Forest Act, and legislation in order to achieve the following:

(i) Put clear terms for benefit sharing.

(ii) Fill gaps of provisions to strengthen community empowerment.

(iii) Clarify responsibility of every PFM stakeholder.

(iv) Clarify means through which every stakeholder will be accountable in the course of undertaking the responsibilities.

Knowledge Contribution of the Paper

This paper shed light on potential of power relations amongst the main stakeholders (community and the government) as determinant factor to success or failure of PFM. The findings contribute to a more informed academic and political sphere on power relations potential barrier to REDD+ programme which is existing in PFM as a framework for REDD+.

Conflict of Interests

The authors declare that there is no conflict of interests regarding the publication of this paper.

Acknowledgments

The authors are highly grateful to the Government of the Kingdom of Norway under Tanzania-Norway Climate Change Partnership Programme for financial support to this study through CCIAM research project. The authors acknowledge the entire Kondoa district staff and AWF management and communities of study villages for their tireless collaboration during data collection.

References

[1] A. Agrawal and C. C. Gibson, "Enchantment and disenchantment: the role of community in natural resource conservation," *World Development*, vol. 27, no. 4, pp. 629–649, 1999.

[2] S. Wiggins, K. Marfo, and V. Anchirinah, "Protecting the forest or the people? Environmental policies and livelihoods in the forest margins of Southern Ghana," *World Development*, vol. 32, no. 11, pp. 1939–1955, 2004.

[3] L. Tole, "Reforms from the ground up: a review of community-based forest management in tropical developing countries," *Environmental Management*, vol. 45, no. 6, pp. 1312–1331, 2010.

[4] D. Ghai and J. M. Vivian, *Grassroots Environmental Action: People's Participation in Sustainable Development*, Routledge, London, UK, 2014.

[5] G. C. Kajembe and G. C. Monela, "Empowering communities to manage natural resources: where does the new power lie? A case study of Duru-Haitemba, Babati, Tanzania," in *Empowering Communities to Manage Natural Resources: Case Studies from Southern Africa*, pp. 151–163, 2000.

[6] D. B. Raik, A. L. Wilson, and D. J. Decker, "Power in natural resources management: an application of theory," *Society and Natural Resources*, vol. 21, no. 8, pp. 729–739, 2008.

[7] L. A. Wily, *Land Tenure Reform and the Balance of Power in Eastern and Southern Africa*, Overseas Development Institute, 2000.

[8] L. A. Wily, "A review of new policy towards participatory forest management in eastern Africa," For CARE Review of Policy Towards Participatory Forest Management in Eastern Africa, 2002.

[9] M. Nuijten, "Power in practice: a force field approach to natural resource management," *The Journal of Transdisciplinary Environmental Studies*, vol. 4, no. 2, pp. 1–14, 2005.

[10] C. Ansell and A. Gash, "Collaborative governance in theory and practice," *Journal of Public Administration Research and Theory*, vol. 18, no. 4, pp. 543–571, 2008.

[11] F. Berkes, "Evolution of co-management: role of knowledge generation, bridging organizations and social learning," *Journal of Environmental Management*, vol. 90, no. 5, pp. 1692–1702, 2009.

[12] H. Zheng and S. Cao, "Threats to China's biodiversity caused by policy contradictions and unexpected consequences," *Ambio*, vol. 44, pp. 23–33, 2014.

[13] S. Cao, L. Chen, and Q. Zhu, "Remembering the ultimate goal of environmental protection: including protection of impoverished citizens in China's environmental policy," *Ambio*, vol. 39, no. 6, pp. 439–442, 2010.

[14] H. Vihemäki, "Participation or further exclusion? Contestations over forest conservation and control in the East Usambara Mountains, Tanzania," 2009.

[15] A. Giddens, *The Constitution of Society: Outline of the Theory of Structuration*, Polity Press, Cambridge, UK, 1984.

[16] F. Cleaver, "Understanding agency in collective action," *Journal of Human Development*, vol. 8, no. 2, pp. 223–244, 2007.

[17] S. Lukes, *Power: A Radical View*, Palgrave MacMillan, New York, NY, USA, 2nd edition, 2005.

[18] A. Benjaminsen Tor and H. Svarstad, *Political Ecology: The Environment, People and Power*, Universitetsforlaget, Oslo, Norway, 2010.

[19] M. Barnett and R. Duvall, "Power in international politics," *International Organization*, vol. 59, no. 1, pp. 39–75, 2005.

[20] United Republic of Tanzania, *The National Forest Policy*, Ministry of Natural Resources and Tourism, Forest and Beekeeping Division, Government Press, Dar es Salaam, Tanzania, 1998.

[21] United Republic of Tanzania, *The New Forest Act. No. 14 of 7th June 2002*, Government Printer, Dar es Salaam, Tanzania, 2002.

[22] United Republic of Tanzania (URT), *The Local Authorities Act no. 7 (District and Urban Authorities)*, Government Printer, Dar es Salaam, Tanzania, 1982.

[23] United Republic of Tanzania, *The Land Act 1999. No. 4 of 15th May 1999*, Government Printer, Dar es Salaam, Tanzania, 1999.

[24] United Republic of Tanzania, *The Village Land Act (No. 5 of 1999)*, Government Printer, Dar es Salaam, Tanzania, 1999.

[25] A. Pallotti, "Tanzania: decentralising power or spreading poverty?" *Review of African Political Economy*, vol. 35, no. 2, pp. 221–235, 2008.

[26] R. Bullock, *Rhetoric versus reality in participatory forest management in East Usambaras, Tanzania [Doctoral dissertation]*, University of Florida, 2010.

[27] T. Blomley and S. Iddi, *Participatory Forest Management in Tanzania: 1993–2009, Lessons Learned and Experiences to Date*, Ministry of Natural Resources and Tourism, 2009.

[28] A. Agrawal and E. Ostrom, "Collective action, property rights, and decentralization in resource use in India and Nepal," *Politics & Society*, vol. 29, no. 4, pp. 485–514, 2001.

[29] J. Phelps, E. L. Webb, and A. Agrawal, "Does REDD+ threaten to recentralize forest governance?" *Science*, vol. 328, no. 5976, pp. 312–313, 2010.

[30] I. Mustalahti, A. Bolin, E. Boyd, and J. Paavola, "Can REDD+ reconcile local priorities and needs with global mitigation benefits? Lessons from Angai Forest, Tanzania," *Ecology and Society*, vol. 17, no. 1, article 16, 2012.

[31] T. Blomley and H. Ramadhani, "Going to scale with participatory forest management: early lessons from Tanzania," *International Forestry Review*, vol. 8, no. 1, pp. 93–100, 2006.

[32] D. Brockington, "Forests, community conservation, and local government performance: the village forest reserves of Tanzania," *Society and Natural Resources*, vol. 20, no. 9, pp. 835–848, 2007.

[33] I. Mustalahti, *Handling the stick: practices and impacts of participation in forest management [Ph.D. thesis]*, Danish Centre for Forest, Landscape and Planning, Faculty of Life Sciences, University of Copenhagen, Copenhagen, Denmark, 2007.

[34] H. Sjoholm and S. Luono, "Traditional pastoral communities securing green pastures through participatory forest management: a case study from Kiteto District, United Republic of Tanzania," in *Proceedings of the 2nd International Conference on Participatory Forest Management in Africa*, February 2002.

[35] C. A. Mwakasendo, *Forest income and rural livelihoods under Suledo community based forest management in Kiteto district [Ph.D. dissertation]*, Sokoine University of Agriculture, 2009.

[36] MOAC, "Status and causes of land degradation in Mbulu, Kondoa and Singida districts, Tanzania and strategies for conservation and rehabilitation," Tech. Rep., Ministry of Agriculture and Co-Operatives, FAO, 1996.

[37] E. J. M. Shirima, "Benefits from dual purpose goats for crop and livestock production under small-scale peasant systems in Kondoa eroded areas, Tanzania," *Livestock Research for Rural Development*, vol. 17, no. 12, 2005.

[38] KDC, *Kondoa District Socio-Economic Profile*, Kondoa District Council, Kondoa, Tanzania, 2011.

[39] K. John, D. S. A. Silayo, and A. Vatn, "The cost of managing forest carbon under REDD+ initiatives: a case of Kolo hills forests in Kondoa District, Dodoma, Tanzania," *International Journal of Forestry Research*, vol. 2014, Article ID 920964, 12 pages, 2014.

[40] G. J. Casimir and H. Tobi, "Defining and using the concept of household: a systematic review," *International Journal of Consumer Studies*, vol. 35, no. 5, pp. 498–506, 2011.

[41] C. Frankfort-Nachmias and D. Nachmias, *Research Methods in the Social Sciences*, St. Martin's Press, New York, NY, USA, 5th edition, 1996.

[42] B. Cooksey, L. Anthony, J. Egoe et al., *Management of Natural Resources Programme, Tanzania TAN-0092*, Final Evaluation

Report, Ministry of Natural Resources and Tourism and the Royal Norwegian Embassy, Dar es Salaam, Tanzania, 2006.

[43] J. C. Ribot, A. Agrawal, and A. M. Larson, "Recentralizing while decentralizing: how national governments reappropriate forest resources," *World Development*, vol. 34, no. 11, pp. 1864–1886, 2006.

[44] A. Scheba and I. Mustalahti, "Rethinking 'expert' knowledge in community forest management in Tanzania," *Forest Policy and Economics*, 2015.

[45] I. Mustalahti and J. F. Lund, "Where and how can participatory forest management succeed? Learning from Tanzania, Mozambique, and Laos," *Society and Natural Resources*, vol. 23, no. 1, pp. 31–44, 2010.

12

Tree Species Diversity, Richness, and Similarity in Intact and Degraded Forest in the Tropical Rainforest of the Congo Basin: Case of the Forest of Likouala in the Republic of Congo

Suspense Averti Ifo,[1] Jean-Marie Moutsambote,[2] Félix Koubouana,[2] Joseph Yoka,[3] Saint Fédriche Ndzai,[2] Leslie Nucia Orcellie Bouetou-Kadilamio,[3] Helischa Mampouya,[2] Charlotte Jourdain,[4] Yannick Bocko,[3] Alima Brigitte Mantota,[2] Mackline Mbemba,[2] Dulsaint Mouanga-Sokath,[2] Roland Odende,[2] Lenguiya Romarick Mondzali,[2] Yeto Emmanuel Mampouya Wenina,[2] Brice Chérubins Ouissika,[2] and Loumeto Jean Joel[3]

[1] ENS, Département de Sciences et Vie de la terre, Université Marien Ngouabi, BP 69, Brazzaville, Congo
[2] ENSAF, Laboratoire d'Ecologie Appliquée Université Marien Ngouabi, BP 69, Brazzaville, Congo
[3] Faculté des Sciences, Département de Biologie et Physiologie Végétales, Université Marien Ngouabi, Brazzaville, Congo
[4] Via Costantino Beltrami 2, 00154 Roma, Italy

Correspondence should be addressed to Suspense Averti Ifo; ifo.suspense@hotmail.fr

Academic Editor: Timothy Martin

Trees species diversity, richness, and similarity were studied in fifteen plots of the tropical rainforests in the northeast of the Republic of Congo, based on trees inventories conducted on fifteen 0.25 ha plots installed along different types of forests developed on terra firma, seasonally flooded, and on flooded terra. In all of the plots installed, all trees with diameter at breast height, DBH ≥ 5 cm, were measured. The Shannon diversity index, species richness, equitability, and species dominance were computed to see the variation in tree community among plots but also between primary forest and secondary forest. A total of 1611 trees representing 114 species and 35 families were recorded from a total area of 3.75 ha. Euphorbiaceae was the dominant family in the forest with 12 species, followed by Fabaceae-Mimosoideae (10 species) and Phyllanthaceae (6 species) and Guttiferae (6 species). The biodiversity did not vary greatly from plot to plot on the whole of the study area (3.75 ha). The low value of Shannon index was obtained in plot 11 ($H' = 0.75$) whereas the highest value was obtained in plot 12 ($H' = 4.46$). The values of this index vary from 0.23 to 0.95 in plots P11 and P15, respectively. Results obtained revealed high biodiversity of trees of the forest of Impfondo-Dongou. The information on tree species structure and function can provide baseline information for conservation of the biodiversity of the tropical forest in this area.

1. Introduction

Tropical forests are the subject of several studies to better understand the role they could play in sustainable development, climate change, and floristic biodiversity [1, 2]. Tropical forests provide many goods and ecosystem services, such as prevention of soil erosion and preservation of habitats for plants and animals [3]. Globally, 52% of the total forests are in tropical regions and they are known to be the most important areas in terms of biodiversity [2, 4]. This diversity is an indicator that allows appreciating links between the richness and the abundance of individuals' trees; it reflects the degree of heterogeneity or stability of vegetation [5]. In the Republic of Congo (RoC), according to the definition of the forest, forests cover 69% of the territory [6]. Sustainable management of these forests requires a good knowledge of all the natural forest resource; this knowledge could be reliable only through studies of the forest environment.

Vegetation's studies led to either conducting a physiognomic research of the architectural type or identifying a number of representative reporting vegetation parameters, allowing defining simply, in order to compare it to other vegetation (Lescure, 1985). For the present study, the second approach was used, that of the floristic and structural parameters.

Many tropical forests are under great anthropogenic pressure and require management interventions to maintain the overall biodiversity, productivity, and sustainability [7]. Understanding tree composition and structure of forest is a vital instrument in assessing the sustainability of the forest, species conservation, and management of forest ecosystems [8]. Long-term biodiversity conservation depends basically on the knowledge of the structure, species richness, and the ecological characteristics of vegetation.

Some studies on the knowledge of the plant resource were conducted in Republic of Congo ([9–14] for the massif of Mayombe, [15]), but these studies remained generally piecemeal and predominantly localized inner protected areas and logging forest concessions. These studies were related to the ethnobotanical aspects and general knowledge of the flora of the Republic of Congo. And most of these studies were done essentially in the south of Republic of Congo and just one in the centre-west of our country. This work will provide more information on the tropical forest of Likouala, RoC. The aims of this research paper are to identify and quantify tree forest species of the tropical rainforest of Likouaka and specific objectives are (i) a floristic analysis of the forest of the axis Impfondo-Dongou, Likouala; (ii) analysis of floristic heterogeneity between interforest plots.

The study area is located within the Likouala department, which is of the most important forest regions in Republic of Congo.

2. Material and Methods

2.1. Study Area. The study was carried out within the tropical rainforest of the North of Congo Brazzaville in the department of Likouala (Figure 1). The zone of study covers a total surface of 155274 ha. It lies between $1°27'52,85''$ and $2°6'55,76''$ of northern latitude and between $17°52'35,04''$ and $18°04'32,65''$ of longitude.

The climate of the study area is of equatorial type. Mean rainfall is of 1760 mm y^{-1}, with a dry season from December to January and a long wet season from March to November (Figure 2). In the Dongou district, the soil cover is of tertiary clay sandy formation and a quaternary alluvial formation to the east. The soils derived from there are impoverished ferrilitic brown-red clay-sand soils on the Western plateau, ferralitic/hydromorphic alluvial soils on alluvial terraces, and waterlogged peat soils in flooded areas. This area has one of the very low densities of human population (0.93 km^{-2}) of the Republic of Congo. The forest of Likouala contains a high diversity of trees and plants [16]. In the Dongou district, the forests of the study area are rainforest. The principal vegetation types are partially deciduous dense rainforests of Ulmaceae and Sterculiaceae, swampy flooded forest of *Uapaca heudelotii*, and forest of *Guibourtia demeusei* [17]. Tree canopy closure of the forest varies from 93% to 100%

TABLE 1: Distribution of number of plots inside each type of forest.

Type of forest	Plots	Number of trees	Density (n/ha)	G (m^2/ha)
DF[1]	P1	67	268	6.75
	P2	73	312	9.11
	P3	212	848	34.24
	P11	217	868	26.73
PF[2]	P5	115	460	23.06
	P6	103	412	25.88
	P7	64	256	36.37
	P8	70	280	30.80
	P9	36	144	16.01
	P10	115	460	29.20
	P4	133	532	36,38
	P12	132	528	29.51
	P13	162	648	35.54
	P14	121	484	21.60
AF[3]	P15	51	204	36.52

1: degraded forest; 2: primary forest; 3: agroforestry.

while the tree height varies from 30 m to above 45 m (own data).

2.2. Data Collection. The tree sampling for the data collection was performed in 15 plots of 50 m × 50 m each placed in different forest strata of the study area: primary forest, secondary forest, and a mosaic of primary and secondary forest (Figure 3). Table 1 indicates the distribution of plots on the extent of the zone of study. The plot of intact forests and degraded forests inventoried was selected after image processing Landsat (OLI 8) of the study area. Coordinates GPS of the zones chosen on the satellite images were recorded in a GPS and on the ground we used the function Goto to go towards the points selected for the installation of the plot of inventories. The ground data allowed validating the classification of different type of forest in primary forest and secondary or degraded forest but also of the forest agro plot. Four plots fell into the zone from forest degraded, ten plots fell in the primary forest, and 1 plot fell into an agro drill forest.

GPS points of all plots were recorded, and inside each plot all living trees with diameter at breast height (DBH) ≥ 5 cm were recorded by species using latest botanical classification. All tree species were assigned to families and relative diversity (number of species in a family) was obtained for tree species diversity classification.

2.3. Measuring Biodiversity. We apply the Shannon diversity index (H') as a measure of species abundance and richness to quantify diversity of the woody species. This index takes both species abundance and species richness into account:

$$H' = -\sum_{i=1}^{s} p_i \ln p_i, \tag{1}$$

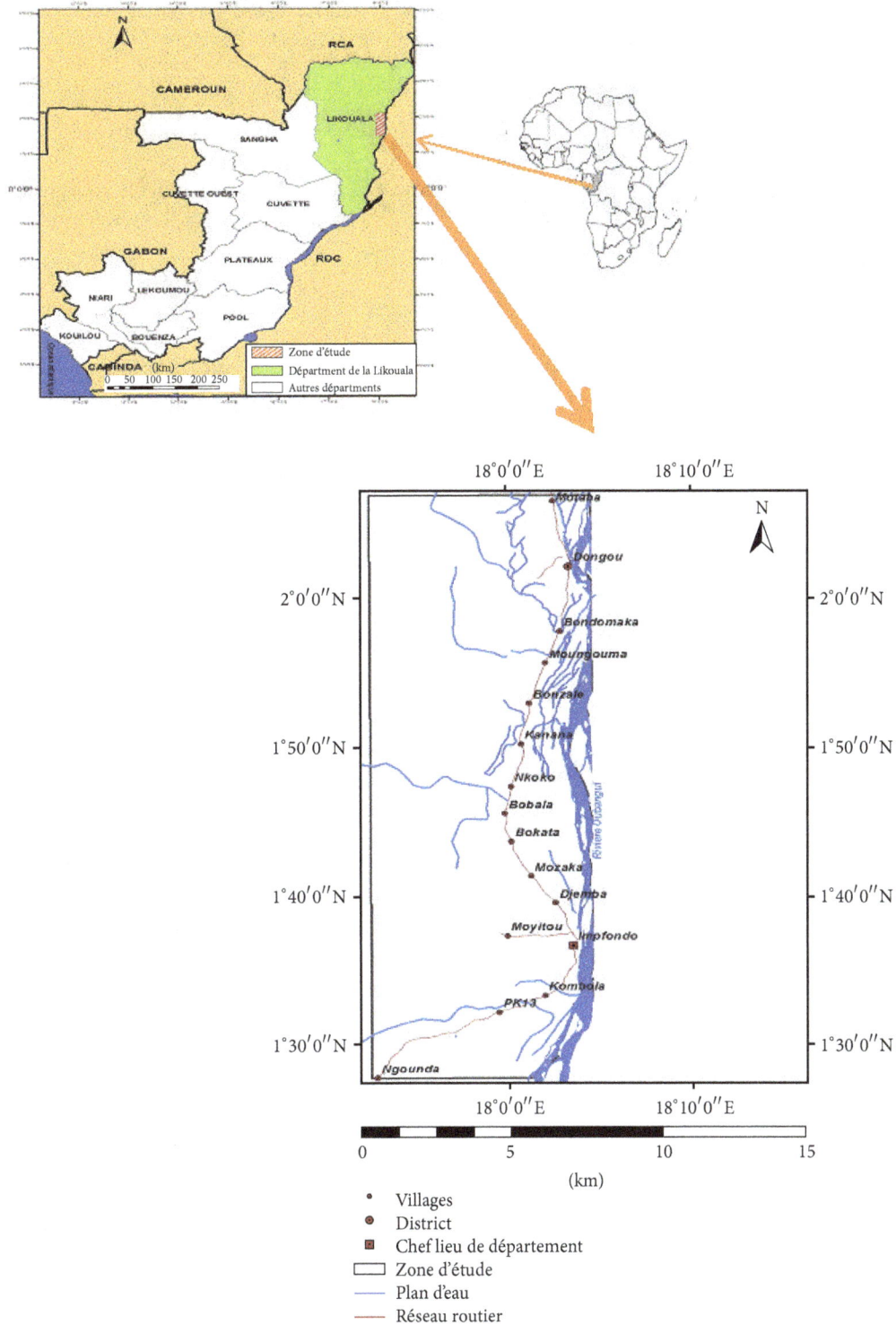

FIGURE 1: Localization of the department of Likouala, Congo Brazzaville.

where s equals the number of species and p_i equals the ratio of individuals of species i divided by all individuals N of all species. The Shannon diversity index ranges typically from 1.5 to 3.5 and rarely reaches 4.5 [18].

The variance of H' is calculated by

$$\operatorname{var} H' = \frac{\sum p_i \left(\ln p_i\right)^2 - \left(\sum p_i \ln p_i\right)^2}{N} + \frac{s-1}{2N^2} \qquad (2)$$

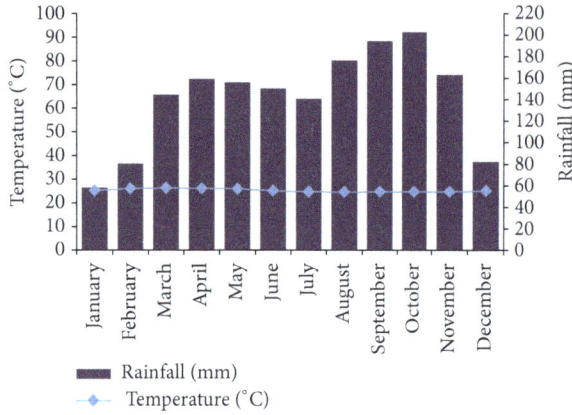

FIGURE 2: Ombrothermic diagram of Likouala (data from 1932 to 2015), ANAC Congo.

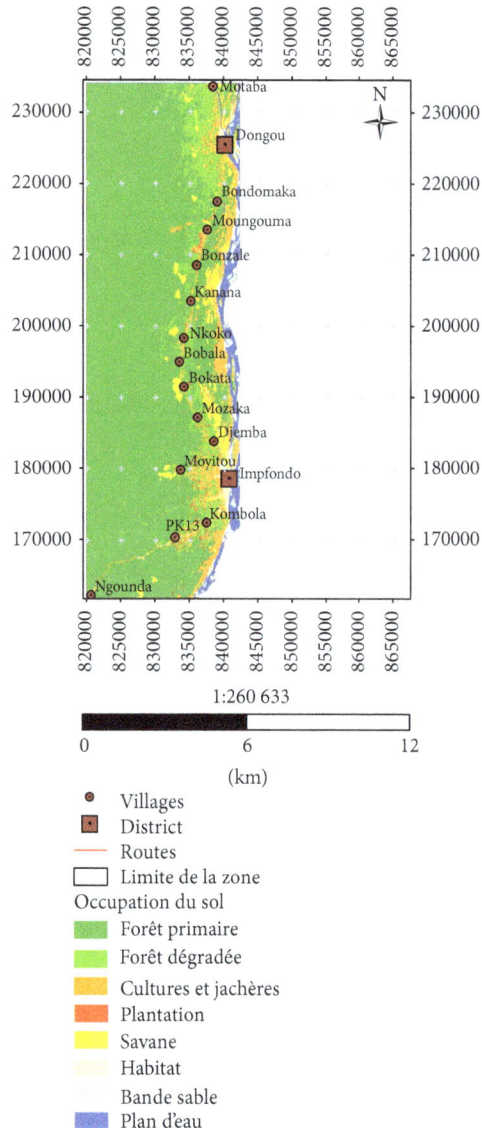

FIGURE 3: Cartography of land use change inside study's area.

and a t-statistic to test the significant differences between two plots or samples as

$$t = \frac{H_1' - H_2'}{\sqrt{\mathrm{var}\, H_1' + \mathrm{var}\, H_2'}}, \tag{3}$$

where H' is the Shannon diversity index of sample j.

Degrees of freedom for this test are equal to

$$\mathrm{d.f.} = \frac{\left(\mathrm{var}\, H_1' + \mathrm{var}\, H_2'\right)^2}{\left(\mathrm{var}\, H_1'\right)^2 / N_1 + \left(\mathrm{var}\, H_2'\right)^2 / N_2}, \tag{4}$$

where N_1 and N_2 are the number of individuals in samples 1 and 2, respectively [19]. We have also considered the Simpson index (D), a measure of species dominance, and the Shannon diversity index (E), a measure of evenness of spread.

The Simpson index is defined as

$$D = \sum_{i=1}^{s} \left(\frac{n_i (n_i - 1)}{N (N - 1)} \right), \tag{5}$$

where n_i is the number of individuals in the ith species and N equals the total number of individuals. As biodiversity increases, the Simpson index decreases. Therefore to get a clear picture of species dominance, we used $D' = 1 - D$.

The Shannon-Wiener index is defined as

$$E = \frac{H'}{H_{\max}} = \frac{-\sum_{i=1}^{s} p_i \ln p_i}{\ln s}. \tag{6}$$

$H'_{\max}$ is the natural logarithm of the total number of species. A value for evenness approaching zero reflects large differences in abundance of species, whereas an evenness of one means all species are equally abundant:

$$\mathrm{Margalef'\ Index}\ (d) = \frac{(S - 1)}{\ln (N)}, \tag{7}$$

where S is the total number of species, "N" is the number of individuals, and "ln" is the natural logarithm.

2.4. Similarity. The Jaccard index was used to calculate similarities of species between the forest types in different forest fragments. These coefficients are used to measure the association between samples. The similarity of two samples (floristic sample) is based on the presence or absence of certain species in the two samples [20]. To study the similarity of our different floristic samples, we used two binary factors excluding the double zeros, that is, the coefficient of Sorensen (K) and the coefficient of Jaccard (S). The Sorensen coefficient provides a twice higher weight to double presence; we can consider the presence of a more informative than this absence [20]:

$$S (\%) = \frac{(a \times 100)}{(a + b + c)}$$

$$K (\%) = \frac{(2a \times 100)}{(2a + b + c)} \tag{8}$$

TABLE 2: Floristic lists and their frequencies of the study area.

Family	Scientific name	Number of species	Number of trees
Achariaceae	*Caloncoba welwitschii* (Oliv.) Gilg.	1	9
Anacardiaceae	*Pseudospondias microcarpa* (A. Rich.) Engl. *Trichoscypha acuminata* Engl.	2	14
Annonaceae	*Anonidium mannii* (Oliv.) Engl. & Diels *Monodora angolensis* Welw.	2	19
Apocynaceae	*Alstonia boonei* De Wild.	1	1
Aptandraceae	*Ongokea gore* (Hua) Pierre	1	1
Bignoniaceae	*Markhamia tomentosa* (Benth.) K.	1	1
Burseraceae	*Dacryodes pubescens* (Verm.) Lam.	1	3
Cannabaceae	*Celtis adolfi-friderici* Engl.	1	12
Chrysobalanaceae	*Parinari congensis* F. Didr. *Parinari congolana* T. Durand et H. Durand *Parinari excelsa* Sabine *Maranthes glabra* (Oliv.)	4	32
Combretaceae	*Terminalia superba* Engl. et Diels.	1	3
Ebenaceae	*Diospyros crassiflora* Hiern *Diospyros ituriensis* (Gùrke) R. Let et F. White	2	47
Euphorbiaceae	*Cleistanthus itsogohensis* Pellegr. *Croton haumanianus* J. Léonard *Dichostemma glaucescens* Pierre *Grossera macrantha* Pax *Macaranga barteri* Mull.-Arg. *Macaranga monandra* Mull.-Arg. *Macaranga schweinfurthii* Pax *Macaranga spinosa* Mull.-Arg. *Plagiostyles africana* (Mull.-Arg.) Prain *Ricinodendron heudelotii* (Baill.) Pierre ex Pax *Sapium ellipticum* (Hochst.) Pax *Tetrorchidium didymostemom* (Baill.) Pax & K. Hoffm.	12	239
Fabaceae-Caesalpinioideae	*Copaifera salikounda* Heckel *Daniellia pynaertii* De Wild. *Dialum pachyphyllum* Harms *Guibourtia demeusei* (Harms) Léon. *Swartzia Bobgunnia fistuloides* (Harms) G.H Kirkpr.	5	116
Fabaceae-Faboideae	*Angylocalyx pynaertii* De Wild. *Baphia dewevrei* De Wild. *Millettia sanagana* Harms *Pterocarpus soyauxii* Taub.	4	123
Fabaceae-Mimosoideae	*Albizia ferruginea* (Guill. & Perr.) Benth. *Albizia laurentii* De Wild. *Cathormion rhombifolium* (Benth.) Hutch. & Dandy (syn: *Albizia rhombifolia* Benth.) *Albizia zygia* (DC) J. F. Macbr. *Newtonia devredii* G. C. C. Gilbert	10	58

TABLE 2: Continued.

Family	Scientific name	Number of species	Number of trees
	Parkia filicoidea Welw. ex Oliv.		
	Parkia bicolor A. Chev.		
	Pentaclethra macrophylla Benth.		
	Piptadeniastrum africanum (Hook. F.) Bren.		
	Tetrapleura tetraptera (Schum. & Thonn.) Taub.		
Guttifereae	*Allanblackia floribunda* Oliv.	6	55
	Garcinia punctata Oliv.		
	Garcinia ovalifolia Oliv.		
	Mammea africana Sabine		
	Symphonia globulifera L. f.		
	Garcinia smeathmannii Oliv.		
Irvingiaceae	*Irvingia excelsa*	3	16
	Irvingia grandifolia (Engl.) Engl.		
	Klainedoxa gabonensis Pierre ex Engl.		
Lamiaceae-Viticoideae	*Vitex pachyphylla* Bak.	1	8
Lauraceae	*Persea americana* L.	1	1
Lecythidaceae	*Petersianthus macrocarpus* (P. Beauv.) Liben.	2	63
	Brazzeia congensis Baill.		
Malvaceae-Sterculioideae	*Cola nitida* (Vent.) Schott & Endl.	1	2
Malvaceae-Tilioideae	*Duboscia macrocarpa* Brocq.	1	8
Meliaceae	*Carapa procera* var. *palustre* DC	5	46
	Carapa procera var. *procera* DC		
	Entandrophragma cylindricum (Sprague) Sprague		
	Trichilia monadelpha (Thonn.) J. J. De Wild.		
	Trichilia tessmannii Harms		
Moraceae	*Antiaris toxicaria* var. *welwitschii* Lesch.	5	17
	Ficus exasperata Vahl.		
	Ficus vogeliana (Miq.) Miq.		
	Milicia excelsa (Welw.) C. C. Berg		
	Trilepisium madagascariense DC.		
Myristicaceae	*Coelocaryon preussii* Warb.	3	226
	Pycnanthus angolensis (Welv.) Exell		
	Staudtia kamerounensis Warb. var. *gabonensis* Fouilloy		
Ochnaceae	*Lophira alata* Banks ex Gaertn.	2	14
	Rhabdophyllum welwitschii Van Tiegh.		
Olacaceae	*Heisteria parvifolia* Smith	3	52
	Strombosia grandifolia Hoof. F.		
	Strombosia pustulata Oliv.		
Pandaceae	*Panda oleosa* Pierre	1	6
Passifloraceae	*Barteria fistulosa* Mast.	1	1
Putranjivaceae	*Drypetes pellegrini* Léandri	2	7
	Drypetes leonensis (Pax) Pax et K. Hoffm.		

TABLE 2: Continued.

Family	Scientific name	Number of species	Number of trees
Phyllanthaceae	*Cleistanthus mildbraedii* Jabl. *Hymenocardia ripicola* J. Léonard *Hymenocardia ulmoides* Oliv. *Maesobotrya dusenii* (Pax) Hutch. *Uapaca guineensis* Mull.-Arg. *Uapaca heudelotii* Baill.	6	37
Rubiaceae	*Aidia micrantha* (K. Schum.) F. White *Colleactina papalis* N. Hallé *Massularia acuminata* (G. Don) Bullock ex Hoyle *Morelia senegalensis* A. Rich. *Morinda pynaertii* Benth. *Oxyanthus schumannianus* De Wild. et Th. Dur *Psydrax subcordata* DC *Psydrax arnoldiana* (De Wild.)	5	44
Rutaceae	*Zanthoxylum heitzii* (Aubrév. & Pellegr.) P. G. Waterman	1	1
Sapindaceae	*Blighia welwitschii* (Hiern) Radlk. *Eriocoelum microspermum* Radlk. *Lecaniodiscus cupanioides* Planch. ex Benth. *Pancovia pedicellaris* Radlk. & Gilg	4	55
Sapotaceae	*Chrysophyllum beguei* Aubrév. *Synsepalum brevipes* (Baker) TD Penn *Tridesmostemom omphalocarpoides* Engl. *Manilkara* sp. *Manilkara fouilloyana* Aubr. et Pellegr.	4	18
Thomandersiaceae	*Thomandersia hensii* De Wild.	1	21
Urticaceae	*Musanga cecropioides* R. Br. *Myrianthus arboreus* P. Beauv.	2	235
	Total	114	1611

with a = number of common presences for both floristic samples, b = number of presences in the first floristic sample, c = number of presences in the second floristic sample, and d = number of species absent in both floristic samples.

According to L. Legendre and P. Legendre [20], the Sorensen coefficient is fully compared with the Jaccard coefficient; that is, if the similarity of a pair of objects computed by the Jaccard coefficient is higher than the similarity of another pair of objects, it will also be higher if we use the coefficient of Sorensen for the calculation of similarity.

3. Results

3.1. Floristic Composition and Species Richness. A total of 1611 trees representing 114 species and 35 families were identified from the total area (3.75 ha). Euphorbiaceae was the dominant family in the forest with 12 species, followed by *Fabaceae Mimosoideae* with 10 species. In terms of the number of trees individuals per family, Euphorbiaceae was the dominant in the whole forest with 239 trees, followed by Urticaceae with 235 trees (Table 2).

In terms of characterization of forest type, this inventory allowed distinguishing several forest types like *Lophira alata, Uapaca heudelotii, Guibourtia demeusei,* and *Celtis adolfi-friderici.* Inventories have revealed the existence of three vertical strata, whose upper stratum is dominated by species referred to above.

The biodiversity did not vary greatly from plot to plot on the whole of the study area (3.75 ha). A low Shannon diversity index value was obtained in plot 11 ($H' = 0.75$) whereas the highest value was obtained in plot 12 ($H' = 4.46$). A statistical analysis made by launched ANOVA revealed that plot 11 was significantly different to the other plots ($\alpha = 0.05$). A great difference was also noted in biodiversity between secondary plots and primary plots (Table 3). The evenness index was calculated. The values varied from 0.23 in plot P11 to 0.95 in plot P15.

The evenness index E was calculated for each plot. The value of equitability varied from 0 to 1. It is equal to 1 when all

TABLE 3: Biodiversity values by biodiversities index and static parameters.

Plots	Type of forest	Total individual	S	Shannon diversity index (H)	Fisher's α	Simpson index	Evenness index = H'_{max}	Variances (H)	Ecartype
P1	Degraded forest	67	26	4.33	15.76	0.05	3.06	0.28	0.53
P2	Degraded forest	79	22	3.54	13.99	0.14	2.64	0.18	0.43
P3	Degraded forest	212	38	4.13	23.23	0.08	2.62	0.09	0.3
P4	Primary forest	132	26	3.57	14.92	0.14	2.52	0.11	0.34
P5	Primary forest	111	24	3.97	11.3	0.08	2.88	0.15	0.39
P6	Primary forest	102	29	4.14	17.39	0.07	2.83	0.18	0.42
P7	Primary forest	52	15	3.36	8.39	0.11	2.86	0.23	0.48
P8	Primary forest	61	9	2.46	3.81	0.22	2.58	0.11	0.33
P9	Primary forest	34	12	2.93	8.13	0.16	2.72	0.28	0.53
P10	Primary forest	106	27	3.94	16.32	0.09	2.76	0.16	0.4
P11	Degraded forest	217	10	0.75	5.85	0.05	0.75	0.04	0.2
P12	Primary forest	126	33	4.47	17.76	0.05	2.94	0.16	0.41
P13	Primary forest	153	31	3.89	19.03	0.1	2.61	0.11	0.33
P14	Primary forest	109	31	4.38	16.54	0.06	2.93	0.18	0.43
P15	Agroforestry	47	20	4.12	11.37	0.04	3.17	0.36	0.6

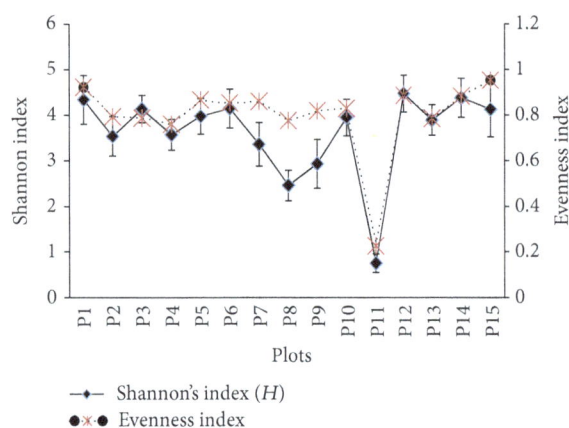

FIGURE 4: Shannon diversity index and evenness index trends in all the study areas.

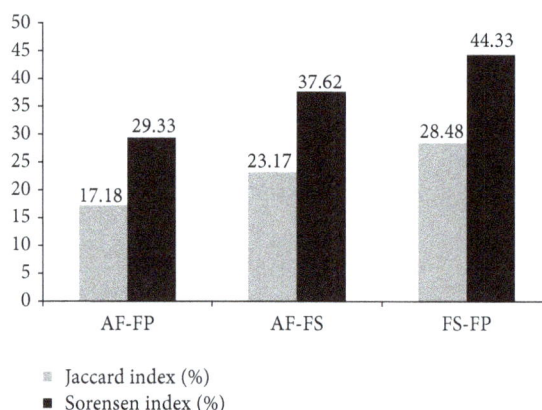

FIGURE 5: Similarity index between two types of forest. AF = agroforestry, FP = primary forest, and FS = secondary forest.

the species have same abundance and tend towards 0 when the near total of flora is concentrated on only one species. The values of this index varied from 0.23 to 0.95 in plots P11 and P15, respectively (Figure 4). The value of plot 11 confirms well conducted survey in the plot which is dominated by one species, *Musanga cecropioides*. The E value obtained in plot P11 is the one with a value inferior to 0.5 out of the entire results. Two plots have value of E superior to 0.9 (plots P1 and P15). Twelve plots have a value of E varying between 0.7 and 0.89.

3.2. Biodiversities Indexes and Other Parameters.

Other analyses of the biodiversity made by applying the other indices such as the index of Fisher α revealed interesting information. Whereas with the Shannon diversity index, it is in plot 11 that we noted the weakest biodiversity, the application of the index Fisher α (Table 3) showed that the low value of the

biodiversity was obtained in plot 08 which is a primary forest plot whereas the strongest value of F is observed in P3 plot, which is a mosaic of secondary and primary forest. Plot 11 (monodominant plot of *Musanga cecropioides*) does not have the low value of the biodiversity like Shannon diversity index revealed.

The species richness of 114 species was observed in 3.75 ha of the Likouala Forest Department. *Musanga cecropioides* was the most dominant with 222 trees censured followed by *Staudtia kamerounensis* var. *gabonensis* with 117 trees.

3.3. Similarity: Sorensen (K) and Jaccard (S) Index.

Species similarities between the forest types were studied between primary forest and secondary forest (FS-FP), primary forest and agroforestry land (AF-FP), and agroforestry land and secondary forest (plot 15). We have noted that the lowest Jaccard index value was obtained between AF-FP (Figure 5) (17.18%). The highest value was noted between FS-FP.

TABLE 4: The characteristics of ecological factors in some forests of Republic of Congo.

Name of forest	Rainfall (mm)	Authors	Length of dry season
Forest of Impfondo-Dongou	1800–2000	Our study	2
Forest of Mayombe	1600	Koubouana et al. [21]	4
Forest of centre-west of Congo	2132.6	Our own data	2
Forests of the littoral	1500	Kimpouni [22]	4

The indices of diversity enabled us to conclude that the studied zones are rich in cash. Are various studied forests similar from the floristic composition point of view?

The values of coefficient of similarity vary from 17.18% to 28.48% for the index of Jaccard and 29.33% to 44.33% for the index of Sorensen.

4. Discussion

The analysis of the tree flora of the study area showed that the families of Euphorbiaceae (10.53%) are the most represented, followed by the Fabaceae-Mimosoideae (8.77%), Rubiaceae (7.89%), and the Guttiferae (6.14%). Indeed, the presence of the Euphorbiaceae and Rubiaceae generally represented by species of wood is a character common to all tropical rainforests as noted by Reynal-Roques [29]. However the abundance of the Fabaceae-Mimosoideae and Guttiferae is proof of the old age or maturity of the inventoried forest [11]. From the point of view of physiognomy of forest areas studied, the results show that these are the Euphorbiaceae (14.84%) which are abundant in terms of number of trees from beneath wood and stratum average, followed by the Urticaceae (14.59%), the Myristicaceae (14.03%) of the Fabaceae-Faboideae (7.64%), and Fabaceae-Mimosoideae (7.20%). The abundance of the Urticaceae is explained by the presence of quasi-monospecific stands to *Musanga cecropioides* in degraded forests. However the abundance of the Myristicaceae and Fabaceae is a specific character of the forests studied axis Impfondo-Dongou. Indeed, the results obtained in our study are totally different from those obtained by Kimpouni [22] in Congolese coastal forests, Koubouana et al. [30] in the forest of Western Centre at Mbomo and Kelle. Indeed the work of Kimpouni [22] performed in the littoral showed an abundance of the Fabaceae-Caesalpinioideae, followed by the Rubiaceae and Euphorbiaceae. Those of Koubouana et al. (in press) in the centre-west of the Congo showed an abundance of the Fabaceae-Caesalpinioideae (18.05%), followed by the Meliaceae (7.52%), Fabaceae-Mimosoideae (6.02%), Euphorbiaceae (5.26%), Annonaceae (4.51%), and the Myristicaceae (3.76%). In the study conducted at the Mayombe in the South of Congo by Koubouana et al. [21], the results obtained show an abundance of family Burseraceae (19.17%) followed by the Fabaceae-Caesalpinioideae (16.09%), Myristicaceae (13.18%), Annonaceae (9.49%), Euphorbiaceae (8.32%), and Fabaceae-Mimosoideae (7.32%). This variation of the floristic composition of the different forests studied is explained by the diversity of geological substrate and the diversity of climate (Table 4). Table 5 showed the characteristics of ecological factors in the forests studied.

It is important to note that Brazzaville is in the south of the Republic of Congo and Mbomo-Kelle's locality is in the northwest of Republic of Congo. In comparison with these two localities, our study area is in the extreme northeast. Each of the study areas has a local climatic condition (Table 4).

Table 3 shows the values for the assessment of the biodiversity of trees surveyed in 15 parcels that were the subject of this study on floristic biodiversity on ≥10 cm DBH trees. In this study we wanted to focus our attention on the biodiversity indices to assess the level of biodiversity across the study area, but also the microvariations that would exist between the plots of the study area. Moreover, in the scope of this study, we tested the role that vegetation indexes might play in the evaluation of forest degradation between primary and secondary forests.

Considering the Shannon diversity index, our study showed that plot 15 has a lower Shannon index of 1, while two plots have clues to Shannon between 2 and 3. The rest of the plots with higher values have 3. High species richness is a hallmark of many tropical forests (Gentry et al. 2010).

Our study revealed the existence of variability of biodiversity in the study area. According to Orth and Colette [31] the Shannon diversity index has strong values for species with recoveries of same importance and it takes low values, when some species have strong recoveries.

Low biological diversity noted in plot 11 ($H' = 0.75$) could be explained by the fact that it is dominated by a single species *Musanga cecropioides*. This species contributes nearly 90% of the total number of trees in the plot. In two plots with the highest values, the plots contain more than 30 species of trees with at least two species of codominant trees, but with lower contributions. In parcel 12 ($H' = 4.47$), *Angylocalyx pynaertii* De Wild and *Plagiostyles africana* species each have a 13% contribution. In these same plots the other two species following in terms of specific contribution are *Grossera macrantha* and *Strombosia grandifolia* with, respectively, 7% and 6% of a total of 126 species inventoried in this plot. As shown in Table 1, the application of the other indices of biodiversity gives a different result. The Pioulou biodiversity index (H'_{max}) indicates that plot 15 ($H'_{max} = 3.17$) has the highest biodiversity followed by plot 1 ($H'_{max} = 3.06$) and plot 12 ($H'_{max} = 2.94$). Several causes could explain variations in the degree of biodiversity between the plots of the study area: soil type, rainfall trends, anthropogenic action, land use change, and so forth.

The Shannon diversity index values obtained in this study are lower than those obtained in other studies both in the Republic of Congo and in other tropical forests in the Congo basin compared to other tropical countries. For

TABLE 5: The characteristics of ecological factors in some forests of Republic of Congo.

Type of vegetation	Minimum tree DBH	Study area (ha)	Countries	Rainfall mean (mm/year)	Shannon-Wiener index (bit)	Number of families	Number of genera	Number of trees	Number of species	Authors
	DBH ≤ 10 cm	1.5	Southwest of the Republic of Congo	1200–1500	1.9 ± 0.5	47	120		153	Kimpouni et al. [23]
Forest	DBH ≥ 20 cm	35		1600	3.75	41		5076		Koubouana et al. [21]
Shrub savannah		400	Plateaux Teke, Republic of Congo	1600–2100	2.16	15	16	3075	25	Mampouya Wenina [24]
Fabaceae-Caesalpinioideae	DBH ≥ 20 cm	88.5	Northwest of Republic of Congo	1900	5.3	31	107	11012	133	Koubouana et al. (in press)
Forest	DBH ≥ 10 cm		Plateau des Cataractes, Republic of Congo	1400–1600 mm		42	116		153	Kimpouni [22]
Forest (monodominant *Aucoumea klaineana*)	DBH ≥ 10 cm		Youbi	1200				1186	71	Kimpouni et al. [25]
Mosaics of natural forest and grassland	(DBH) ≥ 10 cm	0.72	Uganda, Youbi, Republic of Congo (southwest)	1397–1500 mm	4.02	26			93	Nangendo et al. [26]
	DBH ≥ 15 cm	1		1300	3.55, 3.47, 3.48, and 3.32	40	73	1789	92	Premavani et al. [27]
Forest	DBH ≥ 10 cm	1.96		2500–3000	3.795			808	72	Aigbe and Omokhua [28]

instance, in the forest of centre-west of Republic of Congo in Mbomo-Kelle (Republic of Congo), Shannon diversity index varies from 5.91 to 5.95 in bloc 4 and bloc 9, respectively (Koubouana et al. in press). In the southwest of the Republic of Congo, studies were conducted by Kimpouni et al. [23] and Koubouana et al. [21] and revealed different H' values. Kimpouni et al. [23] in a degraded forest in Brazzaville obtained for woody species a Shannon diversity index of 1.9 bits. But in the tropical forest of southwest of the Republic of Congo, Koubouana et al. (in press) noted an old secondary forest that the Shannon diversity index was about 3.08.

Regarding heterogeneity, many authors think that the structural heterogeneity of the forests and their high species richness are often interpreted in terms of forest dynamics and relationship with the resulting phenomena of succession [5, 32]. In this work, we have mainly focused on the study of the biodiversity of trees to make a comparison between the degraded forest areas and nondegraded forest areas.

Several factors could explain the variations of biodiversity in our study: the topography of the area [33, 34] or edaphic factors [35, 36] to explain the issue of floral heterogeneity of tropical forests. In our study area, three forest types were identified: flooded forests, solid ground forest, and partially flooded forest. In the context of this work we have studied the differences that exist between forest types through the Jaccard and of Sorensen similarity indexes. Moreover, (K) and Jaccard (S) index give a very good idea of the presence or absence of species in the different transects of the inventory. The range of this coefficient is between 0 and 1. Interpretation of the CSJ values is as follows: 1: both survey sites have only common species; 0: both survey sites have only singular species; 1/2: the two survey sites have as many common species as the sum of singular species at each survey site; [0, 1/2]: the similarity in terms of species diversity between both survey sites is rather low; [1/2.0]: the similarity in terms of species diversity between both survey sites is rather high.

In our case, it ranges from 17.18 to 23.48%, taking into account the three combinations (AF-FP, AF-FS, and FS-FP). This means the similarity in terms of species diversity between both survey sites is rather low. The results showed that there is a high tree biodiversity in our study area.

The values of similarity index are lower than 50%, which enables us to conclude that there is obviously a difference in point of floristic composition between the primary forests and the secondary forests, thus confirming the floristic data that we presented above.

4.1. Evaluation of Forest Degradation through the Biodiversity Indexes. Degradation is considered to be a temporal process. There is however a consensual definition that is accepted by the various stakeholders, which is as follows: forest degradation is the reduction of the capacity of the forest to provide goods and services. In the context of REDD+, forest degradation can be defined as the partial loss of biomass due to logging or other causes of removal of wood from biomass [37]. A degraded forest is a secondary forest that has lost, as a result of human activities, the structure, function, composition, or productivity of species normally associated with a natural forest. Thus, a degraded forest offers a supply

of goods and services and has only limited biodiversity. Biological diversity in a degraded forest includes many nontree components which can dominate the understory vegetation cover (CBD (2005; 2001).

To assess the level of forest degradation, a maximum biodiversity index has been applied and presents different values between the secondary and primary forest. These forests are characterized by a very high biodiversity, especially in the case of secondary forests of *Macaranga spinosa* or *Macaranga barteri*. In the case of plots P1 and P2 the objective of this study is to identify settings that allow assessing forest degradation.

5. Conclusion

The study of the biodiversity of the Likouala forests and the Impfondo-Dongou axis revealed a high floral biodiversity of trees considering a diameter of 5 cm at DBH 1.30 m. Tree biodiversity is very important in primary forests. In secondary forests, the biodiversity varies in line with the secondary forest type: secondary forest of *Macaranga spinosa* or secondary forest of *Musanga cecropioides*. Moreover, biodiversity varies according to the nature of the substrate: forest on dry land, forest in partially flooded areas, and flooded forest.

Competing Interests

The authors declare that there is no conflict of interests regarding the publication of this paper.

Acknowledgments

The authors are thankful to GEOFORAFRI for funding the project. The authors thank Benoit Mertens for kind support during the project.

References

[1] S. L. Lewis, G. Lopez-Gonzalez, B. Sonké et al., "Increasing carbon storage in intact African tropical forests," *Nature*, vol. 457, no. 7232, pp. 1003–1006, 2009.

[2] M. N. K. Djuikouo, J.-L. Doucet, C. K. Nguembou, S. L. Lewis, and B. Sonké, "Diversity and aboveground biomass in three tropical forest types in the Dja Biosphere Reserve, Cameroon," *African Journal of Ecology*, vol. 48, no. 4, pp. 1053–1063, 2010.

[3] M. Anbarashan and N. Parthasarathy, "Tree diversity of tropical dry evergreen forests dominated by single or mixed species on the Coromandel coast of India," *Tropical Ecology*, vol. 54, no. 2, pp. 179–190, 2013.

[4] L. R. Holdridge, "Life Zone Ecology, Tropical Science Center, San Jose, Costa Rica," 1967.

[5] V. Trichon, "Hétérogénéité spatiale d'une forêt tropicale humide de Sumatra: effet de la topographie sur la structure floristique," *Annales des Sciences Forestières, INRA/EDP Sciences*, vol. 54, no. 5, pp. 431–446, 1997.

[6] CNIAF, "Carte de Changement de Couverture Forestière en République du Congo pour la Période 2010–2012," 2015.

[7] A. Kumar, B. G. Marcot, and A. Saxena, "Tree species diversity and distribution patterns in tropical forests of Garo Hills," *Current Science*, vol. 91, no. 10, pp. 1370–1381, 2006.

[8] D. S. Kacholi, "Analysis of structure and diversity of the Kilengwe Forest in the Morogoro Region, Tanzania," *International Journal of Biodiversity*, vol. 2014, Article ID 516840, 8 pages, 2014.

[9] J.-M. Moutsambote, *Dynamique de reconstitution de la forêt Yombe (Dimonika, R.P. du Congo) [Ph.D. thesis]*, These de 3e cycle, University of Bordeaux, Bordeaux, France, 1985.

[10] G. Cusset, *La Flore et la Végétation du Mayombe Congolais. État des Connaissances*, Université Pierre et Marie Curie, Paris, France, 1987.

[11] G. Cusset, "La flore et la végétation du Mayombe congolais, état des connaissances," in *Revue des Connaissances sur le Mayombe*, J. Sénéchal et al., Ed., pp. 103–136, Unesco, Paris, France, 1989.

[12] E. J. Adjanohoun, A. M. R. Ahyi, L. AkeAsi et al., *Contribution aux Études Ethnobotaniques et Floristiques en République Populaire du Congo: Médecine Traditionnelle et Pharmacopée*, ACCT, Paris, France, 1988.

[13] V. Kimpouni and F. Koubouana, "Étude ethnobotanique sur les plantes médicinales et alimentaires dans et autour de la réserve de Conkouati," Rapport Final, PROGECAP/GEF-Congo, UICN, 1997.

[14] F. Dowsett-Lemaire, "The vegetation of the Kouilou basin in Congo," in *Flore et Faune du Bassin du Kouilou (Congo) et Leur Exploitation*, R. J. Dowsett and F. Dowsett-Lemaire, Eds., vol. 4, pp. 17–51, Tauraco Research Report, 1991.

[15] F. Koubouana and J. M. Moutsambote, *Etude Préliminaire de la Végétation de l'UFA Letili et Bambama*, Rapport D'étude, Brazzaville, Congo, 2006.

[16] J.-M. Moutsamboté, *Etude écologique, phytogéographique et phytosociologique du Congo septentrional (Plateaux, Cuvettes, Likouala et Sangha) [Thèse de Doctorat d'Etat]*, Faculté des Sciences, Université Marien Ngouabi, Brazzaville, République du Congo, 2012.

[17] IUCN, "La conservation des ecosystèmes forestiers du Congo. Basé sur le travail de Philippe Hecketsweiller. IUCN, Gland, Suisse et Cambridge, Royume uni. 187., illustré," 1989.

[18] W. L. Gaines, J. R. Harrod, and J. F. Lehmkuhl, "Monitoring biodiversity: quantification and interpretation," General Technical Report PNW-GTR-443, USDA Forest Service, Pacific North-West Research Station, 1999.

[19] A. E. Magurran, *Ecological Diversity and Its Measurement*, CroomHelm, London, UK, 1988.

[20] L. Legendre and P. Legendre, *Écologie Numérique, Tome 1: Traitement Multiple des Données Écologiques*, Masson, Paris, France, 2nd edition, 1984.

[21] F. Koubouana, S. A. Ifo, J.-M. Moutsambote et al., "Structure and flora tree biodiversity in congo basin: case of a secondary tropical forest in southwest of congo-brazzaville," *Research in Plant Sciences*, vol. 3, no. 3, pp. 49–60, 2015.

[22] V. Kimpouni, "Contribution to the inventory and analysis of the ligneous flora of the plates of the Cataracts (Congo-Brazzaville)," *Acta Botanica Gallica*, vol. 156, no. 2, pp. 233–244, 2009.

[23] V. Kimpouni, Å. Apani, and M. Motom, "Analyse phytoécologique de la flore ligneuse de la Haute Sangha (République du Congo)," *Adansonia, Série 3*, vol. 35, no. 1, pp. 107–134, 2013.

[24] Y. E. Mampouya Wenina, *Biodiversité et variabilité de la densité du bois des arbustes de savane dans les environs du village Mâh (Plateaux TEKE, République du Congo) [M.S. thesis]*, Université Marien Ngouabi, 2015.

[25] V. Kimpouni, J. Loumeto, and J. Mizingou, "Woody flora and dynamic of *Aucoumea klaineana* forest in the Congolese littoral," *International Journal of Biological and Chemical Sciences*, vol. 8, no. 4, pp. 1393–1410, 2014.

[26] G. Nangendo, A. Stein, M. Gelens, A. de Gier, and R. Albricht, "Quantifying differences in biodiversity between a tropical forest area and a grassland area subject to traditional burning," *Forest Ecology and Management*, vol. 164, no. 1–3, pp. 109–120, 2002.

[27] D. Premavani, M. T. Naidu, and M. Venkaiah, "Tree species diversity and population structure in the Tropical Forests of North Central Eastern Ghats, India," *Notulae Scientia Biologicae*, vol. 6, no. 4, pp. 448–453, 2014.

[28] H. I. Aigbe and G. E. Omokhua, "Tree species composition and diversity in Oban Forest reserve, Nigeria," *Journal of Agricultural Studies*, vol. 3, no. 1, pp. 10–24, 2015.

[29] Reynal-Roques, *La Botanique Redecouverte*, Reynal-Roques, Berlin, Germany, 1994.

[30] F. Koubouana, S. A. Ifo, J.-M. Moutsambote, and R. Mondzali-Lenguiya, "Floristic diversity of forests of the Northwest Republic of the Congo," *Open Journal of Forestry*, In press.

[31] D. Orth and M. G. Colette, "Espèces dominantes et biodiversité: relation avec les conditions édaphiques et les pratiques agricoles pour les prairies des marais du cotentin," *Ecologie*, vol. 27, no. 3, pp. 171–189, 1996.

[32] A. Aubréville, "La forêt coloniale: les forêts de l'afrique occidentale française," *Annales—Académie des Sciences Coloniales*, vol. 9, pp. 1–245, 1938.

[33] F. Kahn, *Architecture comparée de forêts tropicales humides et dynamique de la rhizosphère [Ph.D. thesis]*, USTL, Montpellier, France, 1983.

[34] K. Basnet, "Effect of topography on the pattern of trees in tabonuco (Dacryodes excelsa) dominated rain forest of Puerto Rico," *Biotropica*, vol. 24, no. 1, pp. 31–42, 1992.

[35] J.-P. Lescure and R. Boulet, "Relationships between soil and vegetation in a tropical rain forest in French Guiana," *Biotropica*, vol. 17, no. 2, pp. 155–164, 1985.

[36] J. S. Gartlan, D. M. Newbery, D. W. Thomas, and P. G. Waterman, "The influence of topography and soil phosphorus on the vegetation of Korup Forest Reserve, Cameroun," *Vegetatio*, vol. 65, no. 3, pp. 131–148, 1986.

[37] M. Kanninen, D. Murdiyarso, F. Seymour, A. Angelsen, S. Wunder, and L. German, *Do Trees Grow on Money? The Implications of Deforestation Research for Policies to Promote REDD*, Forest Perspectives no. 4, CIFOR, Bogor, Indonesia, 2007.

13

Analysis of the Distribution of Forest Management Areas by the Forest Environmental Tax in Ishikawa Prefecture, Japan

Yuta Uchiyama and Ryo Kohsaka

Kanazawa University Graduate School of Human and Socio-Environmental Studies, Kanazawa, Japan

Correspondence should be addressed to Yuta Uchiyama; yutanu4@yahoo.co.jp

Academic Editor: Kihachiro Kikuzawa

Forest management approaches vary according to the needs of individual municipalities with unique geographic conditions and local social contexts. Accordingly, there are two types of subsidies: a unified national subsidy and a prefecture-level subsidy, mainly from forest environmental taxes. The latter is a local tax. Our focus is on examining forest management using these two types of taxes (i.e., central and prefecture-level) and their correlations with social and natural environmental factors. In this paper, we examine the spatial distribution of management areas using subsidies from the central government, the Forestry Agency of Japan, and prefectural forest environmental taxes in Ishikawa. In concrete terms, the spatial correlations of the management areas under two tax schemes are compared with the natural hazard areas (as a natural environmental factor) and areas with high aging rates (as a social factor). The results are tested to see whether the correlations of areas with the two factors are significant, to examine whether the taxes are used for areas with natural and social needs. From the result, positive correlations are identified between the distribution of management areas and natural hazard areas and between the distribution of management areas and areas with high aging rates.

1. Introduction

Owing to the trend of the decreasing size of business in the Japanese forestry sector, the management of the forestland has involved serious reduction in forestry workers. Most parts of trees in forest plantations in Japan need to be harvested. Sustainable forest management issues, including tending and periodic thinning, need to be tackled over several generations. In mountainous forestlands and other forest areas that are difficult to access, most forest owners and forest associations are no longer able to maintain forestlands as their own businesses; forest management in those regions depends on support from national and regional governments.

The need for forest management differs among Japanese regions because their physical, geographical, and social contexts differ. Prefectures implement their forest policies in the context of their individual regions. The forest environmental tax is a policy option for sustainable management in individual prefectures. Forest management support from the prefectures depends on the general revenue resources of the prefectures. In addition to general revenue resources, revenue from the forest environmental tax in each prefecture is used for forest management. Kochi Prefecture introduced the first forest environmental tax in 2003, and 33 prefectures have now introduced the tax, not including urban prefectures such as Tokyo. Due to a shortage of revenue resources, it is necessary to use revenue resources effectively to deal with specific regional issues.

The forest environmental tax can be regarded as an institution of the Payment for Ecosystem Services (PES) [1–6]. To acknowledge the ecosystem services from forestlands sustainably, this tax has been introduced in the prefectures. The purpose of the tax is to implement forest management as a charge to the beneficiaries. In terms of PES, revenue from the forest environmental tax should be paid to forest owners and related people who maintain forests, to make them provide ecosystem services. In practice, revenues from the tax have been paid mainly to forest associations. The characteristics of the forest environment tax include the fact that the tax system is at a regional level, and beneficiaries in wider regions are required to pay relatively small amounts for the tax [7].

Because the tax system is at the prefecture level, prefectures are not necessarily uniform in their application of the system. A common feature of the forest environmental tax at the prefectural level is that taxpayers are frequently unaware of the tax, because of the limited amount (frequently 500 yen per annum) and the mode of payment (i.e., it is collected with the water usage fees or residence taxes). Almost all residents within the prefectures are obliged to pay these forest environmental taxes, except for certain urban prefectures such as Tokyo and Osaka, where they do not have such systems.

Because most taxpayers are unaware of their existence, the purpose and operations relating to the use of this tax money are not thoroughly examined from our perspectives based on GIS data. Are the taxes used in areas in urgent need of forest management, consistent with the scheme's original design? If so, which social or natural factors play a larger role?

As with many other public budget programs, management areas are distributed unevenly and without strict rules and standards. The areas are selected through negotiation between forest owners and forest cooperatives. Dialogues are critical because the main management areas under the forest environmental tax are artificial forests owned by private owners. The decisions of negotiations are outcomes of plans, lobbying, and prioritization. As a result, the decisions of management areas may be unequally distributed or even biased within a certain prefecture.

Other factors that may cause unequal distribution of management areas are the characteristics of forest cooperatives with different scales of budget and organizational structures. Negotiations between forest cooperatives and forest owners and budget sizes and organization structures of forest cooperatives are social factors that affect the distribution of forest management areas. Physical environmental factors of forest areas include the densities of trees, slopes, and accessibilities, such as existing roads. Because of these factors, the prefectures and municipalities have difficulty in implementing forest management equally in the forest area of each prefecture.

Under the influence of social and physical environmental factors, forest management areas under the forest environmental tax may be unevenly distributed. An evaluation of the use of the forest environmental tax is not conducted regarding the effective use of the tax for the whole area of a prefecture. The forest management areas that are selected by forest cooperatives in a bottom-up manner can be distributed in high risk areas of natural disaster or areas with high aging rates and few human resources for forestry.

When these areas do not overlap, there is a risk that forest management is implemented in areas that are not consistent with the goals of the introduction of the tax in each prefecture. Consequently, comparisons of the distribution of forest management areas by the tax and the distribution of high risk areas of natural disasters or population can contribute to evaluations of the effectiveness of the use of the tax.

The results of the comparative analysis can be used to examine the directions of forestry policies and the forest environmental tax. Even if it is difficult to evenly allocate

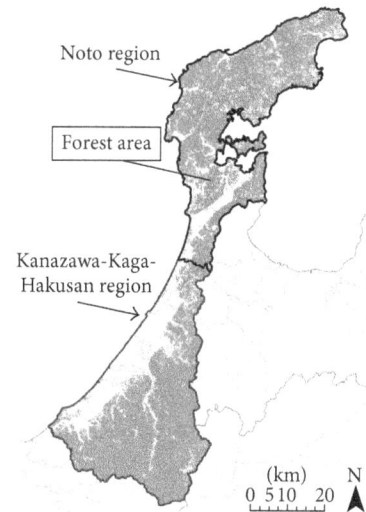

FIGURE 1: Distribution of forest areas in the Ishikawa prefecture.

forest management areas in a prefecture, management areas can be allocated in areas with a high risk of natural disaster or high aging rate and a lack of the required management. Forest management under the forest environmental tax needs to be implemented to sustainably acknowledge services from forestlands, to maximize such services, and to reduce the risk of natural hazard based on scientific evaluation of the use of the tax.

In this research, forest management areas under the forest environmental tax in Ishikawa prefecture are examined. The prefecture includes a peninsula, and mountainous areas are located in the north and southeast part of the prefecture (Figure 1). Forest area covers 73.7% of the land. In the forest area, the rate of coniferous forest is 47%, and the rate of broad-leaved forest is 53%. Aging rates of the mountainous areas are relatively high, and those areas have few human resources in forestry. Through a comparative analysis of the distribution of forest management areas under the tax and distribution of high risk areas of natural hazards and older population, correlations of the distributions of the forest management areas and those specified areas are identified.

2. Methodology

2.1. Research Site and Forest Management Schemes. Since 2007, the forest environmental tax has been used in the Ishikawa prefecture, and forest management and educational activities have been implemented using revenue from the tax. Ishikawa has areas with different aging rates (Table 1), and the proportion of the population over 65 years of age in Noto region, located on the peninsula, is as high as 34% (2010). Compared with the Kanazawa-Kaga-Hakusan region, where 21% of the population is over 65 of age, the elderly proportion of the Noto region is high. Areas with a high risk of natural hazards are almost the same between the former and latter regions. The proportion of forest areas in the Noto region is relatively high, but the difference in these rates is not large.

TABLE 1: Forest area and population of Noto and Kanazawa-Kaga-Hakusan region.

	Forest areas (100 ha)	(a) Forest environment conservation areas (ha)	(b) Management areas by the forest environmental tax (ha)	(a) + (b) (ha)	Disaster risk areas in forest areas (100 ha)	Population (000)	Rate of population over 65 (%)	Rate of forest area (%)
Noto	1,513	1,237	5,988	7,224	373.4	211.3	33.6	76.5
Kanazawa-Kaga-Hakusan	1,572	1,744	3,028	4,772	368.1	956.7	21.3	71.2
Total	3,085	2,981	9,015	11,996	741.5	1,168.0		
Average							23.5	73.7

In the Ishikawa prefecture, there are two types of government supported forest management approaches (Table 2). One is forest environment conservation using subsidies from the central government (i.e., the Forestry Agency of Japan) and the other is forest management funded mainly by the forest environmental tax. The main aim of both types of management is conservation of the multiple functions of forests through revitalization of forestry in the prefecture, and management approaches, including development of roads for forestry, are implemented. The main operation of both types of management is the thinning of trees in the coniferous plantations. The purpose of thinning with high intensity in the management is the induction to mixed forest with broad-leaved trees in the abandoned coniferous plantations.

There are differences between the two types of management. For example, the aims of forest environment conservation operations include cost reduction of forestry, while the aims of forest management using the forest environmental tax include implementing management of disadvantaged forestlands. An example of a disadvantaged forestland is one with a steep gradient where the management cost is relatively high. The organizations primarily responsible for the former operations in the Ishikawa prefecture are the forestry cooperatives and other forestry organizations, and those responsible for the latter management are the forestry cooperatives. Regarding the required intensity of thinning, basically more than 20% of trees in the site are thinned in the forest environment conservation operations, while more than 40% of them are thinned in the forest management using the forest environmental tax. Selections of operations and management areas depend on the decision of the organizations. The prefecture does not select the areas that will be managed by the forestry cooperatives and other forestry organizations in a top-down manner. The management areas are selected based on a point in common between the two types of management.

2.2. Data. We apply the following four data sets for the analysis: (1) areas under forest management operations, (2) forest areas, (3) disaster risk areas, and (4) population age over 65 years. Explanations of the four data sets are given below.

2.2.1. Areas under Forest Management Operation. Data sets for areas under forest management operation were provided by Ishikawa prefecture [8] on the scale of the Oaza. The Oaza unit is a smaller administrative unit than a municipality. It can be regarded as a small-scale district in a municipality. The period was for five fiscal years (2008 to 2013) for forest environment conservation operations and six fiscal years for forest management by the forest environmental tax (2007 to 2013). We use the sum of the two schemes in individual fiscal years. The two schemes are added because in both schemes the decisions for area operations are made by local foresters or forest cooperatives in a bottom-up manner. In the analysis, we apply the polygon data provided by Zenrin Co. for all of the following analysis from forest areas and disaster risk areas at both the Oaza and municipality scale.

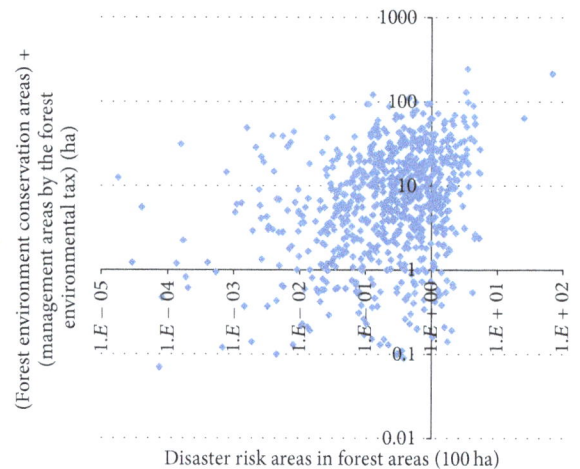

FIGURE 2: Correlation of government supported forest management areas and disaster risk areas (unit: microlevel Oaza).

2.2.2. Forest Areas. To analyze forest areas, the land use grid data provided by the Land, Infrastructure and Transportation Ministry (MLIT) [9] is used. The year of the data is 2009, and the grid resolution is 100 m square. Specifically, the grid data that has the land use category of "forest" is used as the data that shows the distribution of the forest areas. This data is used because it has relatively high grid resolution and it is possible to analyze correspondence with disaster risk areas in detail by aggregating the data in the Oaza unit. Aggregation of the data of forest areas is required to implement this analysis, because the forest areas of the Oaza unit are not disseminated in official statistical data sets.

2.2.3. Disaster Risk Areas. The data for disaster risk areas provided by the MLIT [10] is used. These polygon data show the distribution of areas at high risk for sediment disaster. The proportion of disaster risk area in each forest area of Oaza is calculated. The disaster risk areas are designated by prefecture, and the MLIT collects and disseminates data regarding the designated disaster risk areas.

2.2.4. Population Age over 65 Years. Census data for the year 2010 [11] are used in this analysis. The data is aggregated in the Oaza unit.

3. Results and Discussion

In this section, we review the results from the municipality and smaller microscale (with the unit of Oaza).

3.1. Results from Microscale. We capture the correlation of forest environment conservation operations and forest management areas with the forest environmental tax and disaster risk area. From the results, the areas with the two types of management have correlations of 0.43 with the disaster risks area, which is a relatively weak correlation (Figure 2).

When the individual management areas are examined, the correlation is 0.36 for forest environment conservation

TABLE 2: Forest area and population of the Noto and Kanazawa-Kaga-Hakusan region.

	Main financial resource	Operating entity	Main aims	Obligation to make management plan	Necessity to make agreement with local governments
Forest management mainly with the forest environmental tax	Fund of forest environmental tax	Prefecture, municipality, forest cooperative & NPO	Conservation of multiple functions of forests & management of disadvantaged forests	—	○
Forest environment conservation operation	Subsidies from central government, Forestry Agency	Prefecture, municipality, forest cooperative, NPO, forest owner & contractor of management	Conservation of multiple functions of forest & cost reduction of forestry	○	—

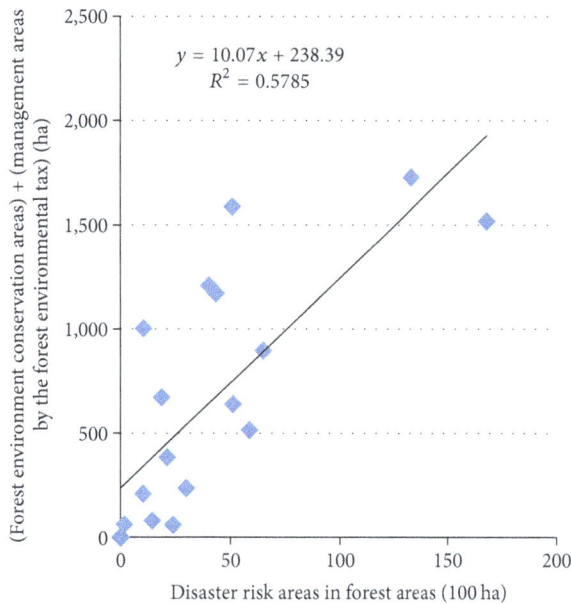

$$y = 10.07x + 238.39$$
$$R^2 = 0.5785$$

FIGURE 3: Correlation of government supported forest management areas and disaster risk areas (unit: municipality).

areas and 0.38 for forest management areas with the forest environmental tax. These correlation coefficients show that the two schemes selected different areas, resulting in a higher correlation of the total areas of both management areas with disaster risk areas.

In this research, we use the data of total forest areas. As the preliminary step, relationships between distribution of the two types of the management areas and forest areas are examined. The results show that the certain amount of forest management areas can be situated in individual Oaza that includes disaster risk areas. However, the main type of trees in individual Oaza is not considered because of the data limitation. In the future research, the main type of trees in individual Oaza will be detected by remotely sensed data with high resolution, and the detailed relationships between distribution of the two types of the management areas and forest areas with each type of trees in microlevel Oaza will be examined.

Next, we examine the ratio of forest areas with the two management schemes and their correlations with the proportion of residents over the age of 65 in each unit of Oaza. The resulting correlation is 0.16. We examine the two schemes individually, but the results indicate lower correlations, below 0.16. The results suggest weaker correlations with residents over the age of 65 in each unit of Oaza and the two schemes.

3.2. Results from Municipality Scale. We apply the same analysis with the same indicators to the scale of municipalities. The correlation with forest management areas under the two schemes with disaster risk areas was 0.76, a high positive correlation (Figure 3). For the individual schemes, the correlation is 0.55 for forest environment conservation operations and 0.74 for forest management areas by the forest environmental tax. By capturing correlation of the disaster

risk area with each management scheme, it becomes clear that stronger correlation is found with the forest management scheme. The results indicate that forest management areas correspond closely to disaster risk areas, which potentially contributes to the prevention of disaster or lowering of the risks.

In the analysis of the microscale, if Oaza has forest areas with disaster risk and neighboring Oaza have the forest management areas, those relationships of the Oaza cannot be reflected to the correlation between the distribution of the high risk areas of natural disasters and the management areas, even if those management areas can contribute to reduction of disaster risk in the Oaza. That is the cause of the stronger correlation in municipality level. The boundaries of municipalities based on the borders of watersheds and parts of the management areas in a municipality can be situated near the high risk areas of natural disasters, and those management areas are not precisely overlapped with the high risk areas but they can contribute to reduction of the disaster risk in the municipality.

In the following step, we examine the ratio of forest areas to the two management schemes and their correlations with the proportion of residents over the age of 65 in each unit of the municipality. The resulting correlation is 0.46, which is a weaker correlation. We examine the areas of the two schemes individually, and the results indicate correlations of −0.14 for forest environment conservation operations and 0.63 for forest management areas with the forest environmental tax (Figure 4). The results suggest higher correlations with forest management areas and weaker correlations with forest environment conservation areas for residents over the age of 65 at the municipality level.

This result implies that forest management areas with the forest environmental tax were implemented with a relatively high concentration in areas with high proportions of the population over the age of 65. The cause of the different correlations of the forest management areas by the tax and the forest environment conservation areas with the proportion of residents over the age of 65 can be related to the difference in the schemes of the two types of the management (Table 2). To implement the former management, making management plan is not needed so that the implementation of the management of disadvantaged forests in the areas with rack of human resources is easier. In addition, the latter management can be implemented in the areas with relatively proactive in forestry, because the aims of the latter management include cost reduction of forestry. In this respect, the latter management areas can be distributed in relatively advantaged areas in terms of social and natural conditions.

We will discuss the reason for larger correlations at the municipality level than at the microlevel of Oaza in Section 4.

4. Discussion

The results for the two schemes showed higher correlations at the municipality level than at the micro-Oaza level for residents over 65. Regarding disaster risk areas, correlations were found at both the micro-Oaza and municipality

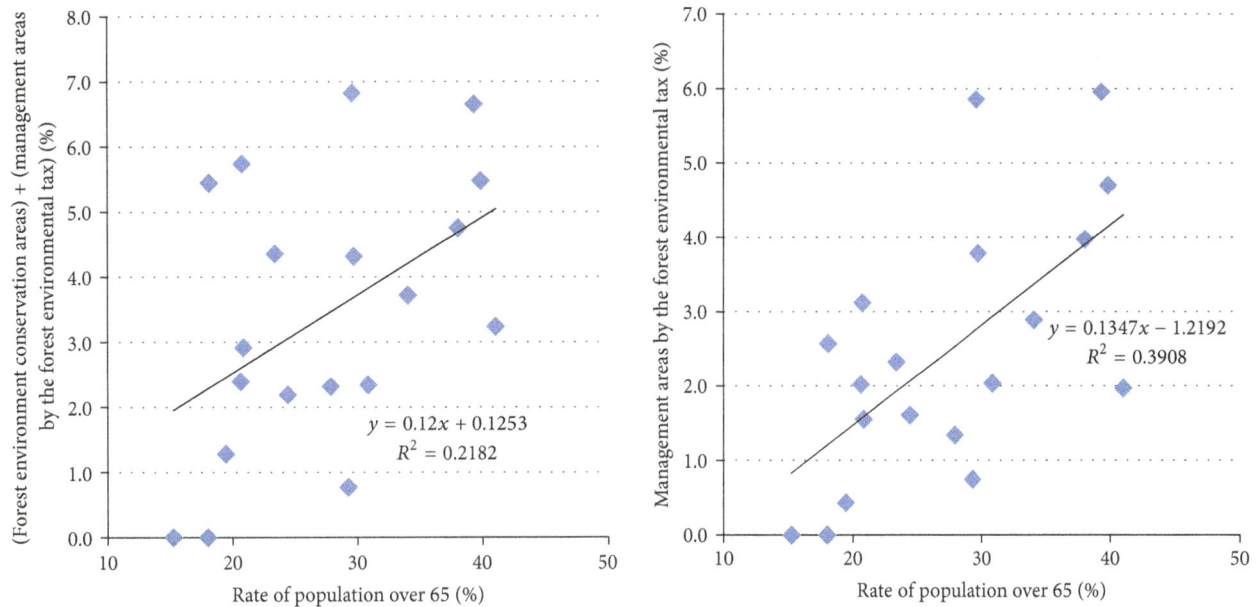

FIGURE 4: Correlation of the proportion of government supported forest management areas and the rate of population over 65 (unit: municipality).

levels. The correlations of both schemes with the disaster risk areas imply that bottom-up decisions in the selection of forest management areas mainly by the forest cooperatives were instrumental in managing areas for disaster prevention.

The municipality level may be the appropriate level to understand the correlation of the rate of forest management area with the rate of people who are related to forest management. Their places of residence often do not correspond to Oaza with the management areas on a one-to-one basis, but they do correspond to municipalities. To analyze the spatial correlation of the distribution of the human resources of forestry and the forest management area with the environmental tax, the municipality level may well be more appropriate than the Oaza level. In this paper, the correlation of the rate of people who are over 65 years with the rate of the forest management area was examined at the municipality level, and a relatively strong and positive correlation was identified.

As a general trend, it was noted that the forest environment management schemes were implemented for areas with a relatively high proportion of the population over 65 years; it is suggested that the scheme potentially contributed to areas with less workforce (and less economic activity) in forestry. In other words, the forest environment conservation scheme was conducted in areas with more social need, in addition to disaster risk areas that need disaster prevention.

5. Future Research

In future research, the specific selection processes for forest areas managed by forestry cooperatives need to be investigated. Through an understanding of these processes, methods

to lead forest cooperatives to select appropriate areas that need management via the environmental tax can be identified. The variables (e.g., ratio of plantation area, tree types, decreasing ratio of population) that can be related to the distribution of the management areas will be identified based on that investigation and multivariate analysis. As mentioned in Section 3.1, types of trees in individual management area need to be considered in the future research. To overcome the limitation of the data of forest area with different type of trees, the analysis to detect the tree types in the management areas by using remotely sensed data with high resolution will be implemented.

Local governments are required to manage forestry cooperatives to effectively use revenue from the forest environmental tax, and they need to know the conditions for appropriate selection of forest cooperatives in management areas.

Competing Interests

The authors declare that there are no competing interests regarding the publication of this paper.

Acknowledgments

This work was supported by the MEXT, KAKENHI [26360062, 15H01597]; Environment Research and Technology Development Fund [1-1303, S-15-2(3)]; Research Institute for Humanity and Nature; Obayashi Foundation; Heiwa Nakajima Foundation; Kurita Water and Environment Foundation; and the Education and Research Base regarding the Ionic Liquid in the Satoyama Biomass Refinery. Thanks are extended to Mr. Shinichi Amada and Dr. Jun Fukuda at the Forest Agency of Japan and Mr. Yuho Hifumi and

Mr. Shota Kimoto at the forest management division of the Ishikawa prefecture.

References

[1] S. Engel, S. Pagiola, and S. Wunder, "Designing payments for environmental services in theory and practice: an overview of the issues," *Ecological Economics*, vol. 65, no. 4, pp. 663–674, 2008.

[2] S. Wunder, S. Engel, and S. Pagiola, "Taking stock: a comparative analysis of payments for environmental services programs in developed and developing countries," *Ecological Economics*, vol. 65, no. 4, pp. 834–852, 2008.

[3] M. Sakagami and K. Kuriyama, *Values of Ecosystem Services: Economical Evaluation*, Koyo-Shobo, Kyoto, Japan, 2009 (Japanese).

[4] K. H. Redford and W. M. Adams, "Payment for ecosystem services and the challenge of saving nature," *Conservation Biology*, vol. 23, no. 4, pp. 785–787, 2009.

[5] R. Muradian, M. Arsel, L. Pellegrini et al., "Payments for ecosystem services and the fatal attraction of win-win solutions," *Conservation Letters*, vol. 6, no. 4, pp. 274–279, 2013.

[6] S. Schomers and B. Matzdorf, "Payments for ecosystem services: a review and comparison of developing and industrialized countries," *Ecosystem Services*, vol. 6, pp. 16–30, 2013.

[7] L. Bespyatko and H. Imura, "Investigation on environmental tax for forestry in Japan as a scheme of payments for environmental services," *Environmental Science*, vol. 21, no. 2, pp. 115–132, 2008 (Japanese).

[8] Ishikawa Prefecture, "Areas of forest environment conservation operations and forest management by the forest environmental tax," 2015.

[9] Land, Infrastructure, and Transportation Ministry (MLIT), Land use distribution, 2016, http://nlftp.mlit.go.jp/ksj/gml/datalist/KsjTmplt-L03-b.html.

[10] Land and Infrastructure and Transportation Ministry (MLIT), "Disaster risk areas," 2016, http://nlftp.mlit.go.jp/ksj/gml/datalist/KsjTmplt-A26.html.

[11] Statistics Bureau of Japan, Census data for the year 2010, 2016, http://www.stat.go.jp/data/kokusei/2010/.

14

Carbon Stocks in the Small Estuarine Mangroves of Geza and Mtimbwani, Tanga, Tanzania

Edmond Alavaisha[1] and Mwita M. Mangora[2]

[1]*Institute of Resource Assessment, University of Dar es Salaam, P.O. Box 35097, Dar es Salaam, Tanzania*
[2]*Institute of Marine Sciences, University of Dar es Salaam, Mizingani Road, P.O. Box 668, Zanzibar, Tanzania*

Correspondence should be addressed to Mwita M. Mangora; mmangora@yahoo.com

Academic Editor: Kurt Johnsen

Mangrove forests offer important ecosystem services, including their high capacity for carbon sequestration and stocking. However, they face rapid degradation and loss of ecological resilience particularly at local scales due to human pressure. We conducted inventory of mangrove forests to characterise forest stand structure and estimate carbon stocks in the small estuarine mangroves of Geza and Mtimbwani in Tanga, Tanzania. Forest structure, above-ground carbon (AGC), and below-ground carbon (BGC) were characterised. Soil carbon was estimated to 1 m depth using loss on ignition procedure. Six common mangrove species were identified dominated by *Avicennia marina* (Forsk.) Vierh. and *Rhizophora mucronata* Lamarck. Forest stand density and basal area were 1740 stems ha^{-1} and 17.2 m^2 ha^{-1} for Geza and 2334 stems ha^{-1} and 30.3 m^2 ha^{-1} for Mtimbwani. Total ecosystem carbon stocks were 414.6 Mg C ha^{-1} for Geza and 684.9 Mg C ha^{-1} for Mtimbwani. Soil carbon contributed over 65% of these stocks, decreasing with depth. Mid zones of the mangrove stands had highest carbon stocks. These data demonstrate that studied mangroves are potential for carbon projects and provide the baseline for monitoring, reporting, and verification (MRV) to support the projects.

1. Introduction

Mangrove forests occur in fragmented stands along almost the entire coastline of Tanzania mainland. Mangroves flourish in river estuaries and deltas and in enclosed bays, lagoons, and tidal creeks. Despite representing only about 0.3% (108,000–115,000 ha) of the total forest area in the country [1, 2], these forests provide numerous ecosystem services including wood and nonwood products; buffering lands and coastal properties [1, 3]; and supporting fisheries as they serve as nursery, feeding, and shelter grounds for numerous commercially important fishes and invertebrates [4–6]. In global terms, mangroves are reported to be up to five times efficient carbon sinks compared to other forms of terrestrial forests [7–9]. However, there exists large nonlinearity in ecosystem services provided by different mangrove formations [8, 10] that hinder generalised estimates of mangrove forest ecosystems' carbon sequestration and stocks which is complicated by local conditions such as climate and soil factors; forest age, growth, and structure; utilisation

and management regime of respective forests [3, 7, 11–13]. As such, there is a considerable knowledge gap and uncertainty at local levels regarding the carbon pool size and variability of carbon sequestration within mangrove forests [14, 15]. Nonetheless, the reported high rate of mangrove carbon sequestration provides global benefits in mitigating the effects of climate change [9, 13, 14] and demonstrates potential for livelihood enhancement through community carbon market schemes for sustainable conservation [16, 17].

Like in many other countries where mangroves occur, overexploitation and conversions to other land uses like solar salt pans, aquaculture, agriculture, and urban expansion are major threats to mangroves of Tanzania [1, 18]. While the global rate of mangrove loss is approximated at 1-2% per year [19–21], there are variations at regional, national, and local scales. In Tanzania, FAO [22] estimated the rate of mangrove loss at about 0.7% per year. In most cases there are no reliable records at local scales, and where they exist, they are potentially out dated [2]. In general, mangrove degradation jeopardizes their ability to provide ecosystem goods and

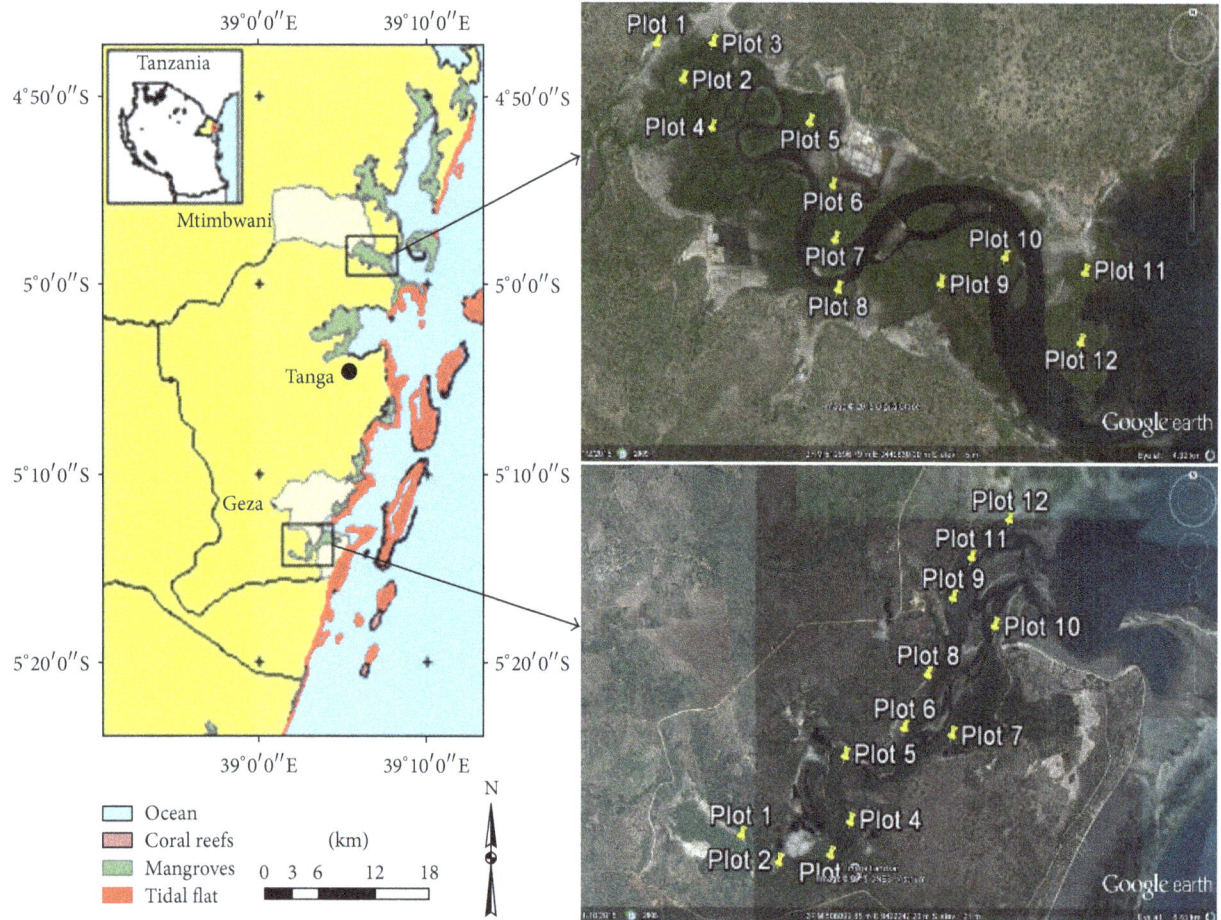

FIGURE 1: Map and Google Earth snapshots showing study sites and sampling plot locations.

services [7], but the full implication of the loss is not well understood and appreciated. This calls for urgent actions to design and implement new effective conservation strategies to protect and restore mangroves to sustain the ecosystem services that they provide [21, 23, 24].

Over the recent past, international climate change agreements have highlighted Reduced Emissions from Deforestation and Degradation (REDD) as one of the possible cost-effective strategies for mitigating the effects of climate change [7, 14, 25]. Reducing emissions aim at maintaining forest carbon stores through financial incentives for forest conservation, for example, carbon credits [26]. With the emerging market-based conservation strategies involving carbon credits in schemes like REDD+ (Reducing Emissions from Deforestation and Forest Degradation, Sustainable Management of Forests, and Enhancement of Forest Carbon Stocks in Developing Countries) and PES (Payment for Ecosystem Services), mangroves are now increasingly recognized as promising natural solutions for carbon capture and long-term storage putting them on top of the agenda in the debates on climate change adaptation mechanisms due to their high rates of carbon sequestration [7, 8, 14]. This warrants for site-specific estimations of the carbon stocks in mangrove forests. In this paper, we report on results of the inventory

in the mangrove forests of Geza and Mtimbwani villages in the north-eastern region of Tanga, Tanzania, that aimed to (i) characterise the tree composition and structure; (ii) estimate the carbon stocks; and (iii) demonstrate the potential for community based carbon market schemes.

2. Materials and Methods

2.1. Description of Study Sites. Geza village is about 26 km south of the city of Tanga whereas Mtimbwani village is 16 km north of the city on the north coast of Tanzania (Figure 1). Mangroves of Geza are approximately 84 ha and are fed seasonally by Bongoa River while those of Mtimbwani are about 326 ha and receive freshwater discharge throughout the year from Ngole River. Common species in both forests are *Avicennia marina* (Forsk.) Vierh., *Rhizophora mucronata* Lamarck, *Ceriops tagal* (Perr.) C. B. Robinson, *Sonneratia alba* J. Smith, *Bruguiera gymnorrhiza* (L.) Lamarck, and *Xylocarpus granatum* König. The forests receive normal tidal inundation twice a day though the extent of tidal influence in upstream varies during rain and dry season. The average annual precipitation is between 1100 mm and 1900 mm, relative humidity of about 76% during daytime and 96% at night, and the average range of atmospheric temperature is from 18°C to 35°C.

2.2. Sampling Plots.
Each mangrove site was divided into three zones along the river gradient from the sea landward using tidal range as criteria: lower, mid, and upper zone. In each zone, four circular nested plots of 10 m radius [27] spaced at approximately 200 m were established, giving a total of twelve sampling plots per each site (Figure 1). Plot locations were predetermined on Google Earth (GE), coordinates recorded and reached during field work by aid of the handheld GPS receiver (Garmin GPSMAP 64s).

2.3. Vegetation Inventory.
Mangrove forest inventory protocols described by Kauffman and Donato [27] were adopted and modified accordingly to suite local conditions. In each plot, all trees with <2.5 cm stem diameter were counted and grouped as seedlings by species. Trees with ≥2.5 cm stem diameter at breast height (DBH, 1.3 m above the ground) were identified by species and measured for DBH (cm) and height (m). For the stilt rooted *R. mucronata*, DBH were measured at 30 cm above the highest root [28]. Standing dead wood trees were practically defined according to their death status. Trees with branches and twigs, which resemble a live tree, were categorized as "status 1," trees with large branches only as "status 2," and trees with bole (trunk) only, no branches, as "status 3." For dead and downed wood, a line-intersect method was used for counting and measuring intersections of woody pieces with diameter >7.6 cm along a vertical sampling plane (transect). For the purpose of degradation, stumps were counted regardless of their species [27].

2.4. Soil Sampling for Carbon.
Soil cores were retrieved from the centre of each sampling plot up to 1 m deep using a multistage sediment core sampler (AMS, Inc., American Falls, USA). The full length of core sampler was steadily inserted vertically into the soil at each depth stage. Soil samples were retrieved in divided segment depths of 0–30 cm, 30–60 cm, and 60–100 cm. Each retrieved core was subdivided in 5 cm and mean values of each segment depth were used for analysis.

2.5. Data Analyses

2.5.1. Forest Structure and Composition.
Forest inventory data were processed using standard analysis procedures as described by Cintron and Novelli [29] to derive forest stand characteristics: stand frequency distribution, density (stems ha^{-1}), basal area (m^2 ha^{-1}), relative frequency (1), relative density (2), and relative dominance (3). Ecological importance values (IV) of each species were determined by summing the respective relative frequency, relative density, and relative dominance. Importance value measures relative dominance of species by criteria of how often it occurred, number of species, and area it occupies in a community.

The species that attained the highest IV was considered the principal species:

$$\text{Relative frequency} = \frac{100Fi}{\sum_{i=1}^{m} Fi}, \tag{1}$$

$$\text{Relative density} = \frac{100ni}{\sum_{i=1}^{m} ni}, \tag{2}$$

$$\text{Relative dominance} = \frac{100Bai}{Ba}, \tag{3}$$

where ni is the number of trees sampled for species i; m is the number of species; Fi is number of plots in which species i occurred, multiplied by 100.

2.5.2. Above- and below-Ground Biomass and Carbon Pools.
Estimation of above-ground biomass (AGB) and below-ground biomass (BGB) for roots in live trees used general allometric equations (4) and (5), respectively, developed by Komiyama et al. [28, 30]:

$$AGB = 0.251 \times \rho \times (DBH)^{2.46}, \tag{4}$$

$$BGB = 0.199 \times \rho^{0.899} \times (DBH)^{2.22}, \tag{5}$$

where AGB and BGB are biomass (kg), DBH is diameter at breast height (cm), and ρ is average general wood density (0.752 g cm^{-3}). These general equations and wood density were used because local or regional species-specific models and respective wood density for all the species have not been developed and established.

Biomass of standing dead wood was obtained by subtraction of a percentage factor from the supposed biomass of live tree derived by a general formula to account for the loss of leaf and twigs biomass. For standing dead trees of status 1 and status 2, factors of 2.5% and 10% biomass of live tree were subtracted, respectively. For status 3, standing dead trees and downed dead wood biomass were estimated by factoring average wood density (0.752 g cm^{-3}) to the volume of dead wood. Volume of downed dead wood was calculated using (6) by Brown [31]:

$$\text{Volume}\left(\text{m}^3\,\text{h}^{-1}\right) = \frac{\pi^2\left(d_1^2 + d_2^2 + d_3^2 + \cdots + d_n^2\right)}{8L}. \tag{6}$$

π is constant; $d_1, d_2, \ldots, d_n$ are diameters of intersecting pieces of large dead wood (cm), and L is the length of the transect line for large size class (m).

To obtain above-ground carbon (AGC) and below-ground carbon (BGC) density, default values of carbon concentration were used: 0.48 for above-ground live trees, 0.5 for dead and downed trees, and 0.39 for below-ground roots as suggested by Kauffman and Donato [27] and Howard et al. [32].

2.5.3. Soil Carbon Pool.
Dry combustion method was used to analyse soil carbon. Soil samples were dried at 60°C to constant weight for at least 72 hours and then allowed to cool to room temperature before weighing. Cooled samples

were weighed and used for bulk density determination. Bulk density $(g\,cm^{-3})$ was determined by dividing the oven-dry soil sample mass (g) by the volume (cm^{-3}) of the sample:

$$\text{Soil bulk density}\left(g\,cm^{-3}\right) = \frac{\text{oven-dry sample mass}\,(g)}{\text{sample volume}\,(cm^3)}. \quad (7)$$

Twenty grams of oven dried subsample from each sample was ignited in a muffle furnace (AAF 11/7, Wolf Laboratories Limited, UK) at 540°C for 5 hours. The differences obtained between the dry weight and ignited weight (loss on ignition, LOI) indicated organic matter content expressed in percentage (% LOI) as

$$\% \text{ LOI} = \left[\frac{(\text{dry mass before combustion} - \text{dry mass after combustion})}{\text{dry mass before combustion}}\right] \times 100. \quad (8)$$

Soil carbon density was determined as

$$\text{Soil carbon}\left(Mg\,ha^{-1}\right) = \text{bulk density}\left(g\,cm^{-3}\right) \times \text{soil depth interval}\,(cm) \times \%\,C. \quad (9)$$

To obtain organic carbon concentration (% C), an equation suggested by Kauffman and Donato [27] was adopted:

$$\text{Organic carbon concentration}\,(\%\,C) = 0.415 \times \%\,\text{LOI} + 2.89. \quad (10)$$

2.5.4. Total Ecosystem Carbon Stock. Estimation of ecosystem carbon density was done by summing all of the component carbon stocks of above-ground, below-ground, and soil.

2.5.5. Statistical Tests. Statistical tests were performed for descriptive and inferential statistics on forest inventory variables. Mean values were subjected to one-way Analysis of Variance (ANOVA) at significance level of 0.05 to find the differences of mean forest density, basal area, stumps, regeneration, and carbon stocks within and between sites.

3. Results

3.1. Forests Structure and Composition. Summary forest structural stand characteristics are presented in Table 1. Six species of mangroves were found in both sites. *A. marina* was the most frequently observed and dominant species in both sites. The least observed species was *B. gymnorrhiza*, while *S. alba* was the least dominant species. In terms of density, *R. mucronata* had high relative stem density in Geza (58.9%) and Mtimbwani (41.8%) whereas *S. alba* was the least contributing species to stem density in both Geza (0.8%) and Mtimbwani

(1.1%). Based on their IVs, *R. mucronata* was the principal species in Geza and *A. marina* in Mtimbwani. Seedlings and saplings count indicated *R. mucronata* and *A. marina* as the most abundant in both sites.

Mangroves of Mtimbwani had high mean stem density across the three zones compared to Geza mangroves (Figure 2(a)). In both sites, mid zones had the highest mean stem density (Geza = 2252 ± 810 stems ha^{-1} and Mtimbwani = 2944 ± 634 stems ha^{-1}) than other zones but not significantly different both between the sites ($p = 0.075$) and within zones ($p = 0.233$ for Geza and $p = 0.115$ for Mtimbwani). For basal areas, differences were significant ($p = 0.041$) between sites where Mtimbwani site had large mean basal area than that of Geza (Figure 2(b)). The mid zones represented large basal areas in both Geza ($22.25 \pm 7.1\,m^2\,ha^{-1}$) and Mtimbwani ($46.10 \pm 12.1\,m^2\,ha^{-1}$) although the differences were not significant among zones in Geza ($p = 0.397$) and Mtimbwani ($p = 0.114$).

The mean number of stumps was not significantly different ($p = 0.112$) between Geza and Mtimbwani sites (Figure 2(c)). Comparison within zones revealed a significant difference ($p = 0.031$) in mean stumps count between the zones of Geza mangroves with highest number in the upper zone (49 ± 14 stumps ha^{-1}). In Mtimbwani, the difference in mean stumps count was not significant ($p = 0.539$) between the zones although the lower zone had higher count (56 ± 22 stumps ha^{-1}). Fewer stumps were counted in mid zones for both sites. Overall, Geza had significantly higher mean number of seedlings than Mtimbwani ($p = 0.003$; Figure 2(d)). Mean seedlings and saplings count was not significantly different between zones in both Geza ($p = 0.183$) and Mtimbwani ($p = 0.168$) although counts were higher in the mid zone of Geza (1107 ± 437) and lower zone of Mtimbwani (512 ± 91).

3.2. Above- and below-Ground Carbon Stocks. The difference in AGC and BGC density was significantly different between sites ($p = 0.041$ for Geza and $p = 0.046$ for Mtimbwani). There was no significant difference in AGC ($p = 0.214$) and BGC ($p = 0.158$) between zones of Geza but the difference was significant for Mtimbwani AGC ($p = 0.021$) and BGC ($p = 0.038$) where mid zone had higher AGC and BGC compared to lower and upper zones (Figure 3). Standing live trees contributed more than 70% of the AGC (Table 2). On average, BGC in Geza and Mtimbwani represented 24.26% and 21.93%, respectively, of vegetative carbon.

3.3. Mangrove Soil Carbon Stocks

3.3.1. Vertical Profile of Carbon Stocks. The general profile in mean soil bulk density and carbon stock in different depths is summarized in Table 3. The mean bulk density and soil carbon density significantly decreased with depth in both sites, Geza ($p = 0.038$, $p = 0.024$) and Mtimbwani ($p = 0.007$, $p = 0.040$). The top 30 cm had higher carbon density compared to 30–60 cm and 60–100 cm, demonstrating a higher C% in the top layers of the soil.

TABLE 1: Forest stand characteristics for Geza and Mtimbwani mangroves.

Site	Species	Frequency	Density (stems ha^{-1})	Basal area (m^2 ha^{-1})	Seedlings/saplings (counts ha^{-1})	Frequency	Relative Density	Dominance	Importance value
Geza	A. marina	12	406	10.22	778	28.57	23.32	59.37	111.26
	B. gymnorrhiza	2	27	0.24	17	4.76	1.52	1.39	7.67
	C. tagal	10	231	0.43	531	26.19	13.26	2.49	41.94
	R. mucronata	11	1026	5.82	1219	26.19	58.99	33.78	118.97
	S. alba	3	13	0.11	0	7.14	0.76	0.67	8.57
	X. granatum	3	37	0.4	24	7.14	2.13	2.31	11.59
	Total		1740	17.21	2569	100	100	100	300
Mtimbwani	A. marina	12	621	20.78	326	26.09	26.59	68.64	121.31
	B. gymnorrhiza	4	80	0.49	35	8.7	3.41	1.63	13.73
	C. tagal	8	536	3.51	238	17.39	22.95	11.59	51.93
	R. mucronata	11	976	4.49	483	23.91	41.82	14.83	80.56
	S. alba	6	27	0.18	18	13.04	1.14	0.58	14.76
	X. granatum	5	95	0.83	44	10.87	4.09	2.74	17.7
	Total		2334	30.27	1144	100	100	100	300

(a)

(b)

(c)

(d)

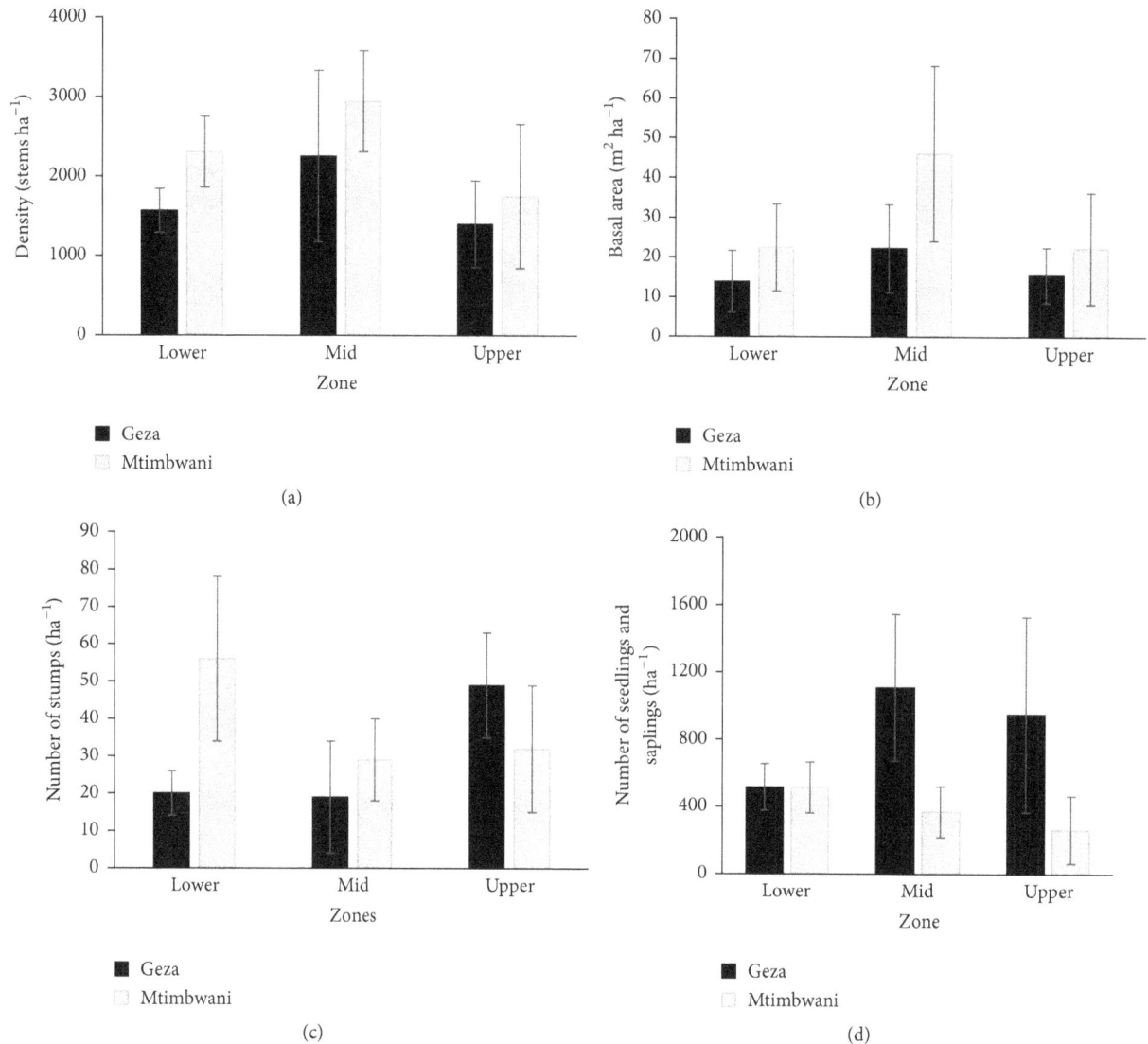

FIGURE 2: Forest structure indicating (a) stems density (b), basal area, (c) number of stumps, and (d) potential regeneration in different zones (mean ± SD).

TABLE 2: Mean AGC density (Mg C ha^{-1}) for standing live, dead and downed trees (mean ± SD).

Site	Component	Zone		
		Low	Mid	Upper
Geza	Standing live tree	54.88 ± 19.16	102.63 ± 30.67	65.66 ± 14.45
	Standing dead trees	9.02 ± 3.06	1.32 ± 0.48	1.16 ± 0.53
	Downed trees	0.23 ± 0.08	0.03 ± 0.02	0.22 ± 0.18
Mtimbwani	Standing live tree	103.44 ± 39.53	167.84 ± 57.11	99.50 ± 45.15
	Standing dead trees	7.69 ± 4.90	133.59 ± 126.43	10.12 ± 4.05
	Downed trees	0.24 ± 0.20	0.10 ± 0.01	0.07 ± 0.03

3.3.2. Horizontal Profile of Soil Carbon Stocks. Generally, mangroves of Mtimbwani had significantly ($p = 0.002$) higher mean soil carbon density in all zones than Geza. The variations of soil carbon density were significantly different within zones in Geza ($p = 0.037$) and Mtimbwani

($p = 0.042$). Mid zones had highest mean soil carbon density than lower and upper zones in both sites (Table 4).

3.3.3. Total Carbon Stocks. The total ecosystem carbon stock for Mtimbwani was 684.99 Mg C ha^{-1} and for Geza was

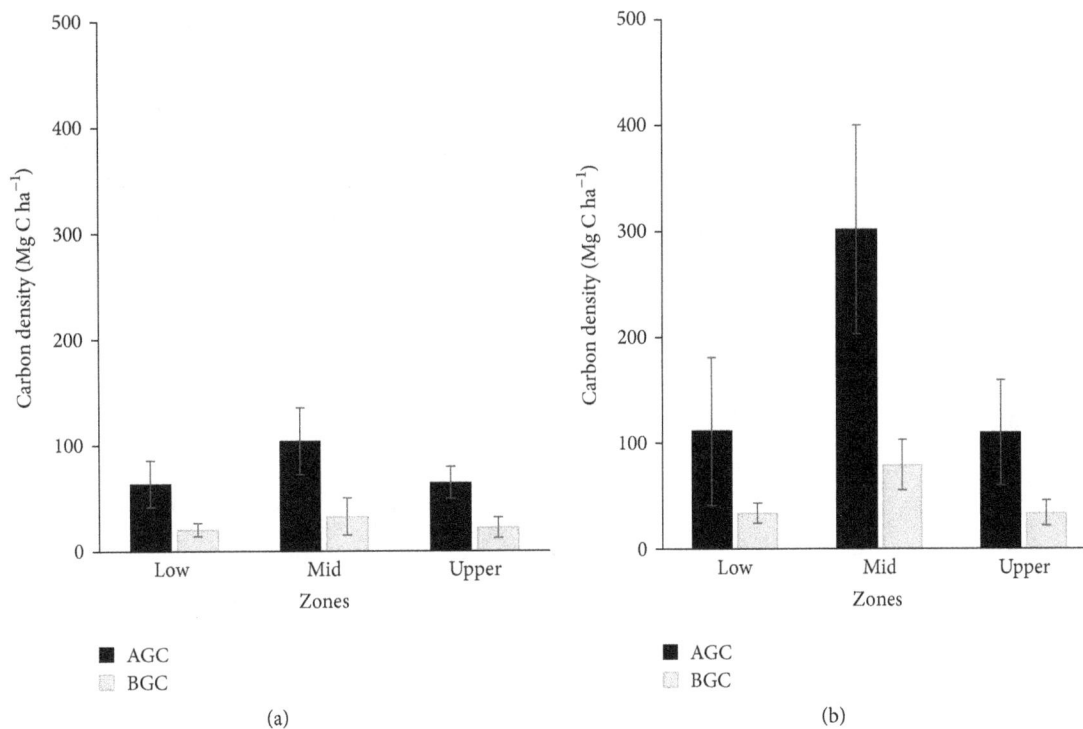

FIGURE 3: Mean AGC and BGC density ($MgCha^{-1}$) in different zones for (a) Geza and (b) Mtimbwani (mean ± SD).

TABLE 3: Vertical profile of mean soil bulk density and carbon stocks (mean ± SD).

Site	Depth (cm)	Mean soil bulk density ($g cm^{-3}$)	Mean soil carbon ($MgCha^{-1}$)
Geza	0–30	$0.62 ± 0.20^a$	$106.17 ± 7.29^a$
	30–60	$0.65 ± 0.25^a$	$104.80 ± 16.14^b$
	60–100	$0.70 ± 0.16^b$	$100.36 ± 8.80^c$
Mtimbwani	0–30	$0.83 ± 0.17^a$	$160.00 ± 8.65^a$
	30–60	$0.92 ± 0.29^b$	$152.59 ± 15.42^{ab}$
	60–100	$0.94 ± 0.12^c$	$149.32 ± 5.82^b$

Corresponding letters a, b, and c denote significant differences at $p < 0.05$.

TABLE 4: Horizontal profile of soil carbon stocks (mean ± SD).

Zones	Mean carbon ($MgCha^{-1}$)	
	Geza	Mtimbwani
Lower	$349.75 ± 11.66^a$	$470.50 ± 21.29^a$
Mid	$400.50 ± 31.90^a$	$497.00 ± 48.91^b$
Upper	$183.75 ± 15.26^b$	$418.25 ± 56.51^a$

Corresponding letters a, b, and c denote significant differences at $p < 0.05$.

$414.64 MgCha^{-1}$ (Figure 4). Soil carbon pool was the main contributor to total carbon stock, accounting for 67.43% and 75.08% for Mtimbwani and Geza, respectively. In both sites BGC (roots) was less than 10% of the total carbon stock.

4. Discussion

4.1. Vegetation Structure and Composition. Six mangrove species were found in the two sites with *A. marina* and *R. mucronata* having high relative frequency compared to other species. According to Ksawani et al. [33] the mixture of different species influences the health of forest and enhancement of carbon storage. High frequency of *A. marina* and *R. mucronata* might be attributed to their high regeneration capacity despite these species having high use preferences, particularly the latter [2]. Hamad et al. [34] reported high frequency of *A. marina* and *R. mucronata* in the mangroves of Pemba despite their high use preference for building poles and firewood. The high seedlings and saplings (2569 counts ha^{-1}) implying high regeneration in Geza forest depicted the mangroves to be under human pressure such as cutting for fuelwood, building poles, and conversion for salt pans construction [2]. The higher regeneration might be contributed by opening of canopy gaps that enhances light availability to support seed germination and seedling growth [34, 35].

The species richness creates stable ecosystem that is more likely to self-sustain in the events of harvesting pressures [12].

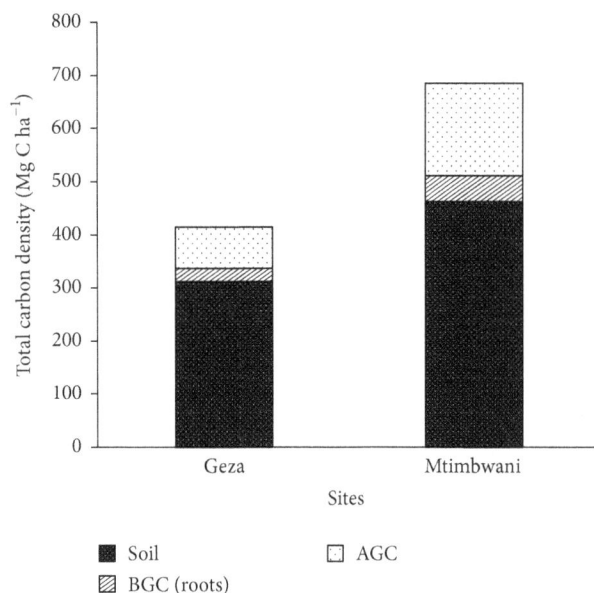

FIGURE 4: Contribution of different carbon pools to the total ecosystem carbon stocks.

Kirui et al. [36] reported that changes in species richness in Gazi Bay were likely to reduce resilience of mangrove ecosystem and make it vulnerable to natural and anthropogenic activities. Results revealed higher stem density and larger basal area in Mtimbwani compared to Geza (Figures 2(a) and 2(b)). But in general, Geza and Mtimbwani mangroves suffer similar pattern of exploitation, largely the conversion for solar salt production. The effect of human pressure is indicated by the low basal area and stem density. Although the observed stem densities and basal areas were low by standards of a healthy forest according to Bundotich et al. [37], they were high compared to reports from other areas in the region. Kairo et al. [38] reporting for Mida Creek indicated tree density to vary from 1197 to 1585 trees ha^{-1}. Recently Trettin et al. [39] reported that overstory trees in the large Zambezi deltaic mangroves averaged to 2000 trees ha^{-1}. Comparing the two study sites, the observed higher stem density and basal area in Mtimbwani might be attributed to the influence of freshwater from the Ngole River, unlike in Geza where Bongoa River is only seasonal. The freshwater inflow brings nutrient from upstream and creates brackish condition, which is an important aspect for mangroves growth.

Mid zones had higher stem density and basal area (Figures 2(a) and 2(b)) which may be attributed to the accessibility as it was similarly reported by Mangora and Shalli [40]. Both ends of the forests are easily accessible from land and sea, with relatively less effort required to harvest the products. Large number of stumps observed in the lower and upper zones (Figure 2(c)) demonstrates the exploitation pressures in these zones. Nonetheless, the observed higher basal areas similar to those reported by Trettin et al. [39] in the well-developed mangroves of Zambezi Delta indicate that mangroves in the study sites are still viable. Mangrove species dominance values indicated A. marina cover large

areas in both sites, despite low stem density (Table 1). This might be attributed to the fact that most of A. marina species were large in size, an indication that the species is less preferred for cutting as compared to the species of the family Rhizophoraceae and therefore has opportunities to grow into large trees. Similar explanation was given for R. mucronata of Mida Creek [38] and Ngomeni forest [37], where R. mucronata was not a dominant species but had high stem density. Importance values indicated A. marina and R. mucronata as the most common species in both sites, suggesting their potential for restoration to maximise carbon stocks and sustain wood stocks.

4.2. Above- and below-Ground (Roots) Carbon Pools. Standing live trees were the major contributor to AGC followed by standing dead trees and downed trees (Table 2). The present reported AGC for Geza and Mtimbwani were comparable to those reported by Kirui et al. [41] and Cohen et al. [42] for Gazi Bay and Vanga which are a little further north of the study sites across the border in Kenya, where carbon density estimate was between 125 and 226 Mg C ha^{-1}, and to those recently reported by Trettin et al. [39] for the large deltaic mangroves of Zambezi in Mozambique which ranged from 111 to 483 Mg ha^{-1} but exceeded those reported by Cohen et al. [42] for Mtwapa Creek (36.4 Mg C ha^{-1}) and Mida Creek (38.5 Mg C ha^{-1}) in Kenya and Sitoe et al. [43] for Sofala Bay (33.3 Mg C ha^{-1}) in Mozambique. The possible explanation for the variation may be the structural characteristics of the forests [33], whereby mangroves of Mtimbwani and Geza had comparably large basal area and high stem density, which also suggest the low level of degradation. The BGC (roots) was the least pool of carbon after soil and AGC components (Figure 4). Nonetheless, the obtained mean estimates of BGC density were above those reported by Sitoe et al. [43] and Fatoyinbo et al. [44] in Mozambique. These differences in the amount of carbon stored in roots between sites and zones might be explained by differences in nutrient cycling, sediment characteristics, and related growth rates [9, 45].

4.3. Soil Carbon Pools. Soil bulk density increased while mean soil carbon decreased with depth (Table 3). Similar trends were reported by Lupembe [46] in the Rufiji Delta (Tanzania), Rahman et al. [47] in the Sundarbans (Bangladesh), Sitoe et al. [43] in Sofala Bay (Mozambique), and Stringer et al. [13] in the Zambezi Delta (Mozambique). Increasing bulk density is associated with the increase of soil compactness as depth increases and decrease in concentration of organic matter [13, 32] which also explains high carbon in upper soil layer as concentration of organic matter triggered by biological activity, particularly litter deposition and decomposition on the soil surface, is high [7, 45].

The average soil carbon density estimated for up to 1 m deep was 311.3 ± 87.2 Mg C ha^{-1} for Geza and 461.9 ± 88.1 Mg C ha^{-1} for Mtimbwani. The estimated soil carbon was comparable to that reported from Madagascar by Jones et al. [16] who found 446.2 Mg C ha^{-1}. The present values of soil carbon density were higher than those reported by Sitoe et al. [43] who found an average of 218.5 Mg C ha^{-1} in Sofala Bay (Mozambique) but low compared to estimates reported by Kauffman et al. [48] for the Mangroves of Yap Micronesia (724.8 Mg C ha^{-1}) and Donato et al. [7] for the coastal fringe mangroves of Republic of Palau (527.7 Mg C ha^{-1}). Again, site-specific difference in forest structure, composition, history of exploitation, and management regimes may explain the variations and warrant site-specific conservation prescriptions.

4.4. Total Ecosystem Carbon Stocks. Total ecosystem carbon stocks were different between sites whereby Geza had 414.6 Mg C ha^{-1} and Mtimbwani had 684.9 Mg C ha^{-1}. These values were within the range of ecosystem carbon stocks reported from other sites in the region [13, 16, 17, 42, 43] and those reported from other regions [32, 49], which ranges from 55 to 800 Mg C ha^{-1}. Higher ecosystem carbon stocks have been reported from the Indo-West Pacific with values ranging from 830 to 1131 Mg C ha^{-1} [7, 48]. In terms of the pattern of carbon density with tidal gradient, mid zones had large stocks. This may be attributed to forest density and basal area, similar to the observation by Stringer et al. [13] in the Zambezi Delta. The observed low density of stumps in the mid zones, implying low deforestation, further supports this. Other reports suggest that differences in mangrove carbon stocks between sites and zones may as well be due to differences in forest age, forest conservation and management status, soil depth, and soil water content [13, 16, 27], factors which need further research for the study sites. The influence of these local factors demonstrates that estimation of carbon stocks in mangroves cannot be generalised, especially for MRV and their potential in climate change mitigation and adaptation.

Despite the differences in mangrove carbon stocks for different ecosystems across the world, the relative contribution of various ecosystem components shows a similar trend. Above-ground carbon density contributed 18.87% and 25.41% of total ecosystem carbon estimates in Geza and Mtimbwani, respectively. Kauffman et al. [48] reported that AGC accounts for about 18 to 25% of total ecosystem carbon. Observations on BGC density were also coherent with values reported from other places, which range from 6 to 23% [7]. Similar to reports from other assessments [7, 13, 43, 48], the soil carbon pool was dominant representing 75.08% and 67.43% of total ecosystem carbon stock for Geza and Mtimbwani mangroves, respectively. Soil carbon values of the present study and those from other areas demonstrate the role of mangrove soil as an important carbon sink. This carbon pool has a significant contribution to primary productivity and health of the forests [9] that can ensure mangroves are able to sustain in providing other ecosystem services that are valuable to local communities.

4.5. Potential for Carbon Markets. Detailed analysis of the socioeconomic factors is not presented here but indicated that communities in Geza and Mtimbwani are substantially dependent on ecosystem services provided by mangroves. Similar to other reports [40], low income and low level of environmental education contributed to rapid degradation of mangrove forests and eventually structure and carbon stocks. Most households in Geza and Mtimbwani are engaged in harvesting and selling of mangrove products to maximise their income in addition to household consumption [50]. Mangroves are widely used as cheap construction materials to build and maintain their homes and a source of household fuelwood. Mangrove areas are also converted to solar salt pans. The observed large amount of carbon stored in the two mangrove communities is vulnerable to loss from such human pressures that are often coupled with the natural phenomena such as sea level rise, disrupted water budgets, and increased salinity. Bhomia et al. [14] suggested that carbon stocks >500 Mg C ha^{-1} confirm the potential of mangroves as sinks and that any loss in such mangrove forests exposes them to significant carbon sources. These data sets provide a clear message that concerted efforts in conservation and management of mangrove areas are warranted to minimize the emissions. The quantification of fairly large carbon stocks in the mangroves of Mtimbwani and Geza provides key baselines for informed policy and planning decisions to safeguard mangroves and restore degraded areas.

Opportunities exist because of the growing realization of the impacts of mangrove degradation coupled with those of climate change that shape the community willingness for conservation initiatives to address mangrove degradation. Emerging market-based conservation through carbon projects (carbon credits) is a potential opportunity to incentivize Geza and Mtimbwani communities, providing them with necessary resources needed to sustain conservation and the carbon sink function of mangroves. This is further supported by the observed vegetative structural indices that proved that mangroves of Geza and Mtimbwani are still viable for self-sustenance given appropriate conservation measures. Incentivized communities would conduct self-enforcement in restoring and protecting the mangroves as they would devise alternative sources of wood products for local consumption through tree planting of fast growing terrestrial based species.

5. Conclusion

This study presented findings that demonstrate that forest structure and ecosystem carbon densities are key elements to voluntary carbon market schemes. Inventory of both mangrove vegetation and soil carbon demonstrated comparatively higher carbon stocks in the study sites and therefore reflecting their potential for carbon markets. Geza and Mtimbwani communities have experience in collaborative arrangements in mangroves management from the former Tanga Coastal Zone Conservation and Development programme [51]. The estimated carbon stocks have provided baseline data for future MRV that is important in supporting carbon projects. Based on present findings and the potential for carbon projects, it is recommended that (a) permanent plots should be established for MRV of additionality in terms of health and carbon stocks; (b) vulnerability assessment and monitoring should be done to assess mangrove cover changes over time and predict extents of human impacts on mangroves; (c) collaborative management should be strengthened by harmonising rules and regulations across stakeholders; (d) detailed studies should be done to investigate whether existing governance mechanism can favour initiation of carbon market schemes.

Competing Interests

The authors declare that there are no competing interests regarding the publication of this paper.

Acknowledgments

This work was funded by a research grant from the Bilateral Marine Science Programme supported by Sida through the Institute of Marine Sciences of the University of Dar es Salaam. The authors appreciate the assistance offered during field work by Mr. M. Yasini, Mr. A. Kurungu, Mr. A. Nyangusi, and Mr. A. Mabwela and the laboratory work by Mr. M. Mwadini.

References

[1] A. K. Semesi, "Developing management plans for the mangrove forest reserves of mainland Tanzania," *Hydrobiologia*, vol. 247, no. 1–3, pp. 1–10, 1992.

[2] Y. Wang, G. Bonynge, J. Nugranad et al., "Remote sensing of Mangrove change along the Tanzania coast," *Marine Geodesy*, vol. 26, no. 1-2, pp. 35–48, 2003.

[3] A. K. Semesi, Y. D. Mgaya, M. H. S. Muruke, J. Francis, M. Mtolera, and G. Msumi, "Coastal resources utilization and conservation issues in Bagamoyo, Tanzania," *Ambio*, vol. 27, no. 8, pp. 635–644, 1998.

[4] B. R. Lugendo, I. Nagelkerken, G. Kruitwagen, G. Van Der Velde, and Y. D. Mgaya, "Relative importance of mangroves as feeding habitats for fishes: a comparison between mangrove habitats with different settings," *Bulletin of Marine Science*, vol. 80, no. 3, pp. 497–512, 2007.

[5] G. Kruitwagen, I. Nagelkerken, B. R. Lugendo, Y. D. Mgaya, and S. E. W. Bonga, "Importance of different carbon sources for macroinvertebrates and fishes of an interlinked mangrove-mudflat ecosystem (Tanzania)," *Estuarine, Coastal and Shelf Science*, vol. 88, no. 4, pp. 464–472, 2010.

[6] I. A. Kimirei, M. M. Igulu, M. Semba, and B. R. Lugendo, "Small estuarine and non-estuarine mangrove ecosystems of Tanzania: overlooked coastal habitats?" in *Estuaries: A Lifeline of Ecosystem Services in the Western Indian Ocean*, S. Diop, P. Scheren, and J. F. Machiwa, Eds., Estuaries of the World, pp. 209–226, Springer, Cham, Switzerland, 2016.

[7] D. C. Donato, J. B. Kauffman, D. Murdiyarso, S. Kurnianto, and M. Stidham, "Mangroves among the most carbon-rich tropical forests and key in land use carbon emissions," *Nature Geosciences*, vol. 4, pp. 293–297, 2011.

[8] D. M. Alongi, "Carbon payments for mangrove conservation: ecosystem constraints and uncertainties of sequestration potential," *Environmental Science and Policy*, vol. 14, no. 4, pp. 462–470, 2011.

[9] D. M. Alongi, "Carbon cycling and storage in mangrove forests," *Annual Review of Marine Science*, vol. 6, pp. 195–219, 2014.

[10] K. C. Ewel, R. R. Twilley, and J. E. Ong, "Different kinds of mangrove forests provide different goods and services," *Global Ecology and Biogeography Letters*, vol. 7, no. 1, pp. 83–94, 1998.

[11] P. A. W. Abuodha and J. G. Kairo, "Human-induced stresses on mangrove swamps along the Kenyan coast," *Hydrobiologia*, vol. 458, pp. 255–265, 2001.

[12] J. A. Okello, N. Schmitz, J. G. Kairo, H. Beeckman, F. Dahdouh-Guebas, and N. Koedam, "Self-sustenance potential of peri-urban mangroves: a case of Mtwapa creek Kenya," *Journal of Environmental Science and Water Resources*, vol. 2, no. 8, pp. 277–289, 2013.

[13] C. E. Stringer, C. C. Trettin, S. J. Zarnoch, and W. Tang, "Carbon stocks of mangroves within the Zambezi River Delta, Mozambique," *Forest Ecology and Management*, vol. 354, pp. 139–148, 2015.

[14] R. K. Bhomia, J. B. Kauffman, and T. N. McFadden, "Ecosystem carbon stocks of mangrove forests along the Pacific and Caribbean coasts of Honduras," *Wetlands Ecology and Management*, vol. 24, no. 2, pp. 187–201, 2016.

[15] M. M. Mangora, M. S. Shalli, I. S. Semesi et al., "Designing a mangrove research and demonstration forest in the rufiji delta, Tanzania," in *Proceedings of the 5th Interagency Conference on Research in the Watersheds*, C. E. Stringer, K. W. Krauss, and J. S. Latimer, Eds., pp. 190–192, U.S. Department of Agriculture Forest Service, Southern Research Station, Asheville, NC, USA.

[16] T. G. Jones, H. R. Ratsimba, L. Ravaoarinorotsihoarana, G. Cripps, and A. Bey, "Ecological segregation of the late jurassic stegosaurian and iguanodontian dinosaurs of the morrison formation in north america: pronounced or subtle?" *Forests*, vol. 5, no. 1, pp. 177–205, 2014.

[17] T. Jones, H. Ratsimba, L. Ravaoarinorotsihoarana et al., "The dynamics, ecological variability and estimated carbon stocks of mangroves in Mahajamba Bay, Madagascar," *Journal of Marine Science and Engineering*, vol. 3, no. 3, pp. 793–820, 2015.

[18] B. Lugendo, "Mangroves, salt marshes and seagrass beds," in *The Regional State of the Coast Report: Western Indian Ocean*, UNEP-Nairobi Convention and WIOMSA, Eds., pp. 53–68, UNEP and WIOMSA, Nairobi, Kenya, 2015.

[19] I. Valiela, J. L. Bowen, and J. K. York, "Mangrove forests: one of the world's threatened major tropical environments," *BioScience*, vol. 51, no. 10, pp. 807–815, 2001.

[20] D. M. Alongi, "Present state and future of the world's mangrove forests," *Environmental Conservation*, vol. 29, no. 3, pp. 331–349, 2002.

[21] N. C. Duke, J.-O. Meynecke, S. Dittmann et al., "A world without mangroves?" *Science*, vol. 317, no. 5834, pp. 41–42, 2007.

[22] FAO, *The World's Mangroves 1980-2005. A Thematic Study Prepared in the Framework of the Global Forest Resources Assessment 2005*, FAO, Rome, Italy, 2007.

[23] P. E. R. Dale, J. M. Knight, and P. G. Dwyer, "Mangrove rehabilitation: a review focusing on ecological and institutional issues," *Wetlands Ecology and Management*, vol. 22, no. 6, pp. 587–604, 2014.

[24] UNEP, *The Importance of Mangroves to People: A Call to Action*, Edited by J. Van Bochove, E. Sullivan and T. Nakamura, United Nations Environment Programme World Conservation Monitoring Centre, Cambridge, UK, 2014.

[25] IPCC, *Climate Change 2014: Synthesis Report. Contribution of Working Groups I, II and III to the Fifth Assessment Report of the Intergovernmental Panel on Climate Change*, IPCC, Geneva, Switzerland, 2014.

[26] J. G. Kairo, C. Wanjiru, and J. Ochiewo, "Net pay: economic analysis of a replanted mangrove plantation in Kenya," *Journal of Sustainable Forestry*, vol. 28, no. 3–5, pp. 395–414, 2009.

[27] J. B. Kauffman and D. C. Donato, "Protocols for the measurement, monitoring and reporting of structure, biomass and carbon stocks in mangrove forests," Working Paper 86, CIFOR, Bogor, Indonesia, 2012.

[28] A. Komiyama, S. Poungparn, and S. Kato, "Common allometric equations for estimating the tree weight of mangroves," *Journal of Tropical Ecology*, vol. 21, no. 4, pp. 471–477, 2005.

[29] G. Cintron and Y. S. Novelli, "Methods for studying mangrove structure," in *The Mangrove Ecosystem: Research Methods*, S. C. Snedaker and J. G. Snedaker, Eds., pp. 91–113, UNESCO, Paris, France, 1984.

[30] A. Komiyama, J. E. Ong, and S. Poungparn, "Allometry, biomass, and productivity of mangrove forests: a review," *Aquatic Botany*, vol. 89, no. 2, pp. 128–137, 2008.

[31] J. K. Brown, "A planar intersects method for sampling fuel volume and surface area," *Forest Science*, vol. 17, pp. 96–102, 1971.

[32] J. Howard, S. Hoyt, K. Isensee, E. Pidgeon, and M. Telszewski, Eds., *Coastal Blue Carbon: Methods for Assessing Carbon Stocks and Emissions Factors in Mangroves, Tidal Salt Marshes, and Seagrass Meadows*, Conservation International, Intergovernmental Oceanographic Commission of UNESCO, International Union for Conservation of Nature, Arlington, Va, USA, 2014.

[33] I. Ksawani, J. Kmarusaman, and M. I. Nurum-Nadhirah, "Biological diversity assessment of Tok Bali mangrove forest, Kelantan, Malaysia," *WSEAS Transactions on Environment and Development*, vol. 3, no. 2, pp. 37–44, 2007.

[34] H. M. Hamad, I. S. Mchenga, and M. I. Hamisi, "Status of exploitation and regeneration of mangrove forests in Pemba Island, Tanzania," *Global Journal of Bio-Science and Biotechnology*, vol. 3, no. 1, pp. 12–18, 2014.

[35] K. Kathiresan and B. L. Bingham, "Biology of mangroves and mangrove ecosystems," *Advances in Marine Biology*, vol. 40, pp. 81–251, 2001.

[36] B. Y. K. Kirui, J. G. Kairo, M. W. Skov, M. Mencuccini, and M. Huxham, "Effects of species richness, identity and environmental variables on growth in planted mangroves in Kenya," *Marine Ecology Progress Series*, vol. 465, pp. 1–10, 2012.

[37] G. Bundotich, M. Karachi, E. Fondo, and J. G. Kairo, "Structural inventory of mangrove forests in Ngomeni," in *Advances in Coastal Ecology: People, Processes and Ecosystems in Kenya*, J. Hoorweg and N. Muthiga, Eds., pp. 111–121, African Studies Centre, Leiden, Netherlands, 2009.

[38] J. G. Kairo, F. Dahdouh-Guebas, P. O. Gwada, C. Ochieng, and N. Koedam, "Regeneration status of mangrove forests in Mida Creek, Kenya: a compromised or secured future?" *Ambio*, vol. 31, no. 7-8, pp. 562–568, 2002.

[39] C. C. Trettin, C. E. Stringer, and S. J. Zarnoch, "Composition, biomass and structure of mangroves within the Zambezi River Delta," *Wetlands Ecology and Management*, vol. 24, no. 2, pp. 173–186, 2016.

[40] M. M. Mangora and M. S. Shalli, "Socio-economic profiles of communities adjacent to Tanga Marine Reserve Systems, Tanzania: key ingredients to general management planning," *Current Research Journal of Social Sciences*, vol. 4, no. 2, pp. 141–149, 2012.

[41] B. Kirui, J. G. Kairo, and M. Karachi, "Allometric equations for estimating above ground biomass of *Rhizophora mucronata*

Lamk. (Rhizophoraceae) mangroves at Gaxi Bay, Kenya," *Western Indian Ocean Journal of Marine Science*, vol. 5, no. 1, pp. 27–34, 2006.

[42] R. Cohen, J. Kaino, J. A. Okello et al., "Propagating uncertainty to estimates of above-ground biomass for Kenyan mangroves: a scaling procedure from tree to landscape level," *Forest Ecology and Management*, vol. 310, pp. 968–982, 2013.

[43] A. A. Sitoe, L. J. C. Mandlate, and B. S. Guedes, "Biomass and carbon stocks of Sofala bay mangrove forests," *Forests*, vol. 5, no. 8, pp. 1967–1981, 2014.

[44] T. E. Fatoyinbo, M. Simard, R. A. Washington-Allen, and H. H. Shugart, "Landscape-scale extent, height, biomass, and carbon estimation of Mozambique's mangrove forests with Landsat ETM+ and Shuttle Radar Topography Mission elevation data," *Journal of Geophysical Research*, vol. 113, pp. 1–13, 2008.

[45] P. Mfilinge, N. Atta, and M. Tsuchiya, "Nutrient dynamics and leaf litter decomposition in a subtropical mangrove forest at Oura Bay, Okinawa, Japan," *Trees*, vol. 16, no. 2-3, pp. 172–180, 2002.

[46] I. B. Lupembe, *Carbon stocks in the mangrove ecosystem of rufiji river delta, Rufiji District, Tanzania [M.S. thesis]*, Sokoine University of Agriculture, Morogoro, Tanzania, 2014.

[47] M. M. Rahman, M. Nabiul Islam Khan, A. K. Fazlul Hoque, and I. Ahmed, "Carbon stock in the Sundarbans mangrove forest: spatial variations in vegetation types and salinity zones," *Wetlands Ecology and Management*, vol. 23, no. 2, pp. 269–283, 2015.

[48] J. B. Kauffman, C. Heider, T. G. Cole, K. A. Dwire, and D. C. Donato, "Ecosystem carbon stocks of Micronesian mangrove forests," *Wetlands*, vol. 31, no. 2, pp. 343–352, 2011.

[49] G. Wang, D. Guan, M. R. Peart, Y. Chen, and Y. Peng, "Ecosystem carbon stocks of mangrove forest in Yingluo Bay, Guangdong Province of South China," *Forest Ecology and Management*, vol. 310, pp. 539–546, 2013.

[50] B. A. Tarimo, *Analysis of local governance in conservation of mangrove forests of Geza and Mtimbwani Coastal Communities [M.S. thesis]*, University of Dar es Salaam, Dar es Salaam, Tanzania, 2015.

[51] S. Wells, M. Samoilys, S. Makoloweka, and H. Kalombo, "Lessons learnt from a collaborative management programme in coastal Tanzania," *Ocean and Coastal Management*, vol. 53, no. 4, pp. 161–168, 2010.

15

Biomass and Soil Carbon Stocks in Wet Montane Forest, Monteverde Region, Costa Rica: Assessments and Challenges for Quantifying Accumulation Rates

Lawrence H. Tanner,[1] Megan T. Wilckens,[1] Morgan A. Nivison,[1] and Katherine M. Johnson[2,3]

[1]*Environmental Science Systems, Le Moyne College, Syracuse, NY 13214, USA*
[2]*Monteverde Institute, Monteverde, Puntarenas, Costa Rica*
[3]*Phinizy Center for Water Sciences, Augusta, GA 30906, USA*

Correspondence should be addressed to Lawrence H. Tanner; tannerlh@lemoyne.edu

Academic Editor: Piermaria Corona

We measured carbon stocks at two forest reserves in the cloud forest region of Monteverde, comparing cleared land, experimental secondary forest plots, and mature forest at each location to assess the effectiveness of reforestation in sequestering biomass and soil carbon. The biomass carbon stock measured in the mature forest at the Monteverde Institute is similar to other measurements of mature tropical montane forest biomass carbon in Costa Rica. Local historical records and the distribution of large trees suggest a mature forest age of greater than 80 years. The forest at La Calandria lacks historical documentation, and dendrochronological dating is not applicable. However, based on the differences in tree size, above-ground biomass carbon, and soil carbon between the Monteverde Institute and La Calandria sites, we estimate an age difference of at least 30 years of the mature forests. Experimental secondary forest plots at both sites have accumulated biomass at lower than expected rates, suggesting local limiting factors, such as nutrient limitation. We find that soil carbon content is primarily a function of time and that altitudinal differences between the study sites do not play a role.

1. Introduction

Modeling of the anthropogenic climate change anticipated in the coming decades requires a thorough understanding of the carbon cycle, in particular, the sources and sinks of carbon that are exchangeable on short (decadal) time scales and the rates of exchange. Multiple studies have estimated both the reservoirs and fluxes of carbon from terrestrial and marine reservoirs [1]. Tropical forests are major sinks for atmospheric carbon, accounting for as much as 37% of the terrestrial carbon sequestered in above-ground biomass and soil [2]. The Amazon forest alone, for example, is estimated to contain from 150 to 200 GtC in biomass and soil [3], equal to approximately one-quarter of the total atmospheric reservoir. Early studies of carbon cycling recognized the importance of land-use changes, deforestation and afforestation in particular, and attendant fluctuations in the above-ground biomass carbon in contributing to variations in the atmospheric reservoir. Disturbance of tropical forests by land-use change and deforestation has been estimated to account for as much as 23% of anthropogenic carbon emissions [4]; most of these emissions (75%) were estimated to derive from the loss of above-ground biomass.

Fewer studies have examined the potential changes in soil carbon that accompany land-use changes. Rhoades et al. [5], for example, estimated that conversion of tropical forest to pasture led to a decrease in soil carbon of 18% to 20%. More recently, Dieleman et al. [6] found the difference in soil carbon between mature tropical forest and pasture to be as much as 40%, but most of the difference was concentrated in the top 10 to 15 cm. Unfortunately, our knowledge of carbon flux rates in tropical systems remains limited as various studies have produced contradictory results; for example, Hughes et al. [7] documented that the carbon stored in above-ground biomass increases predictably with the age of the forest but saw no significant change in soil carbon, a finding echoed by the study of Markewitz et al. [8] but contrary to the results of Bautista-Cruz and del Castillo [9]. Clearly, a more complete understanding of the rates of change of these carbon stocks will be required to track accurately the changes in the global carbon cycle. Additionally, predicting changes in the size of these carbon reservoirs in the future will demand knowledge of how these fluxes will vary as a response to ongoing and anticipated climate change.

This study attempts to resolve some of the issues involved in the study of carbon flux rates in tropical forests. Specifically, by comparing the carbon stocks, both above-ground biomass (AGB_{carbon}) and soil organic carbon, of mature-disturbed forest, experimental secondary forest plots, and cleared land at two forest reserves in the Monteverde cloud forest region, we attempt to resolve the apparent contradictory results in studies of changes in soil carbon with time. Furthermore, one of the challenges in assessing the effectiveness of reforestation at sequestering carbon in tropical forests is that many tropical tree species lack annual growth rings, making precise age measurements problematic in the absence of historical records [10]. We demonstrate herein that carbon-stock data may be useful in approximating the age of tropical forests not otherwise documented.

2. Materials and Methods

2.1. Study Location. Much of Costa Rica's mature forest was cleared during the twentieth century, primarily for agricultural land-use and timber. An estimated 10^6 hectares of forest were cleared just during the 1950s [11], and by 1983 only 26% of the original forest cover of the country remained [12]. Fortuitously, a significant portion of this land has been reforested in recent decades through government intervention, including establishment of an extensive national park system and payments to landowners for environmental services. Consequently, ca. 25% of the total land area of the country is now protected in some way and net forest area is now estimated to increase by ca. 35,000 ha per year [12]. In addition to enhanced ecosystem services (e.g., watershed and biodiversity protection), an important result of this change is that there are now numerous areas of secondary forest of varying age throughout the country available for examination of the rates of forest regrowth in secondary ecological succession.

The Monteverde region is known for the cloud forests on the western slopes of the Cordillera de Tilarán. Leeward cloud forests (*sensu* Lawton and Dryer [14]), are tropical montane evergreen forests that obtain significant moisture from fog drip. In Costa Rica, they occur on the Pacific-facing slopes of the Cordillera de Tilarán where trade winds are forced above the lifting condensation point. The climate of the region is characterized by three seasons: a wet season from May to October; a transitional misty-windy season from

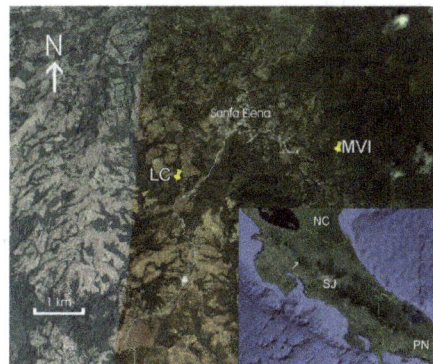

FIGURE 1: Location of the study area in the Monteverde region of Costa Rica. Sampling was conducted at the Monteverde Institute (MVI) and La Calandria (LC). Capitol city of San Jose shown in inset for reference, along with borders of Nicaragua (NC) and Panama (PN). Adapted from Google Earth® imagery (satellite image acquisition 03/2002).

November to January; and the dry season from February to April [14]. Even during the dry season, fog drip keeps the soils continuously moist. In the Monteverde region, these soils, which formed on rhyolites, are classed as Typic Dystrandepts [15].

The Monteverde Institute, a community-based research and educational endeavor that operates in collaboration with a consortium of scientific and educational institutions, is located ca. 2 km southeast of the center of the village of Santa Elena, Puntarenas Province (Figure 1). The Monteverde Institute (MVI) administers 15 hectares in cooperation with the Fundacion Conservacionista Costarricense, and is located within the ca. 9,000 hectare Monteverde Reserve complex. The land cover includes a mix of pasture, secondary forest plots of various ages, and mature forest at a mean elevation of ca. 1490 m in the lower cloud forest, or lower montane wet forest zone [16]. The climate at this location is characterized by a mean annual temperature (MAT) of 18.5°C and ca. 2500 mm mean annual precipitation (MAP) [17]. La Calandria is a private reserve including a lodge and biological field station currently administered by a local hotel (Pension Santa Elena), with conservation efforts coordinated by the Fundacion Conservacionista Costarricense. The reserve is located ca. 1.7 km southwest of the Santa Elena village center and consists of 27 hectares of mixed forest type, at a mean elevation of 1240 m in the premontane wet forest, or rain shadow forest zone [16]. There are no precise climatological data for this location, but based on an adiabatic lapse rate of 5.4°C km^{-1} [18], we estimate a MAT of 19.6°C. MAP at this elevation is estimated at 2100 to 2500 mm yr^{-1} [19]. In both areas, the mature forest is characterized by a discontinuous upper canopy, in which the taller trees are 15 m to 40 m tall, a well-developed sub-canopy, abundant vines and lianas, and epiphytes are common but not abundant [20]. Common species in the forest at the two sites include species include: multiple species of *Cinnamomum*, *Ocotea* and *Persea*; *Hampea appendiculata*, *Citharexylum caudatum*, *Hasseltia floribunda*, *Roupala glaberrima* and *Viburnum costaricanum*

TABLE 1: Summarized data for AGB calculations for Monteverde Institute (MVI) and La Calandria (LC). Measurements of stem dbh calculated as number of stems in size class per hectare. AGB for mature and secondary forest presented as oven-dried biomass calculated by equation of Brown [13] in units of metric tons (=Mg) per hectare (t ha^{-1}). AGB$_{carbon}$ is biomass carbon in t ha^{-1}.

Site	Mature dbh			AGB		AGB$_{carbon}$	
	$\geq$20	$\geq$50	$\geq$70	Mature	Secondary	Mature	Secondary
MVI	340	160	60	457 $\pm$ 108	5.2 $\pm$ 1.7	219.4 $\pm$ 51.8	2.5 $\pm$ 0.8
LC	467	100	0	159 $\pm$ 23.7	23.4 $\pm$ 10.2	76.3 $\pm$ 11.3	11.2 $\pm$ 4.9

(W. Haber and E. Cruz, pers. com.). Both locations exhibited similar numbers of trees per hectare and a mix of large diameter (>50 cm) upper canopy and smaller lower canopy and understory trees. The oldest forest stands are considered disturbed-mature forest indicating a period of uninterrupted growth of at least 30 years. The secondary forests in this study are experimental plots of mixed native species (listed below) planted in grids on cleared plots, with other species allowed to recolonize between the seedlings. Bordering the forests are clear-cut fields and lawns with densely rooted grasses. Grass-cutting (of the lawns) and active grazing by livestock prevent reforestation.

2.2. Study Methods.

The sampling and measurements were conducted during the dry season month of January, 2014. In each of the three sampling environments (mature forest, experimental secondary forest, grassy clearing) we established multiple 100 m^2 (10 m by 10 m) sampling plots (locations recorded by GPS). At MVI, sampling was conducted for five 100 m^2 plots for each land cover type. At La Calandria, each environment was represented by three 100 m^2 plots. Above-ground biomass in each plot was recorded by measuring the breast-height diameter (dbh = 1.3 m above base) of all woody stems 1.0 cm or greater diameter (including vines and lianas), totaling ca. 100 stems per plot (=10^4 stems ha^{-1}). To avoid error induced by uncertainty in height measurement, we calculated biomass with a height-independent algorithm developed by Brown [13] for broadleaf moist tropical forests; AGB $= e^{(-2.134+2.53 \ln(dbh))}$. This equation (Eq. 3.2.4 of [13]) is particularly useful for the plots we sampled because it yields a high r^2 over a wide range of stem diameters. The dbh data for each plot were divided into 5-cm size classes (1–5, 6–10, 11–15, etc.) and the AGB calculated for the midpoint of each class, multiplied by the number of stems in the class. The oven-dried biomass thus calculated was multiplied by 0.48 to convert to carbon weight [21]. The results for the plots for each environment type were averaged to calculate a mean AGB$_{carbon}$ in units of metric tons of elemental carbon per hectare (tC ha^{-1} = Mg C ha^{-1}).

Within each 100 m^2 plot, five soil cores were drilled, one at each corner and one in the center, to a depth of 30 cm using a hand auger with a 2-cm internal diameter. The characteristics (grain size and color using standard Munsell soil color designations) of the core were recorded and the core divided into 10-cm segments (0–10, 11–20, 21–30). Bulk density was calculated from the oven-dried weight of a soil core cylinder of 1 cm diameter by 1 cm height. Aggregate samples from each of the sample intervals were

used to measure mean bulk density values for soils in the forest and clearing locations. In the laboratory, individual soil samples were dried, homogenized, sieved with a 2-mm screen to remove larger rocks and root fragments, and the sub-2 mm fraction pulverized in a ball mill. From each processed sample, 0.1 to 0.125 g was drawn for analysis with a Leco TruSpec CN® by combustion in a pure O$_2$ atmosphere at 950°C. The weight percent carbon was calculated from the composition of the evolved gases as measured by an infrared cell. Soil carbon stocks were calculated using values of 0.7, 0.75 and 0.8 g cm^{-3} for the 0–10, 11–20 and 21–30 cm intervals, respectively, in the forest soils, and 0.9, 0.95 and 1.0 g cm^{-3} for the same intervals in the clearings; these values are consistent with those found in other studies of lower montane forest soils [5]. The statistical significance of the differences in the biomass and soil carbon results between the three environments was tested by one-way ANOVA using SigmaStat software manufactured by Systat Software, Inc. Differences of significance were defined at the level of $p <$ 0.05.

3. Results and Discussion

3.1. Aboveground Biomass

3.1.1. Mature Forest. At both locations, the forest is characterized by a partially open upper canopy, and well-developed lower canopy and understory. The mean number of stems measured (dbh > 1 cm) was similar for both sites, although slightly higher at La Calandria, where the plots averaged 107 stems per 100 m^2, compared to MVI where we measured an average of 91 stems per plot. At MVI, the mature forest plots contained a mean of 1.6 trees with dbh > 50 cm, equivalent to 160 ha^{-1}, and 0.6 trees with dbh > 70 cm, or 60 ha^{-1} (Table 1). At La Calandria, the mature forest plots contained a mean of 1.0 trees per plot with dbh > 50 cm per plot, equivalent to 100 trees ha^{-1}, and no trees with dbh > 70 cm. If all trees of medium size (dbh > 20 cm) or larger size are considered, the MVI plots contained the equivalent of 340 trees ha^{-1}, compared to 467 ha^{-1} in the La Calandria plots. The number of taller trees (>20 m), estimated by inclinometer measurement. was greater at La Calandria where plots averaged almost five trees (ca. 500 ha^{-1}) with height > 20 m, compared to an average of just over one tree (ca. 100 ha^{-1}) per plot at MVI.

The differences in AGB and AGB$_{carbon}$ between MVI and La Calandria are significant ($p <$ 0.05) for both mature and secondary forest sites (Table 1). The mature forest at

TABLE 2: Soil carbon data and carbon stocks for mature and secondary forest and clearings at Monteverde Institute (MVI) and La Calandria (LC), with mean C_{soil}% (first line) and soil carbon stock calculated as described in text in $tC\,ha^{-1}$ (second line) for each 10 cm increment in soil for each site. Bottom line is total carbon stock (soil carbon plus AGB_{carbon}) for each setting $tC\,ha^{-1}$.

Interval	MVI_{mature}	LC_{mature}	$MVI_{secondary}$	$LC_{secondary}$	MVI_{clear}	LC_{clear}
0–10	20.5 ± 2.8	14.8 ± 1.3	10.2 ± 1.2	9.3 ± 0.7	8.2 ± 1.4	7.0 ± 0.4
	143.5 ± 19.6	103.6 ± 9.1	71.4 ± 8.5	65.0 ± 4.9	73.8 ± 12.6	63.0 ± 3.6
10–20	13.0 ± 1.4	11.1 ± 0.8	7.6 ± 2.4	6.3 ± 0.8	5.3 ± 1.4	5.5 ± 0.2
	97.5 ± 10.5	83.3 ± 6.0	57.0 ± 18	47.4 ± 6.0	50.4 ± 13.3	52.3 ± 1.9
20–30	7.7 ± 1.6	6.9 ± 0.9	4.7 ± 0.5	5.5 ± 0.5	4.1 ± 0.2	5.3 ± 2.8
	61.6 ± 12.8	55.2 ± 7.2	37.6 ± 4.0	44.0 ± 4.0	41 ± 2.1	53.0 ± 3.0
Total C_{soil}	302.6 ± 42.9	242.1 ± 22.3	166.0 ± 30.5	156.4 ± 11.0	165.2 ± 27.9	168.3 ± 8.5
Total C	522.0 ± 94.7	318.4 ± 33.6	168.5 ± 31.3	167.6 ± 15.9	165.2 ± 27.9	168.3 ± 8.5

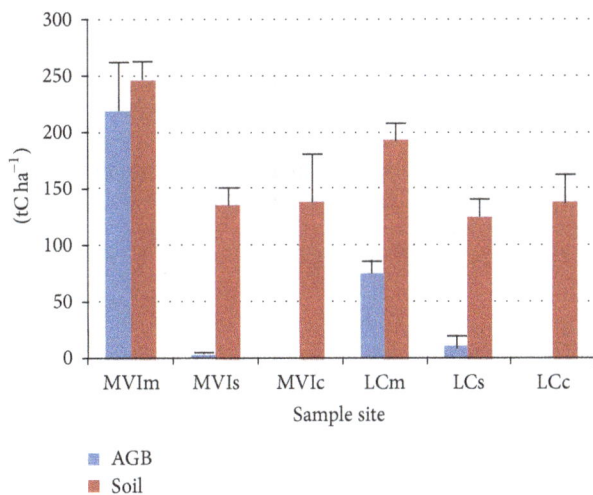

FIGURE 2: Summary plot of carbon stocks measured in this study, expressed in $tC\,ha^{-1}$ for both AGB and C_{soil} with bar for positive standard error. Locations indicated: MVIm = MVI mature forest; MVIs = MVI secondary forest; MVIc = MVI clearing; LCm = La Calandria mature forest; LCs = La Calandria secondary forest; and LCc = La Calandria clearing.

MVI holds an average AGB of 457 ± 108 $t\,ha^{-1}$ (219.4 ± 51.8 $tC\,ha^{-1}$ when converted to AGB_{carbon}), compared to AGB of 159 ± 23.7 $t\,ha^{-1}$ (76.3 ± 11.3 $tC\,ha^{-1}$) at La Calandria (Table 1, Figure 2). The overall difference between these two sites is controlled by the greater abundance of trees with large diameters, due to the exponential relationship between tree diameter and mass. As reflected in the measures of standard error, the variation in AGB between individual plots was much greater at MVI, reflecting the uneven distribution of trunk diameters; that is, large (>50 cm) and very large (>70 cm) diameter trees are very unevenly distributed in the MVI plots compared to the more uniform distribution at La Calandria.

3.1.2. Secondary Forest. The plots studied here are experimental plots of seedling trees planted in a gridded pattern on cleared land that had been forested previously. Other vegetation was allowed to regrow between the planted trees resulting in dense ground layer vegetation. The MVI plots

were established in 2008 as the Rachel Crandell Preserve and measured for this study in 2014. The planted trees in the MVI plots included *Ocotea monteverdensis, Ocotea whitei, Ocotea floribunda, Cinnamomum triplinerve* and *Citharexlum costaricanum*. These had a maximum diameter of 6.5 cm (for an individual stem) and a maximum height of 5 m. Vegetation growing between the planted trees was generally <1 cm dbh and was not measured. The survivorship of one plot was conspicuously low, and no trees exceeded 2.5 m height, whereas the trees in the other plots averaged 2.8 m. The AGB for individual plots ranged from 2.8 kg to 150 kg; the mean AGB for these plots was 5.2 ± 1.7 $t\,ha^{-1}$ (2.5 ± 0.8 $tC\,ha^{-1}$) The plots at La Calandria were planted in 2001; therefore the measurements represent growth after 13 years. The planted species included *C. triplinerve, C. costaricanum, Inga punctate* and *O. whitei*. The trees had a maximum height of 6.2 m, and secondary growth of the understory layer was particularly dense, including abundant *Psidium guajava* trees up to a height of 3.5 m. AGB of individual plots ranged from a minimum of 90.6 kg to 321.4 kg, yielding a mean AGB of 23.4 ± 10.2 $t\,ha^{-1}$ (11.2 ± 4.9 $tC\,ha^{-1}$).

3.2. Soil Carbon. The results of soil analyses (Table 2, Figure 2) present two main trends: (1) The total C_{soil} is higher at MVI than La Calandria for both mature and secondary forest, although the differences are not significant at the $p < 0.05$ level. The C_{soil} for clearings is indistinguishable between locations, however; (2) The soil profiles in all three sample environments (mature forest, secondary forest, clearing) exhibit C_{soil} decreasing from the surface downwards through the 30 cm sample interval.

3.2.1. Mature Forest. Beneath a relatively thick layer of leaf litter (not included in analyses), the mature forest soils typically have a dark brown (Munsell soil color 5YR 2.5/2 to 10YR 3/2), organic rich sandy loam upper layer at least 10 cm thick. The mean carbon content of the soil for this layer is 20.5 ± 2.8%, which equates to a soil carbon stock of 14.4 $kgC\,m^{-2}$, or 143.5 ± 19.6 $tC\,ha^{-1}$ (Table 2, Figure 2). The interval from 10 to 20 cm depth typically forms a transition from the darker, more organic rich layer above, which often continues through all or most of this interval, to an orange brown (10YR 4/4 to 10YR 6/6) layer. The mean C_{soil} for the interval is 13.0 ± 1.4%, which equates to a soil carbon stock of $97.5 \pm 10.5 \, \text{tC ha}^{-1}$. The lowest interval studied, from 20 to 30 cm depth, is silty loam, orange brown in color (10YR 4/4 to 10YR 6/6) with a mean C_{soil} of $7.7 \pm 1.6\%$, equivalent to a soil carbon stock of $61.6 \pm 42.9 \, \text{tC ha}^{-1}$. The top 30 cm of soil at MVI thus represents a carbon stock of $302.6 \pm 42.9 \, \text{tC ha}^{-1}$ (Table 2, Figure 2).

The La Calandria mature forest soils were sampled beneath a thick cover of leaf litter. The upper 10 cm typically consisted of gray-brown (7.5YR 3/3) sandy loam with a mean C_{soil} of $14.8 \pm 1.3\%$, which converts to $103.6 \pm 9.1 \, \text{tC ha}^{-1}$ (Table 2). The interval from 10 to 20 cm is also gray-brown sandy loam with a mean C_{soil} of $11.1 \pm 0.8\%$, equivalent to $83.3 \pm 6.0 \, \text{tC ha}^{-1}$. The 20 to 30 cm interval is orange (10YR 3/2), silty clay loam with C_{soil} of $6.9 \pm 0.9\%$, which equates to $55.2 \pm 7.2 \, \text{tC ha}^{-1}$. The total soil carbon stock for the mature forest is $242.1 \pm 22.3 \, \text{tC ha}^{-1}$ (Table 2, Figure 2).

3.2.2. Secondary Forest. At MVI, the leaf litter thickness varied with the AGB. Beneath this variable litter layer, the uppermost 10 cm of the soil is slightly sandy to slightly clayey fine silt, light brown (10YR 2/2) color. The mean C_{soil} for this layer is $10.2 \pm 1.2\%$, equivalent to a soil carbon stock of $71.4 \pm 8.5 \, \text{tC ha}^{-1}$. The underlying 10 to 20 cm interval is mainly dark brown (10YR 3/3 to 10YR 6/6) sandy loam. C_{soil} for this layer is $7.6 \pm 2.4\%$, equivalent to a soil carbon stock of $57.0 \pm 18 \, \text{tC ha}^{-1}$. The lowest soil interval studied, 20 to 30 cm, is light brown sandy loam (10YR 3/4 to 10YR 5/6). The C_{soil} for this is $4.7 \pm 0.5\%$, which equates to $37.6 \pm 4 \, \text{tC ha}^{-1}$. The soil carbon stock for the entire upper 30 cm interval is $166.0 \pm 30.5 \, \text{tC ha}^{-1}$ (Table 2).

The secondary forest soils at La Calandria were sampled beneath a thin and discontinuous layer of leaf litter. The uppermost 10 cm is brown (10YR 5/2) sandy loam with a mean C_{soil} of $9.3 \pm 0.7\%$, equivalent to $65 \pm 4.9 \, \text{tC ha}^{-1}$. The interval of 10 to 20 cm is transitional to the somewhat lighter colored and loam interval below. The mean C_{soil} for this interval is $6.3 \pm 0.8\%$, which equates to $47.4 \pm 6 \, \text{tC ha}^{-1}$. The 20 to 30 cm interval in the secondary forest soils is orange-brown (10YR 3/3) silty clay loam with a mean C_{soil} of $5.5 \pm 0.5\%$, equivalent to $44 \pm 4 \, \text{tC ha}^{-1}$. The soil carbon stock for the La Calandria experimental forest plots is therefore $156.4 \pm 11.0 \, \text{tC ha}^{-1}$.

3.2.3. Clearings. The clearings at MVI are deforested areas near buildings where natural growth is limited by grass-cutting. The surface is covered mainly by grasses, but also small flowering plants, and locally, planted shrubs. There is no leaf litter, but the uppermost soil is heavily rooted, dark brown (7.5YR 2.5/2) silt. The mean C_{soil} for the uppermost 10 cm is $8.2 \pm 1.4\%$, which equates to $73.8 \pm 12.6 \, \text{tC ha}^{-1}$. The underlying interval of 10 to 20 depth is similarly dark brown loam to sandy loam with a mean C_{soil} of $5.3 \pm 0.4\%$, equivalent to $50.4 \pm 13.3 \, \text{tC ha}^{-1}$. The bottom interval is typically orange-brown (10YR 2/2 to 10YR 6/6) loam with mean C_{soil} of $4.1 \pm 0.2\%$, equating to $41 \pm 2.1 \, \text{tC ha}^{-1}$. The mean soil carbon stock for the upper 30 cm of soil in clearings at MVI is $165.2 \pm 27.9 \, \text{tC ha}^{-1}$ (Table 2).

The sampled clearings at La Calandria Biological Station are mowed lawns, as at MVI, near (but not adjacent to) buildings. The upper 10 cm of the soil is dark brown (10YR 4/3) sandy silt with a mean C_{soil} of $7.0 \pm 0.4\%$, equivalent to $63 \pm 3.6 \, \text{tC ha}^{-1}$. The underlying interval of 10 cm to 20 cm is marked by the transition to orange (5YR 5/6) clay loam. The C_{soil} for this interval is $5.5 \pm 0.2\%$, which equates to $52.3 \pm 1.9 \, \text{tC ha}^{-1}$. The lowermost interval, 20 to 30 cm depth, is continued orange clay loam. The mean C_{soil} for the interval is $5.3 \pm 2.8\%$, equivalent to a carbon stock of $53 \pm 3 \, \text{tC ha}^{-1}$. The soil carbon stock for the 30 cm soil interval in the clearings at La Calandria is $168.3 \pm 8.5 \, \text{tC ha}^{-1}$.

3.3. Discussion

3.3.1. Above-Ground Biomass. One of the limitations of forest studies in tropical forests is the common lack of tree rings, and therefore age control, in many tropical tree species. Hence, determining the rate of carbon accumulation in forests without precise historical records is problematic. Here we explore the potential for using carbon accumulation rates measured in similar environments as a means constraining forest age.

Although the sample size of this study is limited, the mean AGB_{carbon} of $219.4 \, \text{tC ha}^{-1}$ in the mature at MVI compares well with other measurements for mature moist tropical montane forests in Costa Rica. The meta-analysis of Spracklen and Righelato [22] presented values of AGB_{carbon} in Costa Rican tropical montane forests ranging from 145 to $362 \, \text{tC ha}^{-1}$, collected over a wide range of elevations; the mean AGB_{carbon} for the mature forest at MVI accords well with the mean of these data. In particular, the AGB of $457 \, \text{t ha}^{-1}$ at MVI matches well an estimate of $490 \, \text{t ha}^{-1}$ obtained for the mature forest at the nearby Monteverde Cloud Forest Reserve [17]. Oliveras et al. [23] measured an AGB_{carbon} accumulation rate of $4.9 \, \text{tC ha}^{-1} \, \text{yr}^{-1}$ in Andean secondary cloud forests undergoing recovery from pasture useage. If the accumulation rate for that study [23] is applied to the data presented herein, the MVI mature forest minimum age is on the order of 50 years. We note, however, that the MVI forest reserve is part of the local watershed forest that was protected by the Quakers who settled in the region in 1951, and thus has likely been uncut for well over 60 years. Martin et al. [2] noted the log relationship of the carbon pool to age in tropical secondary forests, and concluded that due to a decreasing rate of carbon accumulation, the minimum age of full carbon stock recovery of a forest converted from pasture is at least 80 years. An age of 80 years or greater is consistent with the occurrence of numerous trees with dbh > 50 cm, and a consistent occurrence of trees with dbh > 70 cm [24]. Hence, the age of the MVI forest is very likely greater than the 50 years suggested by strict application of the accumulation rate of $4.9 \, \text{tC ha}^{-1} \, \text{yr}^{-1}$ [23], and potentially greater than 80 years.

The difference in AGB_{carbon} in the mature forests at MVI and La Calandria is striking ($219.4 \, \text{tC ha}^{-1}$ compared to $76.3 \, \text{tC ha}^{-1}$) and is clearly a function of forest age, as demonstrated by the difference in tree size. The age of the mature forest at La Calandria is unknown, but given the frequency of trees with dbh > 50 cm, and the complete lack of trees with dbh > 70 cm, we can state unequivocally that it is younger than the MVI forest, although likely older than 50 years [24]. The younger age explains the lower frequency of large upper canopy trees at La Calandria, with a consequent more open canopy, and the greater frequency of medium size (dbh 20 to 50 cm) trees with greater height than at MVI. This does little to constrain the age of the La Calandria forest more precisely, however.

If we arbitrarily accept a minimum age of 80 years for the mature forest at MVI, the resulting carbon accumulation rate is 2.75 tC ha^{-1} yr^{-1}. Applying this rate to the La Calandria data would suggest a forest age of only 30 years, which is inconsistent with the trunk diameters, as noted above [24]. However, various studies have noted that rates of biomass accumulation during reforestation can vary greatly depending on numerous factors, many related to previous land-use [25]. For example, a study by Uhl et al. [26] found that carbon accumulation rates on reforested pasture varied by more than an order of magnitude depending on the duration and intensity of grazing. As these factors may vary greatly on relatively small spatial scales, AGB accumulation rates from prior studies may not be useful for constraining forest age

There is no ambiguity in regard to the age of the experimental secondary forest plots at the two locations. The plots at La Calandria contained 13-year-old trees at the time of measurement. Thus these plots were just over twice the age of the plots at MVI when measured but contained more than fourfold AGB$_{carbon}$. The MVI secondary plots accrued carbon at a mean rate of just over 0.4 tC ha^{-1} yr^{-1} through the first six years. By contrast, the La Calandria secondary plots added carbon at a mean rate of almost 0.9 tC ha^{-1} yr^{-1}. Both accumulation rates we measured in the secondary forests are much lower than those observed in previous studies of carbon accumulation in tropical secondary forests, for example, the aforementioned rate of 4.9 tC ha^{-1} yr^{-1} measured in Peruvian cloud forests [23], a rate of 5.5 tC ha^{-1} measured in the Amazonian Basin [27], or the hypothetical rate suggested above for the MVI mature forest of 2.75 tC ha^{-1}. We speculate, therefore, that reforestation on these plots is limited by some factor, such as nutrient availability, resulting from prior land-use [25], or nutrient limitation of the rhyolitic material.

3.3.2. Soil Carbon. The soil carbon stock at the MVI mature forest site is higher relative to the La Calandria mature forest (although not at the level of significance of $p < 0.05$), with most of the difference in the upper 10 cm. The difference in soil carbon between the MVI and La Calandria mature forest sites might be expected, given the higher AGB$_{carbon}$ and presumed greater forest age at MVI, if the premise of increasing C$_{soil}$ with time is accepted. Bautista-Cruz and del Castillo [9] measured C$_{soil}$ accumulation rate of 4.3 tC ha^{-1} yr^{-1} in the upper 20 cm during the initial years of secondary succession in a cloud forest in southern Mexico, while Silver et al. [28] measured an increase of just 1.3 tC ha^{-1} yr^{-1} in wet tropical secondary forests of the Amazon Basin. A lower rate, as found in the latter study [28], is more consistent with the observed

difference between the MVI and La Calandria soils if the age assessment based on tree diameter is correct. As mentioned above, however, previous studies [7, 8] have shown little or no increase in soil carbon stocks during reforestation, *contra* the studies cited above. Schedlbauer and Kavanagh [29], for example, concluded from a study of secondary wet forests in Costa Rica that C$_{soil}$ does not increase with age. Similarly, Martin et al. [2] found only a very weak relationship between soil carbon recovery and time in tropical secondary forests. Regardless of age, moreover, a higher soil carbon content might be expected at MVI due to its higher elevation and consequent higher moisture and lower temperature, both of which retard the decomposition of the soil organic matter, as noted by Dieleman et al. [6] and Salinas et al. [30].

We compared C$_{soil}$ of the clearings and the secondary and mature forests at each location in an attempt to discern the dynamics of the soil carbon accumulation at both locations. We found C$_{soil}$ from the clearings at MVI identical to the clearings at La Calandria. Furthermore, at both MVI and La Calandria, clearing C$_{soil}$ is statistically indistinguishable from C$_{soil}$ secondary forest plots. At both MVI and La Calandria, however, the difference in C$_{soil}$ between the secondary plots and the mature forest are statistically significant, with the greatest increase concentrated in the uppermost 10 cm, thus indicating that C$_{soil}$ is increasing with time in the forests at both sites. Moreover, the lack of difference in C$_{soil}$ between MVI and La Calandria for both clearings and secondary forest plots demonstrates that altitude-associated climate differences do not play a significant role here. Thus, we conclude that the difference in C$_{soil}$ in the mature forests at MVI and La Calandria is a function primarily of the difference in age between the sites. At La Calandria, for example, the difference in soil carbon stocks between mature and secondary forest is 86 tC ha^{-1}. If hypothetically this represents 40 to 50 years of accumulation, the accumulation rate is 1.72 to 2.15 tC ha^{-1} yr^{-1}, not an unreasonable range of values in comparison to the values obtained in the studies cited above. Moreover, an accumulation rate within this range would also be consistent with an age for the mature forest at MVI that is several decades older than the La Calandria forest.

4. Conclusions

This study measured above-ground biomass and in soil carbon stocks in clearings, experimental secondary forest plots, and mature forests at two locations in the Monteverde region. At a first approximation, the obvious result is that deforestation results in the loss of hundreds of tons of above-ground biomass carbon per hectare and that complete recovery of the lost carbon requires many decades, as numerous previous studies have shown. Additionally, a substantial portion of the soil carbon is lost during deforestation.

Perhaps more significantly, this study demonstrates some of the challenges and limitations in determining rates of change in tropical forest carbon stocks: in particular, the lack of complete historical data and the inability to apply dendrochronology forces estimation of forest age based on the distribution of large trees. At the Monteverde Institute

(MVI), AGB_{carbon} stock for the mature forest measured in this study is similar to other measurements of mature tropical montane forest biomass carbon in Costa Rica, as well as other wet montane tropical forests. The combination of historical records and the distribution of large trees suggests a mature forest age likely greater than 80 years. Based on the difference in tree size and AGB_{carbon} between MVI and La Calandria, we estimate an age difference of the forests of at least 30 years. The biomass of the secondary forest plots at both locations is anomalously low, however, and indicates very low rates of biomass accumulation in comparison to the rates seen in other studies of tropical montane forests. Hence, we suggest some inhibiting factor is limiting forest regrowth, perhaps soil nutrient content depleted through land-use changes or inherited from a low-nutrient parent. Consequently, the rates of biomass carbon accumulation found in this study may not be widely applicable.

Although the La Calandria site is located at a lower elevation than MVI and experiences a slightly warmer and drier climate, the nearly identical soil carbon stocks in the secondary forest plots and clearings suggests that climate is not sufficiently different between the sites to explain the difference in the mature forest carbon stocks. Rather, continuous accumulation of soil carbon over a period of three to four decades satisfactorily explains the difference in the size of the soil carbon stocks. The rates of soil carbon accumulation observed here are consistent with other studies conducted in montane and tropical forests, in contrast to the above-ground biomass results, and support the conclusion of some studies that soil carbon increases over time during reforestation in wet montane forests.

Conflict of Interests

The authors declare that there is no conflict of interests regarding the publication of this paper.

Acknowledgments

This study was made possible by grants from Le Moyne College's Student Research Fund to Morgan A. Nivison and Megan T. Wilckens and the Faculty Research & Development Fund to Lawrence H. Tanner. The authors are grateful for the support of Debra Hamilton at MVI, which was essential to this project.

References

[1] P. Bousquet, P. Peylin, P. Ciais, C. Le Quere, P. Friedlingstein, and P. P. Tans, "Regional changes in carbon dioxide fluxes of land and oceans since 1980," *Science*, vol. 290, no. 5495, pp. 1342–1346, 2000.

[2] P. A. Martin, A. C. Newton, and J. M. Bullock, "Carbon pools recover more quickly than plant biodiversity in tropical secondary forests," *Proceedings of the Royal Society B*, vol. 280, 2015.

[3] T. R. Feldpausch, J. Lloyd, S. L. Lewis et al., "Tree height integrated into pantropical forest biomass estimates," *Biogeosciences*, vol. 9, no. 8, pp. 3381–3403, 2012.

[4] J. M. Mellilo, R. A. Houghton, D. W. Kicklighter, and A. D. McGuire, "Tropical deforestation and the global carbon budget," *Annual Review of Energy and the Environment*, vol. 21, pp. 293–310, 1996.

[5] C. C. Rhoades, G. E. Eckert, and D. C. Coleman, "Soil carbon differences among forest, agriculture, and secondary vegetation in lower montane Ecuador," *Ecological Applications*, vol. 10, no. 2, pp. 497–505, 2000.

[6] W. I. J. Dieleman, M. Venter, A. Ramachandra, A. K. Krockenberger, and M. I. Bird, "Soil carbon stocks vary predictably with altitude in tropical forests: implications for soil carbon storage," *Geoderma*, vol. 204-205, pp. 59–67, 2013.

[7] R. F. Hughes, J. B. Kauffman, and V. J. Jaramillo, "Biomass, carbon, and nutrient dynamics of secondary forests in a humid tropical region of Mexico," *Ecology*, vol. 80, no. 6, pp. 1892–1907, 1999.

[8] D. Markewitz, E. Davidson, P. Moutinho, and D. Nepstad, "Nutrient loss and redistribution after forest clearing on a highly weathered soil in Amazonia," *Ecological Applications*, vol. 14, no. 4, pp. S177–S199, 2004.

[9] A. Bautista-Cruz and R. F. del Castillo, "Soil changes during secondary succession in a tropical montane cloud forest area," *Soil Science Society of America Journal*, vol. 69, no. 3, pp. 906–914, 2005.

[10] K. J. Anchukaitis and M. N. Evans, "Tropical cloud forest climate variability and the demise of the Monteverde golden toad," *Proceedings of the National Academy of Sciences of the United States of America*, vol. 107, no. 11, pp. 5036–5040, 2010.

[11] A. Ramirez, "Ecological research and the Costa Rican park system," *Ecological Applications*, vol. 14, no. 1, pp. 25–27, 2004.

[12] R. Blasiak, "Ethics and Environmentalism: Costa Rica's Lesson," United Nations University, 2011, http://ourworld.unu.edu/en/ethics-and-environmentalism-costa-ricas-lesson.

[13] S. Brown, "Estimating biomass and biomass change of tropical forests: a primer," FAO Forestry Paper 134, Food and Agriculture Organization of the United Nations, Rome, Italy, 1997, http://www.fao.org/docrep/w4095e/w4095e00.htm.

[14] R. O. Lawton and V. J. Dryer, "The vegetation of the monteverde cloud forest reserve," *Brenesia*, vol. 18, pp. 101–116, 1980.

[15] E. D. Vance and N. M. Nadkarni, "Microbial biomass and activity in canopy organic matter and the forest floor of a tropical cloud forest," *Soil Biology and Biochemistry*, vol. 22, no. 5, pp. 677–684, 1990.

[16] G. V. N. Powell and R. D. Bjork, "Habitat linkages and the conservation of tropical biodiversity as indicated by seasonal migrations of three-wattled bellbirds," *Conservation Biology*, vol. 18, no. 2, pp. 500–509, 2004.

[17] N. Nadkarni and N. Wheelwright, *Monteverde: Ecology and Conservation of a Tropical Cloud Forest*, Oxford University Press, 2000.

[18] M. S. Lachniet and W. P. Patterson, "Oxygen isotope values of precipitation and surface waters in northern Central America (Belize and Guatemala) are dominated by temperature and amount effects," *Earth and Planetary Science Letters*, vol. 284, no. 3-4, pp. 435–446, 2009.

[19] J. Kricher, *Tropical Ecology*, Princeton University Press, 2011.

[20] N. M. Nadkarni, T. J. Matelson, and W. A. Haber, "Structural characteristics and floristic composition of a Neotropical cloud forest, Monteverde, Costa Rica," *Journal of Tropical Ecology*, vol. 11, no. 4, pp. 481–495, 1995.

[21] R. Condit, *Methods for Estimating Above-Ground Biomass of Forest and Replacement Vegetation in the Tropics*, Center for Tropical Forest Science Research Manual, Smithsonian Tropical Research Institute, 2008.

[22] D. V. Spracklen and R. Righelato, "Tropical montane forests are a larger than expected global carbon store," *Biogeosciences*, vol. 11, no. 10, pp. 2741–2754, 2014.

[23] I. Oliveras, C. Girardin, C. E. Doughty et al., "Andean grasslands are as productive as tropical cloud forests," *Environmental Research Letters*, vol. 9, no. 11, 2014, http://iopscience.iop.org/article/10.1088/1748-9326/9/11/115011.

[24] S. T. O'Brien, S. P. Hubbell, P. Spiro, R. Condit, and R. B. Foster, "Diameter, height, crown, and age relationships in eight neotropical tree species," *Ecology*, vol. 76, no. 6, pp. 1926–1939, 1995.

[25] K. D. Holl and R. A. Zahawi, "Factors explaining variability in woody above-ground biomass accumulation in restored tropical forest," *Forest Ecology and Management*, vol. 319, pp. 36–43, 2014.

[26] C. Uhl, R. Buschbacher, and E. A. S. Serrao, "Abandoned pastures in eastern Amazonia. I. Patterns of plant succession," *Journal of Ecology*, vol. 76, no. 3, pp. 663–681, 1988.

[27] T. R. Feldpausch, M. A. Rondon, E. C. M. Fernandes, S. J. Riha, and E. Wandelli, "Carbon and nutrient accumulation in secondary forests regenerating on pastures in central Amazonia," *Ecological Applications*, vol. 14, no. 4, pp. S164–S176, 2004.

[28] W. L. Silver, R. Ostertag, and A. E. Lugo, "The potential for carbon sequestration through reforestation of abandoned tropical agricultural and pasture lands," *Restoration Ecology*, vol. 8, no. 4, pp. 394–407, 2000.

[29] J. L. Schedlbauer and K. L. Kavanagh, "Soil carbon dynamics in a chronosequence of secondary forests in northeastern Costa Rica," *Forest Ecology and Management*, vol. 255, no. 3-4, pp. 1326–1335, 2008.

[30] N. Salinas, Y. Malhi, P. Meir et al., "The sensitivity of tropical leaf litter decomposition to temperature: results from a large-scale leaf translocation experiment along an elevation gradient in Peruvian forests," *The New Phytologist*, vol. 189, no. 4, pp. 967–977, 2011.

16

Comparison of T-Square, Point Centered Quarter, and N-Tree Sampling Methods in *Pittosporum undulatum* Invaded Woodlands

Lurdes Borges Silva,[1,2,3] **Mário Alves,**[2] **Rui Bento Elias,**[4] **and Luís Silva**[1]

[1]*InBIO, Rede de Investigação em Biodiversidade e Biologia Evolutiva, Laboratório Associado, CIBIO-Açores, Universidade dos Açores, Açores, 9501-801 Ponta Delgada, Portugal*
[2]*NATURALREASON, LDA, Caminho do Meio Velho, 5-B, Açores, 9760-114 Cabo da Praia, Portugal*
[3]*3CBIO, Universidade dos Açores, Faculdade de Ciências e Tecnologia, Ponta Delgada, Portugal*
[4]*Centre for Ecology, Evolution and Environmental Changes (CE3C), Azorean Biodiversity Group and Universidade dos Açores, Faculdade de Ciências Agrárias e do Ambiente, Açores, 9700-042 Angra do Heroísmo, Portugal*

Correspondence should be addressed to Lurdes Borges Silva; lurdesborgesilva@gmail.com

Academic Editor: Guy R. Larocque

Tree density is an important parameter affecting ecosystems functions and management decisions, while tree distribution patterns affect sampling design. *Pittosporum undulatum* stands in the Azores are being targeted with a biomass valorization program, for which efficient tree density estimators are required. We compared T-Square sampling, Point Centered Quarter Method (PCQM), and N-tree sampling with benchmark quadrat (QD) sampling in six 900 m² plots established at *P. undulatum* stands in São Miguel Island. A total of 15 estimators were tested using a data resampling approach. The estimated density range (344–5056 trees/ha) was found to agree with previous studies using PCQM only. Although with a tendency to underestimate tree density (in comparison with QD), overall, T-Square sampling appeared to be the most accurate and precise method, followed by PCQM. Tree distribution pattern was found to be slightly aggregated in 4 of the 6 stands. Considering (1) the low level of bias and high precision, (2) the consistency among three estimators, (3) the possibility of use with aggregated patterns, and (4) the possibility of obtaining a larger number of independent tree parameter estimates, we recommend the use of T-Square sampling in *P. undulatum* stands within the framework of a biomass valorization program.

1. Introduction

In ecological research, the basic objective of sampling is to obtain a descriptive estimate of some attribute of a plant population. This estimate should be a relatively accurate representation to allow detection of real differences among plant populations. Quantitative data are essential to adequately characterize the woody component of forest communities [1]. Some form of sampling is required because total counts of individuals in naturally occurring plant populations are generally impractical without an exhaustive expenditure of energy and resources [2]. A number of sampling techniques are available to quantify forest community traits. These techniques vary in their underlying assumptions (e.g., random distribution of target population), equipment required, and time necessary to obtain an adequate sample for statistical analysis [3].

One of the most commonly sampled parameters is density [4]. Stand density or tree density, expressed as the number of trees per unit area, is an important forest management parameter. Together with other forest structure parameters such as crown closure and crown diameter, it is used by foresters to evaluate regeneration, to assess the effect of forest management measures or as an proxy variable for other stand parameters such as age, basal area, and volume [5]. Thus, density is considered as one of the most important numerical indices to explain quantitative values of tree and shrub communities. By affecting many aspects of the ecosystem, it can be used in a wide variety of situations.

One of the oldest techniques of data collecting is quadrat counting [6] but this analysis may be insufficient to distinguish certain point patterns and the size of quadrats may affect the results [7]. Obtaining adequate information with minimum effort and time is a major concern when sampling vegetation [3]. In sampling for indices of forest structure, distance sampling (plotless sampling) can be very efficient [8]. Continued technological advances in range measurements and field computing add to the attraction of distance sampling because measurements can be completed ever faster. Also, in natural populations characterized by an irregular, possibly clustered, distribution of trees, the precision of a density estimate from distance sampling can be better than the precision obtained with fixed area plot sampling [9]. Distance sampling has attracted the attention of researchers over the past 50 years as a means of estimating density. Its main attraction is that it is fast and easy to use, and one or more distances are always recorded at each sample point. In contrast, plot sampling can sometimes be a very time consuming process, boundary trees may be overlooked, and some plots may have no tallies [10]. Distance sampling has a long history in forest inventories where the distance and attributes of interest are measured on the k-trees closest to a sample point [9]. When quadrat sampling is difficult or too costly (*e.g.*, in low density populations or mountain areas) distance sampling is favored [11]. Distance sampling is often faster than fixed area plot sampling, since the number of measurements to take at each location can be fixed in advance. This is because there is no need to sample all the trees present in a plot and therefore measurement efforts are independent of local tree density. Expediency that is faster data collection in stand-wise forest management inventories has been another motivating factor [12].

A variety of methods using distance measures have been developed [1]: *T*-Square sampling [13, 14], Point Centered Quarter Method [15], *N*-tree sampling [16], random pairs [17], variable area transect [18], closest individual [15], ordered distance [19], compound estimators [20, 21], corrected point-distance [22], nearest neighbor [15], quartered neighbor [23], and angle order [24]. All these methods are essentially based on the distance measures from event to event or from point to event [1, 25].

A variety of density estimators have been proposed for different spatial patterns and modifications to improve their robustness, especially when nonrandom spatial patterns are assumed [15, 20, 24, 27]. Some of the distance methods can be used to determine the general spatial pattern or distribution of a population [13, 28] and are useful to determine inter- and intraspecific relationships in plant communities [29]. Advantages arise from ease of use calculation and interpretation by foresters [25]. Their great disadvantage is connected with the loss of detailed information about spatial patterns at different spatial scales [30], describing the spatial structure only at the fine scale of nearest neighbors [25].

Forest stand structure is a key element in understanding forest ecosystems and many methods have been developed for its study [31]. One of the major components of forest stand structure is the spatial arrangement of tree positions [8]. Spatial information for individual trees is increasingly sought by forest managers and modelers as a means to improve the spatial resolution and accuracy of forest models and management scenario [32]. There are three general categories of two-dimensional point patterns, uniform (hyperdispersed), random, and clumped (aggregated). Random type of spatial distribution means that trees are distributed independently of each other and the probability of finding trees in the whole population is the same. In aggregated populations, individuals occur in clumps of different densities and sizes, and in a regular distribution objects are evenly spaced in a population over a given area [25]. We can assume that the spatial pattern reflects the major trends in stand dynamics. For example, regular spatial structures indicate high competition, whereas aggregate patterns indicate massive regeneration without subsequent strong self-thinning [29].

According to Elias et al. [33], in Azorean native mountain forests, the spatial pattern of tree species is largely explained by disturbance regimes and the regeneration strategies of each species. However, factors such as habitat related patchiness, competition, and dispersion limitation may also explain many of the observed patterns. Meanwhile, spatial patterns in Azorean exotic woodland have not been studied. Invasive alien plants are one of the most relevant threats to the long term integrity of biodiversity, affecting the Azores archipelago, with a relatively high rate of nonindigenous species, among which are several top invaders [34–36]. *Pittosporum undulatum* Vent. (Pittosporaceae) is one of the top invasive trees in the Azores, affecting protected areas and indigenous species, and was targeted by the Azorean Regional Program for Control and Eradication of Invasive Plants in Sensitive Areas (PRECEFIAS) [36–40]. Accurate data about abundance and distribution of *P. undulatum* is needed for its control and management, including the evaluation of the energetic potential of its biomass [38, 41]. This information will be crucial for the implementation of a large scale survey of *P. undulatum* stands in the Azores.

Therefore, the aims of this study are (i) to evaluate the most accurate and precise method to estimate tree density in Azorean exotic woodland; (ii) to test plotless density estimators that have been recently validated for other species and regions; (iii) to compare results with those from quadrat counts; (iv) to determine tree distribution pattern in exotic woodland.

2. Material and Methods

2.1. Study Area. We investigated *P. undulatum* dominated forests located in São Miguel Island (Azores) situated in the North Atlantic Ocean, 1500 km west from Portugal (Figure 1). The archipelago of the Azores, scattered across 615 km on a WNW–ESE alignment, covering a total of 2323 km^2, comprises nine volcanic islands. The climate is temperate oceanic with a mean annual temperature of 17°C at sea level. Relative humidity is high while rainfall ranges from 1500 to 3000 mm, depending on the altitude and increasing from east to west [42]. The topography of the island is characterized by volcanic craters, large catchments, ravines, and seasonal water streams

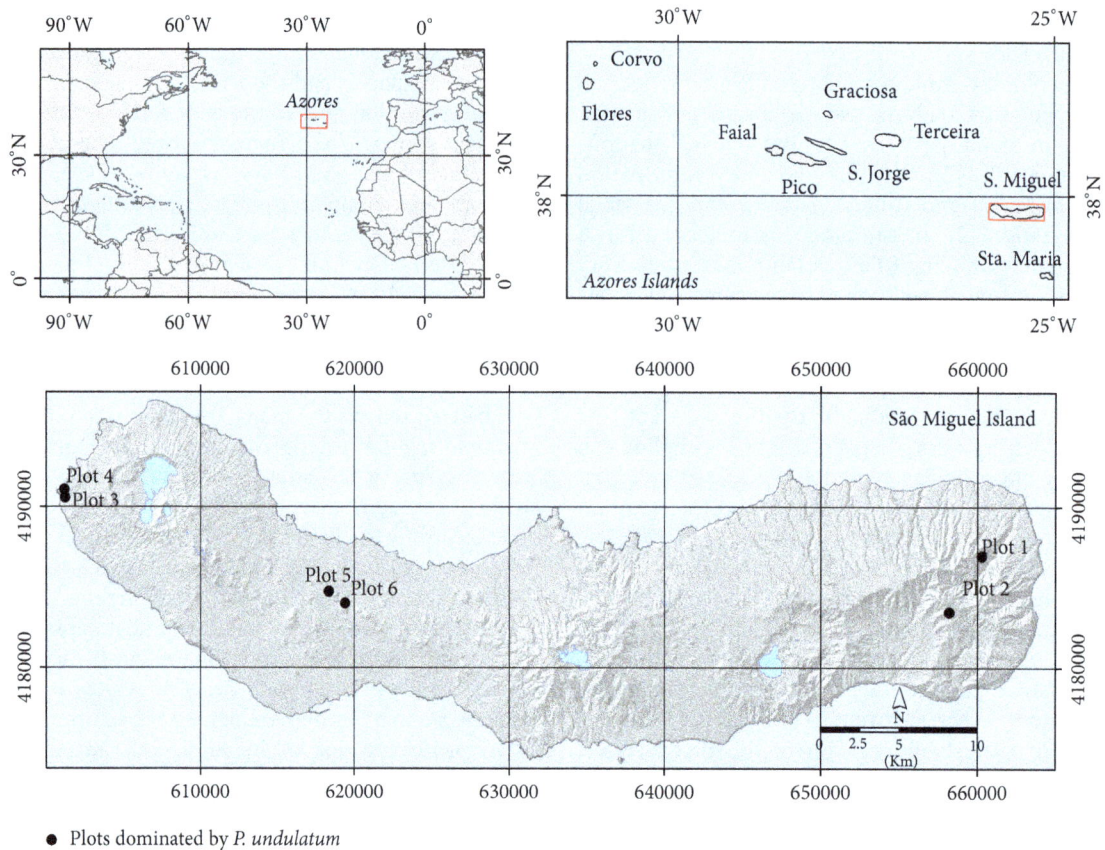

FIGURE 1: Study sites location in São Miguel Island, Azores archipelago, Portugal. Plot 1, Ribeira do Guilherme, Nordeste; Plot 2, Mata dos Bispos, Povoação; Plot 3, Pico das Camarinhas, west slope; Plot 4, Pico das Camarinhas, Caldera; Plot 5, Pico da Furna, São Vicente Ferreira; Plot 6, Picos de Lima, Fenais da Luz.

TABLE 1: Main features of the studied *Pittosporum undulatum* stands.

Plot number	Location	Maximum elevation (m a.s.l.)	Area (ha)	Other species
1	Ribeira do Guilherme	700	120.0	*Laurus azorica, Morella faya*
2	Mata dos Bispos	500	2.5	*Laurus azorica, Erica azorica, Clethra arborea*
3[*] 4[**]	Pico das Camarinhas	250	10.0	*Morella faya*
5	Pico da Furna	311	2.2	*Acacia melanoxylon*
6	Picos de Lima, Fenais da Luz	312	2.3	*Acacia melanoxylon*

[*]West slope, outside volcanic crater, and [**]Caldera, inside volcanic crater.

with a maximum altitude of 1105 m above sea level and a surface area of 745 km^2.

2.2. Stand Characterization.

The studied stands (Figure 1) are located within geomorphological formations [43], namely, Nordeste Volcanic System and Povoação Volcano (Plots 1 and 2), Sete Cidades Volcano (Plots 3 and 4), and Picos Fissural Volcanic System (Plots 5 and 6). Owing to their volcanic origin, soils are generally young Andosols, formed under a humid climate on relatively recent lava flows and pyroclastic deposits. Moreover, Ribeira do Guilherme is a protection area for the management of habitats or species and Pico das Camarinhas is a geological monument [44]. The characterization of each plot is presented in Table 1.

2.3. Plot Sampling.

The field work was carried out in May and June 2014. At each stand we marked a plot with 30 × 30 m (900 m^2). We divided each plot into 36 5 × 5 m quadrats. At each quadrat, the total number of trees was thoroughly counted. Saplings less than 2 cm of diameter at breast height were excluded because most likely they would not survive. All woody species that were within the plots were included in the measurements.

Regarding plant density obtained with plot counts we estimated the confidence interval using the usual formula for the mean:

$$\lambda \pm z_{\alpha/2}SE_\lambda, \tag{1}$$

where λ is the mean density and SE_λ is the respective standard error.

2.4. Plotless Sampling. Thirty random points were selected at each plot. Distances from point-tree and tree-neighbor were measured at 1.30 m high, at the center of each tree. Three methods were used.

In *T*-Square [1, 13, 14, 20, 26], Figure 2(a), a random point is chosen and the distance to nearest tree is measured. A second distance is measured from the nearest tree to its nearest neighbor constrained to be in the hemisphere to the left of the dashed line. Four density estimators were used (Table 2). For *T*-Square sampling the 95% confidence interval for the reciprocal of tree density is calculated as

$$\frac{1}{\lambda_2} \pm t_\alpha \, [\text{SE}]$$

$$\text{SE} = \sqrt{\frac{8\left(\bar{t}^2 S_r^2 + 2\bar{r}\bar{t}S_{rt} + \bar{r}^2 S_t^2\right)}{n}}, \tag{2}$$

where SE is the standard error given by Diggle [31], λ_2 is the density estimator for *T*-Square (Table 2), $\bar{r}$ and $\bar{t}$ are the respective distance means, S_r^2 is the variance of r, S_t^2 is the variance of t, S_{rt} is the covariance of r and t, and t_α is the value taken from a Student distribution with $n - 1$ degrees of freedom.

In Point Centered Quarter Method (PCQM; [15, 19, 24]; Figure 2(b)), the area around the random point is divided into four 90° quadrants and the distance to the nearest tree is measured in each quadrant. Three density estimators were used (Table 2). For PCQM confidence limits are given by Seber [45]:

$$\sqrt{\lambda_5} = \frac{\sqrt{16n - 1} \pm z_{\alpha/2}}{\sqrt{\pi \sum \left(r_{ij}^2\right)}}, \tag{3}$$

where λ_5 is the density estimator for PCQM (Table 2), r_{ij} is the distance from random point i to closest individual in quarter j ($j = 1, 2, 3, 4$; $i = 1, \ldots, n$), and $z_{\alpha/2}$ is the value taken from the standard normal distribution, for an error α.

In *N*-tree sampling (*K*-tree sampling; [16, 27]; Figure 2(c)), a number of trees, *N*, closest to a random point are selected and the respective distances measured. Plot shape is circular and is based on a radius from the sampling point to the centre of the *N*th tree closest to the point. In our study we considered a value of $N = 6$ (see Table 2 for estimators).

2.5. Spatial Pattern Analyses. Since the 1950s, several spatial methods of analysis have been developed and modified to improve our ability to detect and characterize spatial patterns [46]. We used five indexes of spatial pattern, applied to our data from *T*-Square [47, 48], PCQM [28, 49], and quadrat sampling [50].

For *T*-Square we used the index recommended by Ludwig and Reynolds [48]:

$$C = \frac{\sum \left[r_i^2 / \left(r_i^2 + (1/2)\,t_i^2\right)\right]}{n},$$

$$z = \frac{C - 0.5}{\sqrt{1/(12n)}}. \tag{4}$$

The value of *C* is approximately 0.5 for random patterns, significantly less than 0.5 for uniform patterns, and significantly greater than 0.5 for clumped patterns. To test the significance of any departure of *C* from randomness, we compute a value for *z* (see [48], pp. 57-58 for details). We also used the statistic recommended by Hines and O'Hara Hines [47]:

$$H_t = \frac{2n\left[2\sum\left(r_i^2\right) + \sum\left(t_i^2\right)\right]}{\left[\left(\sqrt{2}\sum r_i\right) + \sum t_i\right]^2}. \tag{5}$$

This test statistic is compared with reference values (see [6], pp. 210 for details). Randomness corresponds to 1.27; smaller values indicate a uniform pattern while larger values indicate clumping.

We used our data from PCQM by selecting the tree closest to the sample point [6, 25, 48] and using the index of dispersion described by Johnson and Zimmer [49]:

$$I = (n + 1)\frac{\sum\left(r_i^2\right)^2}{\left[\sum\left(r_i^2\right)\right]^2},$$

$$z = \frac{I - 2}{\sqrt{4(n-1)/(n+2)(n+3)}}. \tag{6}$$

The results are compared with the value defined for a random pattern, 2, with lower values suggesting a uniform pattern and larger values suggesting a clumped pattern (see [48], pp. 58-59 for details); and *z* is compared to a table of critical values for the standard normal distribution to detect significant departures from randomness [48]. The second procedure used for PCQM was suggested by Eberhardt [50] and analyzed further by Hines and O'Hara Hines, [47]:

$$I_E = \left(\frac{s}{\bar{r}}\right)^2 + 1, \tag{7}$$

where s and $\bar{r}$ are, respectively, the standard deviation and mean of the point to nearest individual distances. Critical values have been computed by Hines and O'Hara Hines (see [6], p. 210 for details). The expected value in a random population is 1.27, lower values suggest a regular pattern, and larger values indicate clumping.

Regarding quadrat counts, the data comes from 5 × 5 m quadrats. Morisita's index of dispersion, I_d, has been extensively used to evaluate the degree of dispersion/aggregation of spatial point patterns [51]. It is based on random or regular quadrat counts and is closely related to the simplest and oldest measures of spatial pattern, the variance to mean ratio [6, 30]. We used a standardized version (I_p) of the index which is

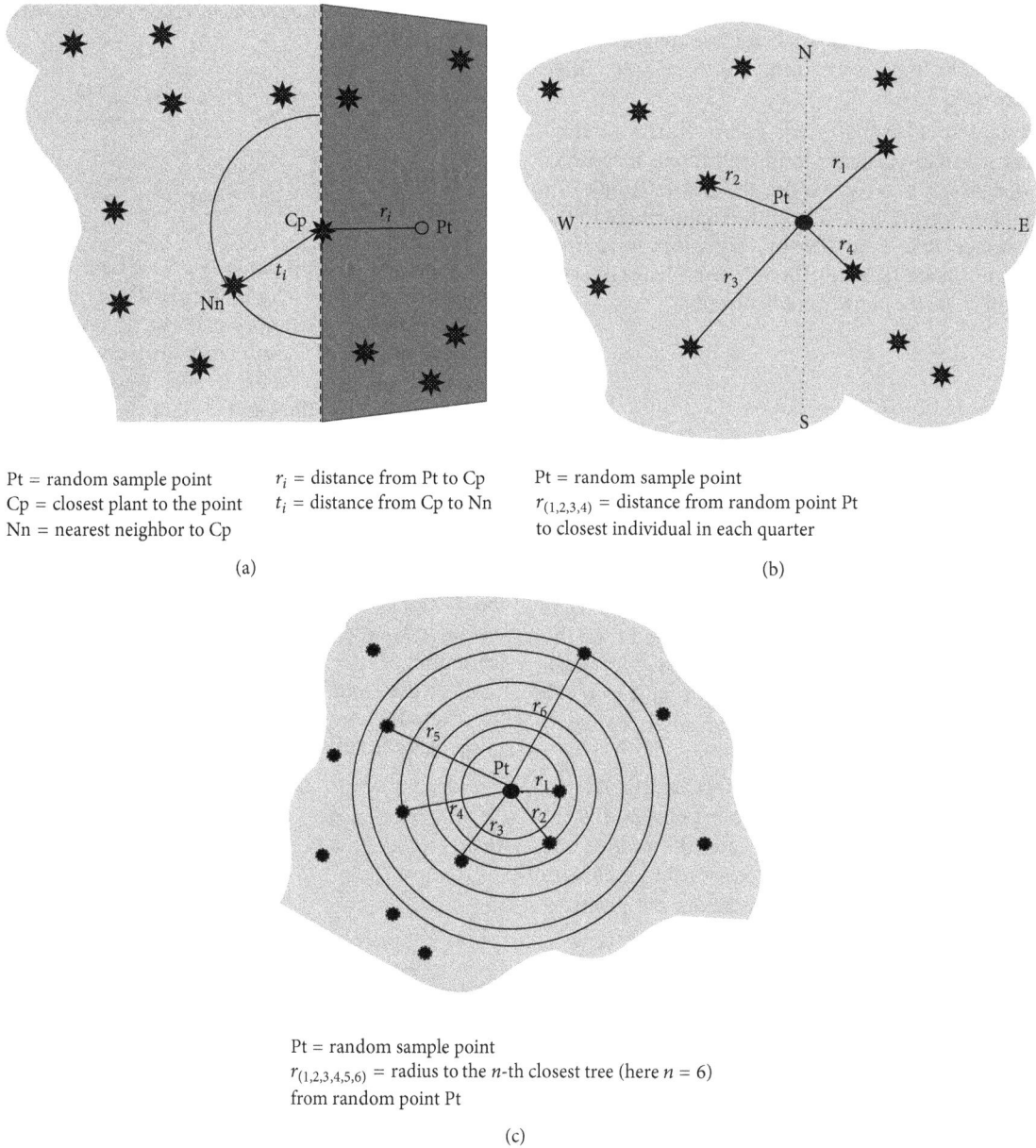

Pt = random sample point r_i = distance from Pt to Cp   Pt = random sample point
Cp = closest plant to the point t_i = distance from Cp to Nn $r_{(1,2,3,4)}$ = distance from random point Pt
Nn = nearest neighbor to Cp   to closest individual in each quarter

(a)   (b)

Pt = random sample point
$r_{(1,2,3,4,5,6)}$ = radius to the n-th closest tree (here n = 6)
from random point Pt

(c)

FIGURE 2: Application of T-Square sampling (a); Point Centered Quarter Method (b); and N-tree sampling method (c).

robust to variation in sample number and size [51] and is obtained by first calculating the raw index:

$$I_d = n \left[\frac{\sum x_i{}^2 - \sum x_i}{\left(\sum x_i\right)^2 - \sum x_i} \right], \qquad (8)$$

where n is the number of samples and x_i is the number of individuals per sample. The standardized Morisita index of dispersion (I_p) ranges from −1 to +1, with 95% confidence limits at +0.5 and −0.5. Random patterns give an I_p of zero, clumped patterns above zero, and uniform patterns below zero (see [6], p. 217 for details). The second index used for quadrat counts is the variance to mean ratio, where a random

pattern, described by the Poisson distribution, corresponds to the value 1:

$$I_c = \left(\frac{s^2}{\bar{x}} \right). \qquad (9)$$

If the variance is larger than the mean, distribution is clumped; if it is smaller than the mean, distribution is regular [6]. To test significant departures from the random expectation, confidence envelopes using the χ^2 test for $n-1$ degrees of freedom are calculated [30].

2.6. *Statistical Analyses.* To compare the different sampling methods, we used a resampling approach [52, 53]: (i) 25 sample points were randomly selected from the 30 available

TABLE 2: Different sampling methods and estimators used in this study (adapted from [1]).

Sampling methods	Estimators		Characteristic
T-Square	$\lambda_1 = \dfrac{2n}{\pi \sum (t_i^2)}$	[13]	λ = estimated density
	$\lambda_2 = \dfrac{n^2}{[2 \sum (r_i)(\sqrt{2} \sum t_i)]}$	[14]	n = sample size t_i = distance from closest individual to nearest neighbor
	$\lambda_3 = \dfrac{2n}{[\pi \sum (r_i^2) + 0.5(\pi \sum (t_i^2))]}$	[20]	r_i = distance from random point to closest individual
	$\lambda_4 = \dfrac{n}{[\pi(\sum(r_i^2) + 0.5 \sum(t_i^2))1/2]}$	[26]	
Point Centered Quarter	$\lambda_5 = \dfrac{4(4n-1)}{\pi \sum (r_{ij}^2)}$	[15]	λ = estimated density n = sample size
	$\lambda_6 = \dfrac{1}{[\sum (r_{ij})/4n]}$	[3]	g = number of individuals to be located in each sector of the area around the random sampling point
	$\lambda_7 = \left(\dfrac{12}{\pi n}\right) \sum \left[\dfrac{1}{\sum \left(r_{ij}^2\right)}\right]$ for $g=1$	[24]	r_{ij} = distance from random point i to closest individual in quarter j
N-tree	$\lambda_8 = \dfrac{1}{n[\sum(2.5/A_i)]}$	For 3-tree	
	$\lambda_9 = \dfrac{1}{n[\sum(3/A_i)]}$	3-tree adjusted	
	$\lambda_{10} = \dfrac{1}{n[\sum(3.5/A_i)]}$	For 4-tree	n = sample size λ = estimated density
	$\lambda_{11} = \dfrac{1}{n[\sum(4/A_i)]}$	4-tree adjusted	A_i = plot area in each plot is calculated separately using the following equation: $A_i = \pi r_i^2$
	$\lambda_{12} = \dfrac{1}{n[\sum(4.5/A_i)]}$	For 5-tree	r_i = plot radius and in normal method prolongs to mid diameter of nth tree but in adjusted methods prolongs between n and $n+1$ trees.
	$\lambda_{13} = \dfrac{1}{n[\sum(5/A_i)]}$	5-tree adjusted	
	$\lambda_{14} = \dfrac{1}{n[\sum(5.5/A_i)]}$	For 6-tree	
	$\lambda_{15} = \dfrac{1}{n[\sum(6/A_i)]}$	6-tree adjusted [16]	

for each plot; (ii) the different estimators were used to calculate tree density; (iii) this procedure was repeated 100 times; (iv) a similar approach was followed to estimate tree density using quadrat counts, where 30 out of 36 were used at each iteration. This procedure allowed simulating the variability in tree density estimates obtained using the different methods. Using this simulated dataset of tree densities, we compared all the estimators using a generalized linear model (ANOVA). Standard contrasts were then used to compare tree densities obtained using quadrat counts (the benchmark method) with all the other methods/estimators (λ_1–λ_{15}). In order to further compare the departure of each estimator from the benchmark method, we calculated the relative mean difference (for each of the six plots) between tree densities obtained by each estimator and the quadrat method and represented those values using a bar plot. In the same sense, the estimators were ranked, according to their distance to the benchmark method, and the ranks for the six plots were represented using a box-plot. To analyze estimator precision, we calculated the coefficient of variation (CV) of tree density for each estimator and divided it by the CV obtained using quadrats. Statistical analyses were all performed within the R software, version 3.2.3 (R Development Core Team, 2014) [54].

3. Results

3.1. Estimated Tree Densities. Tree densities ranged from 344 to 5056 trees/ha and *P. undulatum* tree density ranged from 267 to 4233 tree/ha (Table 3). According to the general results, *T*-Square sampling methods tended to underestimate the tree densities obtained with quadrats (Figure 3).

3.2. Comparison of Sampling Methods. The results of ANOVA, based on 100 simulations, revealed significant differences for the estimation of tree density among the different estimators: plot 1, $F = 8963$ and $P < 0.001$; plot 2, $F = 794$ and $P < 0.001$; plot 3, $F = 3040$ and $P < 0.001$; plot 4, $F = 3653$ and $P < 0.01$; plot 5, $F = 3037$ and

TABLE 3: Calculation of tree density at six *Pittosporum undulatum* dominated stands in São Miguel Island, sampled using a total of 36 5 × 5 m quadrats, corresponding to six 900 m² plots.

Plot	Total number of trees	*P. undulatum* number (%)		Total density (trees/ha)	*P. undulatum* (trees/ha)
1	31	24	77.4	344	267
2	124	84	67.7	1378	933
3	388	381	98.2	4311	4233
4	81	76	93.8	900	844
5	455	366	80.4	5056	4067
6	351	266	75.8	3900	2956

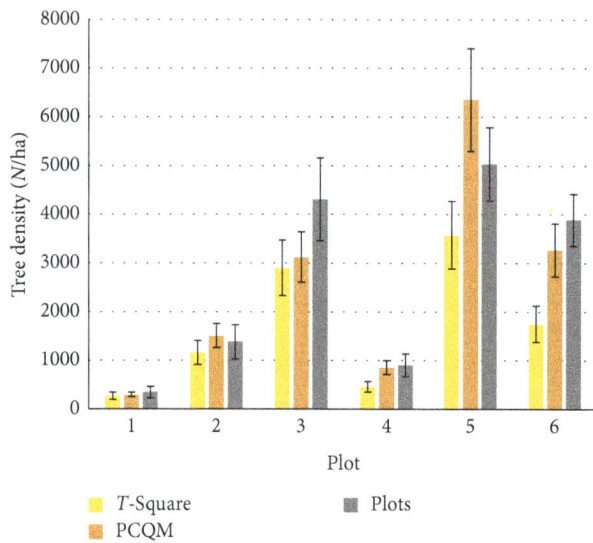

FIGURE 3: Tree density estimates and 95% confidence intervals obtained using the data collected at each plot, for *T*-Square sampling (λ_2), Point Centered Quarter Method (λ_5), and quadrat counts.

$P < 0.001$; and plot 6, $F = 2904$ and $P < 0.001$. In a more detailed analysis, comparing each distance based estimator with the quadrat estimates, *T*-Square and PCQM were consistently more accurate than *N*-tree (Figures 4–6). *T*-Square estimators λ_1, λ_2, and λ_3 were much more accurate and precise than estimator λ_4 that greatly overestimated tree density (Figures 4–7).

The PCQM estimators performed with mix results. In general λ_5 and λ_7 produced more accurate estimates than λ_6 (Figure 4). Overall, the best estimators, in terms of originating a consistently low difference towards the benchmark, include λ_1 and λ_5, (Figure 4). Also, the estimators that showed the lowest differences towards the benchmark, based on the results for the six plots, were λ_1 to λ_3, λ_5, and λ_7 (Figure 5). In the same sense, the estimators that scored the best approximations towards the benchmark more often, based on the six plots, were again λ_1 to λ_3 and λ_5 (Figure 6).

Regarding precision, the more accurate estimators, mentioned above, were also among those with lower CV as compared to the values obtained with quadrat counts (Figure 7).

3.3. Spatial Pattern Analysis.

Regarding *T*-Square sampling, the test of significance for *C* was above the critical value (1.96) for plot 2 only. For H_t, according to critical values for this test statistic, plots 3 and 5 showed regular patterns, plots 1 and 4 showed aggregation, and plots 2 and 6 showed a random pattern (Table 4). Regarding PCQM, for plots 1 and 4, *z* was above the critical value (1.96). For I_e, according to critical values for this test statistic, plots 5 and 6 showed regular pattern, plots 1 and 4 an aggregated pattern, and plots 2 and 3 a random pattern (Table 4).

For quadrat counts, I_p indicated random patterns for all plots, although with several values close to the limits of the 95% confidence interval (i.e., −0.5 to 0.5), for which I_c already indicated a significant departure from randomness (Table 4).

4. Discussion

4.1. Tree Densities.

The tree density values (trees/ha of *P. undulatum*) obtained in the present research are within the range of the results obtained for *P. undulatum* stands in Graciosa Island in a previous survey using PCQM [41]. This confirms the wide variation in tree density that can be found in the Azorean exotic woodland dominated by *P. undulatum* and will have implications in estimating and managing biomass production [38, 41]. Such density variations are dictated by the underlying physical environment and past disturbance history. The results also confirm the potential for the biomass valorization of this species, which together with other uses, such as honey and compost production [55, 56] and reforestation measures, could be included in a broader management program for this invasive plant.

4.2. Sampling Methods.

In terms of accuracy of results, when compared with quadrat counts, the *T*-Square and PCQM estimators provided the best results. *T*-Square estimators evolved as methods to remove some bias due to nonrandomness associated with the nearest neighbor distance measurement [13]. The *T*-Square estimators performed generally slightly better and globally more consistently than the other estimators. The usual *T*-Square estimator, λ_2, showed a similar performance regarding λ_1 and λ_3. However, in studies performed by Engeman et al. [21], λ_2 performed slightly better than λ_3 and λ_4. In our study λ_4 appeared to be inaccurate and should therefore not be used in Azorean forests dominated by *P. undulatum*. The two compound estimators λ_2 and λ_3 suggested by Byth [14] and Diggle [20], respectively, were developed to create a more robust estimator using *T*-Square distance measurements that would perform better in a variety of spatial patterns, relative to more simple estimators utilizing only one distance measurement [13]. Both estimators showed a similar performance in our study, while some authors argue in favor of λ_2 [14]. *T*-Square measurements are essentially nearest neighbor measurements modified only by restricting the direction in which measurements are made but have been shown to be accurate in many situations [29]. However, in our study *T*-Square estimates tended to underestimate the results obtained with the quadrat method (used as a benchmark). This should be further analyzed and taken into account in future studies, although *T*-Square sampling is simple to

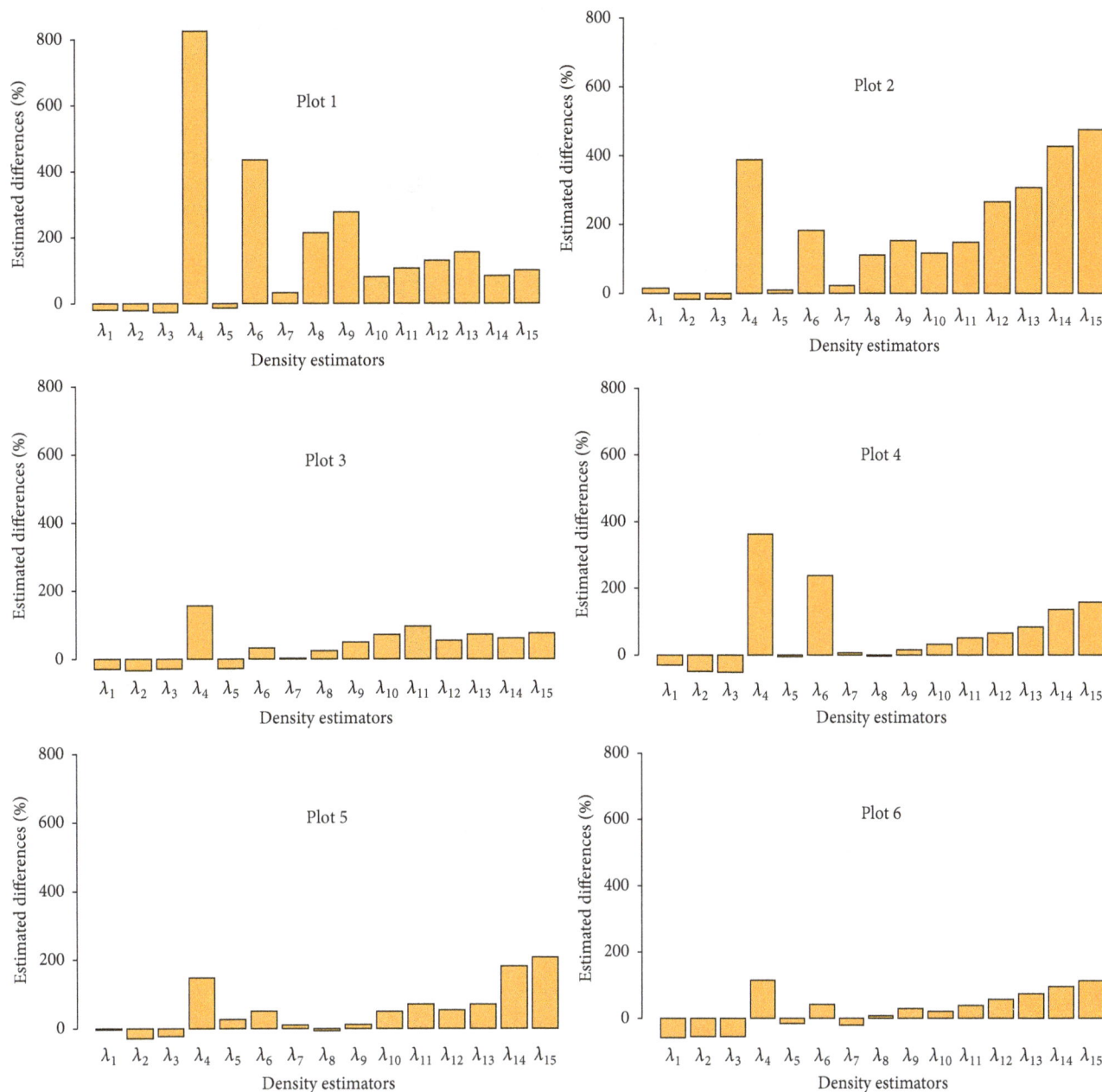

FIGURE 4: Comparison of distance methods and respective estimators (T-Square, λ_1–λ_4; PCQM, λ_5–λ_7; N-tree, λ_8–λ_{15}) with quadrat counts, for tree density estimates in all six *Pittosporum undulatum* stands. The bars represent the estimated relative difference between each distance method and the estimated tree density using quadrat counts, as obtained from contrast analysis applied after ANOVA (divided by tree density estimated by quadrat counts and expressed as a percentage).

TABLE 4: Summary of the analysis performed regarding the spatial distribution patterns in the six plots sampled at *Pittosporum undulatum* stands. Significant deviations from a random pattern are indicated in bold (aggregated) or in italic (regular).

Plots	T-Square			PCQM			Quadrats			
	C	z	H_t	I	z	I_e	I_d	I_p	I_c	χ^2
Plot 1	0.57	1.34	**1.37**	**3.12**	**3.38**	**1.61**	0.93	−0.07	0.94	32.87
Plot 2	**0.62**	**2.29**	1.23	2.13	0.39	1.26	1.34	0.50	**2.20**	**76.90**
Plot 3	0.51	0.10	*1.19*	2.27	0.81	1.26	1.27	0.50	**3.94**	**137.89**
Plot 4	0.58	1.55	**1.31**	**2.98**	**2.97**	**1.39**	1.18	−0.49	1.41	49.22
Plot 5	0.59	1.64	*1.15*	1.35	−1.96	*1.08*	1.13	0.50	**2.66**	**93.04**
Plot 6	0.55	0.96	1.28	1.39	−1.83	*1.07*	1.07	0.50	**1.73**	**60.40**

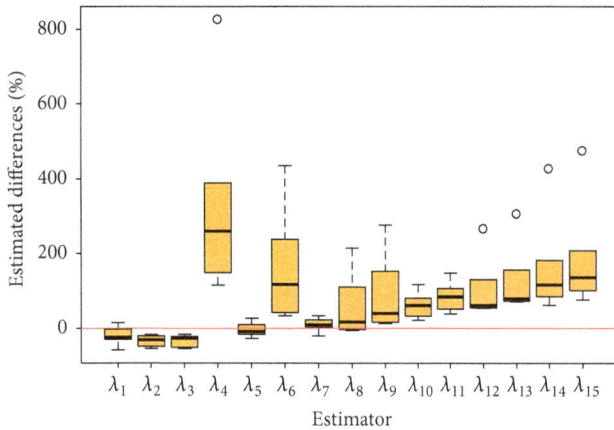

FIGURE 5: Box-plot representing the distribution of mean relative differences in tree density estimation, between each estimator and the quadrat counts, based on the data from the six *Pittosporum undulatum* plots.

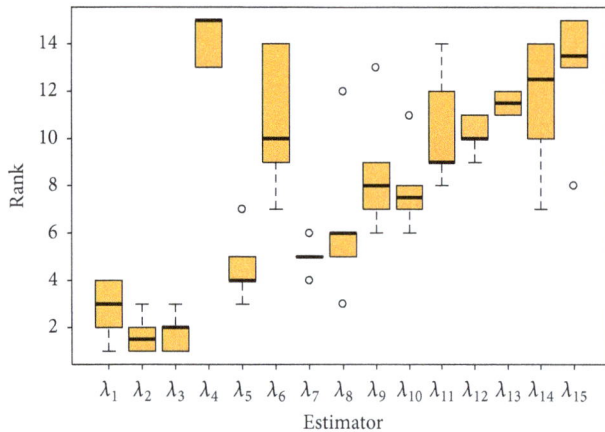

FIGURE 6: Box-plot representing the ranks of each estimator, based on the results for the six *Pittosporum undulatum* plots. A high rank (i.e., a maximum of 1) indicates that a particular estimator was closest to the benchmark more often.

implement in the field, and this simplicity is often preferred by field workers.

Point Centered Quarter Method is often used in forestry, and we found that the estimator λ_5, given by Pollard [15], was among the most accurate among those tested. The advantage of this estimator is that it is very efficient in field sampling. Previous work [2, 3] concluded that PCQM was capable of yielding accurate results and was least susceptible to subjective bias than other methods [1]. However, it has the disadvantage of being very susceptible to bias whenever the spatial pattern is not random [15].

Our results showed that N-tree method performed badly in all plots, showing a considerable overestimation of tree density. Klein and Vilkco [57] proposed two new estimators but it seems that these estimators still need more field work validation. Zobeiry [58] showed that N-tree sampling techniques are not suitable for the natural forest of northern Iran because they consistently underestimated the true population

parameters, a clumped pattern being the most reasonable cause of this behavior. However, N-tree sampling performed well in others studies [1].

4.3. Spatial Pattern. In temperate forests, tree species can have random, aggregated, and even regular spatial patterns, to different degrees, and the spatial pattern of a given species may vary from place to place [33]. The major weakness of density estimators is that their bias is dependent on the spatial distribution of the population [10]. Some distance estimators have been shown to be unbiased over a wide range of spatial patterns if the estimators are adjusted according to the spatial pattern. However, this adjustment would need additional tests to determine a population's spatial distribution before estimating density [10]. T-Square sampling has been largely recommended for spatial pattern analysis based on both simulated patterns and mapped field data [26, 29, 31, 48]. However, the statistics used in our study showed somewhat irregular results. This might be explained if *P. undulatum* stands correspond to a spatial distribution with a slight deviation from randomness, that is, a small degree of aggregation. This was evident in the results obtained for the parameters used for spatial pattern analysis from quadrat counts, with several values very close to the limit for randomness in the standardized Morisita index, which already corresponded to a significant deviation from randomness when using the variance to mean ratio. Therefore, the studied plots are more likely to correspond to random or slightly aggregated distribution patterns. Among Azorean native trees, the aggregate pattern is predominant in saplings but adults are mostly randomly distributed [33]. Aggregate patterns of distribution are reported as the most commonly observed in nature [6, 59], caused by environmental heterogeneity [6] and nonlinearity of biological processes [60] and density-dependent mortality/self-thinning [33]. The main reasons leading to a clustered pattern in a population are the behavioral characteristics of the species and intra- and interspecific relationships [61]. Intraspecific aggregation should limit the mean species richness of the communities to a level less than what it would be if individuals were randomly distributed among the communities [59]. Spatial structure of forest is a complicated characteristic due to the complex factors influencing it [25]: climate, micro-site mosaic, regeneration methods, natural mortality of individuals, biological and ecological characteristics of organisms, natural disturbance (e.g., fire, wind, and landslides), and human activities.

Human activity is an important factor in managed stands, such as the case of *P. undulatum* that was widely cultivated throughout the world, including Atlantic (e.g., Azores, Jamaica) and Pacific islands (e.g., Hawaii) and South Africa [62]. It has become dominant in moist disturbed secondary forests and some primary forests from low to middle elevations [36–39]. Perhaps the main effect of *P. undulatum* comes from its ability to penetrate natural ecosystems by the competitive exclusion of native species [63], regenerating readily and competitively in forest gaps [64] and having dense shade tolerant seedlings that outcompete native seedlings, which affects its distribution pattern [65]. Seedling and sapling

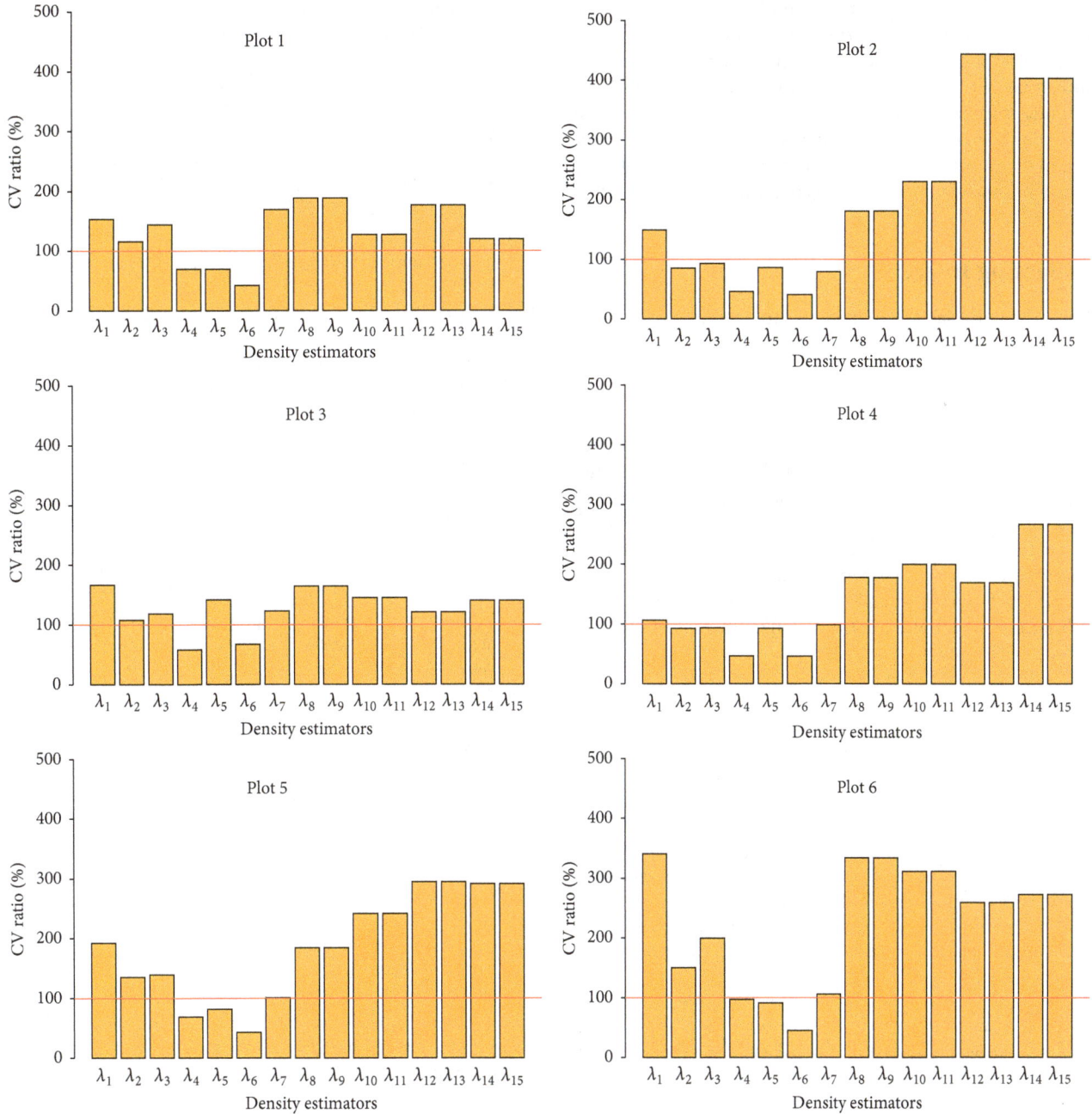

FIGURE 7: Comparison of the precision obtained when using distance methods or quadrat counts to estimate tree density. The bars represent the ratio between the coefficient of variation (CV) obtained for each estimator and the CV obtained when using quadrat counts. The 100% red line corresponds to the same CV for both the distance estimator and the quadrats.

aggregation may also result from the type of dispersion and seed rain [66]. Seed and seedling concentration may occur, for example, in places where animals frequently defecate [67] which is especially important since most Azorean tree species are dispersed by birds [68].

5. Conclusion

Distance methods also known as plotless sampling techniques were introduced because of the practical difficulties

sometimes raised by quadrat sampling [69]. Plotless sampling is considerably more efficient than quadrat sampling, since searching and counting individuals in a large quadrat are very time consuming. However, accuracy and precision are important factors in determining an appropriate sampling method. This is because bias or imprecision in initial density estimates will propagate through derived metrics, such as basal area, biomass, and carbon storage. Among the tested methods, T-Square sampling ranked more often as the closest estimator to the benchmark method. It is recommended for aggregated spatial patterns (a limitation for the use of PCQM). According to our results aggregation might occur in some of the stands. Moreover, T-Square sampling provides more independent tree measures than PCQM (only 2 per sampling station, instead of 4), which might be useful when other parameters are also being sampled. Therefore, we suggest that T-Square sampling might be used to estimate tree density in management programs dedicated to *P. undulatum* in the Azores.

Competing Interests

The authors declare that there is no conflict of interests regarding the publication of this paper.

Acknowledgments

The first author would like to acknowledge NATURAL-REASON, LDA, a company devoted to biofuels production, for supporting her research grant and for its financial contribution to the ongoing research project. This work is also funded by FEDER funds through the Operational Programme for Competitiveness Factors (COMPETE) and by National Funds through FCT (Foundation for Science and Technology) under the UID/BIA/50027/2013 and POCI-01-0145-FEDER-006821.

References

[1] B. Kiani, A. Fallah, M. Tabari, and S. M. Hosseini, "A comparison of distance samling methods in Saxaul (*Halloxylon Ammodendron* C.A. Mey Bunge) shrub-lands," *Polish Journal of Ecology*, vol. 61, pp. 207–219, 2013.

[2] S. L. Beasom and H. H. Haucke, "A comparison of four distance sampling techniques in South Texas Live oak mottes," *Journal of Range Management*, vol. 28, no. 2, pp. 142–144, 1975.

[3] G. Cottam and J. T. Curtis, "The use of distance measures in phytosociological sampling," *Ecology*, vol. 37, no. 3, pp. 451–460, 1956.

[4] Y. Askari, E. S. Kafash, and D. Rezaei, "Evaluation of different sampling methods to study of shrub density in Zagros forest," *European Journal of Experimental Biology*, vol. 3, pp. 121–128, 2013.

[5] V. Lieven, F. Vancoillie, and R. D. Wulf, "Object based forest stand density estimation from very high resolution optical imagery using wavelet based texture measures," *International Archives of Photogrammetry, Remote Sensing and Spatial Information Sciences*, vol. 36, no. 4-C42, article 13, 2006.

[6] C. J. Krebs, *Ecological Methodology*, Addison Wesley Longman, New York, NY, USA, 2nd edition, 1999.

[7] K. Mitchell, *Quantitive Analysis by the Point-Centered Quarter Method*, Department of Mathematics and Computer Science Hobart and William Smith Colleges, Geneva, Switzerland, 2007.

[8] V. Kint, D. W. Robert, and L. Noël, "Evaluation of sampling methods for the estimation of structural indices in forest stands," *Ecological Modelling*, vol. 180, no. 4, pp. 461–476, 2004.

[9] S. Magnussen, C. Kleinn, and N. Picard, "Two new density estimators for distance sampling," *European Journal of Forest Research*, vol. 127, no. 3, pp. 213–224, 2008.

[10] D. Unger, J. P. Stovall, B. P. Oswald, D. Kulhavy, and I.-K. Hung, "A test of the mean distance method for forest regeneration

assessment," *Mathematical and Computational Forestry and Natural-Resource Sciences*, vol. 6, no. 2, pp. 54–61, 2014.

[11] N. Picard, A. M. Kouyaté, and H. Dessard, "Tree density estimations using a distance method in Mali Savanna," *Forest Science*, vol. 51, no. 1, pp. 7–18, 2005.

[12] V. C. Lessard, D. D. Reed, and N. Monkevich, "Comparing n-tree distance sampling with point and plot sampling in northern Michigan forest types," *Northern Journal of Applied Forestry*, vol. 11, pp. 12–16, 1994.

[13] J. E. Besag and J. T. Gleaves, "On the detection of spatial pattern in plant communities," *Bulletin of the International Statistical Institute*, vol. 45, pp. 153–158, 1973.

[14] K. Byth, "On robust distance-based intensity estimators," *Biometrics*, vol. 38, no. 1, pp. 127–135, 1982.

[15] J. H. Pollard, "On distance estimators of density in randomly distributed forests," *Biometrics*, vol. 27, no. 4, pp. 991–1002, 1971.

[16] M. Prodan, "Punktstichprobe für die forsteinrichtung für die forsteinrichtung. (A point sample for forest management planning)," *Forst Holzwirt*, vol. 23, pp. 225–226, 1968.

[17] G. Cottam and J. T. Curtis, "A method for making rapid surveys of woodlands by means of pairs of randomly selected trees," *Ecology*, vol. 30, no. 1, pp. 101–104, 1949.

[18] K. R. Parker, "Density estimation by variable area transect," *The Journal of Wildlife Management*, vol. 43, no. 2, p. 484, 1979.

[19] M. Morisita, "Estimation of population density by a spacing method," *Memoirs of the Faculty of Science, Kyushu University*, vol. 1, pp. 187–197, 1954.

[20] P. J. Diggle, "Robust density estimation using distance methods," *Biometrika*, vol. 62, pp. 39–48, 1975.

[21] R. M. Engeman, R. T. Sugihara, L. F. Pank, and W. E. Dusenberry, "A comparison of plotless density estimators using Monte Carlo simulation," *Ecology*, vol. 75, no. 6, pp. 1769–1779, 1994.

[22] W. G. Warren and C. L. Batcheler, "The density of spatial patterns: robust estimation through distance methods," in *Spatial and Temporal Analysis in Ecology*, R. M. Cormack and J. K. Ord, Eds., pp. 247–270, 1979.

[23] X. Zhu and J. Zhang, "Quartered neighbor method: a new distance method for density estimation," *Frontiers of Biology in China*, vol. 4, no. 4, pp. 574–578, 2009.

[24] M. Morisita, "A new method for the estimation of density by spacing method applicable to non-randomly distributed populations," *Physiology and Ecology*, vol. 7, pp. 134–144, 1957.

[25] J. Szmyt, "Spatial statistics in ecological analysis: from indices to functions," *Silva Fennica*, vol. 48, no. 1, pp. 1–31, 2014.

[26] P. J. Diggle, "Note on the Clark and Evans test of spatial randomness," *Appendix to Hodder and Orton*, pp. 246–248, 1976.

[27] C. Kleinn and F. Vilčko, "Design-unbiased estimation for point-to-tree distance sampling," *Canadian Journal of Forest Research*, vol. 36, no. 6, pp. 1407–1414, 2006.

[28] P. J. Clark and F. C. Evans, "Distance to nearest neighbor as a measure of spatial relationships in populations," *Ecology*, vol. 35, no. 4, pp. 445–453, 1954.

[29] H. B. Özel, "The spatial pattern analyses of natural pure oriental beech (*Fagus orientalis* Lipsky.) stands in the bartin-yenihan district in Turkey," *Biyoloji Bilimleri Araştırma Dergisi*, vol. 7, pp. 34–37, 2014.

[30] M. R. T. Dale, P. Dixon, M.-J. Fortin, P. Legendre, D. E. Myers, and M. S. Rosenberg, "Conceptual and mathematical relationships among methods for spatial analysis," *Ecography*, vol. 25, no. 5, pp. 558–577, 2002.

[31] P. J. Diggle, *Statistical Analysis of Spatial Point Patterns*, Mathematics in Biology, Academic Press, London, UK, 1983.

[32] M. A. Wulder, J. C. White, K. O. Niemann, and T. Nelson, "Comparison of airborne and satellite high spatial resolution data for the identification of individual trees with local maxima filtering," *International Journal of Remote Sensing*, vol. 25, no. 11, pp. 2225–2232, 2004.

[33] R. B. Elias, E. Dias, and F. Pereira, "Disturbance, regeneration and the spatial pattern of tree species in Azorean mountain forests," *Community Ecology*, vol. 12, no. 1, pp. 23–30, 2011.

[34] L. Silva and C. W. Smith, "A quantitative approach to the study of non-indigenous plants: an example from the Azores Archipelago," *Biodiversity and Conservation*, vol. 15, no. 5, pp. 1661–1679, 2006.

[35] L. Silva, E. Ojeda-Land, and J. L. Rodríguez-Luengo, *Invasive Terrestrial Flora and Fauna of Macaronesia. Top 100 in Azores, Madeira and Canaries*, ARENA, Ponta Delgada, Portugal, 2008.

[36] H. Costa, V. Medeiros, E. B. Azevedo, and L. Silva, "Evaluating ecological-niche factor analysis as a modelling tool for environmental weed management in island systems," *Weed Research*, vol. 53, no. 3, pp. 221–230, 2013.

[37] J. Hortal, P. A. V. Borges, A. Jiménez-Valverde, E. B. de Azevedo, and L. Silva, "Assessing the areas under risk of invasion within islands through potential distribution modelling: the case of *Pittosporum undulatum* in São Miguel, Azores," *Journal for Nature Conservation*, vol. 18, no. 4, pp. 247–257, 2010.

[38] P. Lourenço, V. Medeiros, A. Gil, and L. Silva, "Distribution, habitat and biomass of *Pittosporum undulatum*, the most important woody plant invader in the Azores Archipelago," *Forest Ecology and Management*, vol. 262, no. 2, pp. 178–187, 2011.

[39] H. Costa, S. C. Aranda, P. Lourenço, V. Medeiros, E. B. D. Azevedo, and L. Silva, "Predicting successful replacement of forest invaders by native species using species distribution models: the case of *Pittosporum undulatum* and *Morella faya* in the Azores," *Forest Ecology and Management*, vol. 279, pp. 90–96, 2012.

[40] H. Costa, M. J. Bettencourt, C. M. N. Silva, J. Teodósio, A. Gil, and L. Silva, "Invasive alien plants in the azorean protected areas: invasion status and mitigation actions," in *Plant Invasions in Protected Areas: Patterns, Problems and Challenges*, vol. 7 of *Invading Nature—Springer Series in Invasion Ecology*, pp. 375–394, Springer, Amsterdam, The Netherlands, 2013.

[41] A. Teixeira, C. Mir, L. Borges Silva, and L. Hahndorf, "Invasive woodland resources in the azores: biomass availability for 100% renewable energy supply in graciosa Island," in *Proceedings of the 23rd European Biomass Conference and Exhibition*, pp. 14–23, Vienna, Austria, 2015.

[42] E. B. Azevedo, "Condicionantes Dinâmicas do Clima do Arquipélago dos Açores. Elementos para o seu AÇOREANA," *Boletim da Sociedade de Estudos Açoreanos Afonso Chaves*, vol. 9, no. 3, pp. 309–317, 2001.

[43] J. L. Gaspar, J. E. Guest, G. Queiroz et al., "Eruptive frequency and volcanic hazards zonation in São Miguel Island, Azores," in *Volcanic Geology of São Miguel Island (Azores Archipelago)*, J. L. Gaspar, J. E. Guest, A. M. Duncan, F. J. A. Barriga, and D. K. Chester, Eds., vol. 44 of *Geological Society, London, Memoirs*, pp. 155–166, The Geological Society, 2015.

[44] E. A. Lima, M. Machado, and J. Ponte, "Geological heritage management: monitoring the Azores Geopark geosites," *Comunicações Geológicas*, vol. 101, no. 3, pp. 1295–1298, 2014.

[45] G. A. F. Seber, *The Estimation of Animal Abundance*, Charles Griffin and Company, London, UK, 2nd edition, 1982.

[46] M. J. Fortin, M. R. T. Dale, and J. V. Hoef, "Spatial analysis in ecology," in *Encyclopedia of Environmetrics*, vol. 4, pp. 2051–2058, John Wiley & Sons, Chichester, UK, 2002.

[47] W. G. S. Hines and R. J. O'Hara Hines, "The Eberhardt statistic and the detection of nonrandomness of spatial point distributions," *Biometrika*, vol. 66, no. 1, pp. 73–79, 1979.

[48] J. A. Ludwig and J. F. Reynolds, *Statistical Ecology: A Primer on Methods and Computing*, John Wiley & Sons, New York, NY, USA, 1988.

[49] R. B. Johnson and W. J. Zimmer, "A more powerful test for dispersion using distance measurements," *Ecology*, vol. 66, no. 5, pp. 1669–1675, 1985.

[50] L. L. Eberhardt, "Some developments in 'distance sampling'," *Biometrics*, vol. 23, no. 2, pp. 207–216, 1967.

[51] M. Morisita, "I_σ-index, a measure of dispersion of individuals," *Researches on Population Ecology*, vol. 4, no. 1, pp. 1–7, 1962.

[52] B. Efron and R. J. Tibshirani, *An Introduction to the Bootstrap*, vol. 57 of *Monographs on Statistics and Applied Probability*, Chapman & Hall, New York, NY, USA, 1993.

[53] A. C. Davison and D. Kuonen, "An introduction to the bootsrap with applications in R," *Statistical Computing & Statistical Graphics Newsletter*, vol. 13, pp. 6–11, 2002.

[54] R Core Team, *R: A Language and Environment for Statistical Computing*, R Foundation, 2014.

[55] J. Tavares and J. Baptista, "Cultura do Ananás em Estufa Ilha S. Miguel—Açores (Portugal)," in *Ponta Delgada: Produtos—Cooperativa de Produtores de Frutas; Produtos Hortícolas e Florícolas de S. Miguel*, p. 195, 2004.

[56] Azores Government, "Caderno Especificações DOP Mel dos Açores," 2016, https://www.azores.gov.pt/NR/rdonlyres/10DBC106-63C8-4ED8-BD69-9FB765D835BB/452723/Caderno-EspecificaesDOPMeldosAores4.pdf.

[57] C. H. Klein and F. Vilkco, "On estimation in *k*-tree sampling," in *Proceedings of the 7th Annual Forest Inventory and Analysis Symposium*, pp. 203–208, Portland, Ore, USA, 2005.

[58] Y. M. Zobeiry, "Tree sampling in natural forests of northern Iran," *The Forestry Chronicle*, vol. 54, no. 3, pp. 171–172, 1978.

[59] J. A. Veech, "Analyzing patterns of species diversity as departures from random expectations," *Oikos*, vol. 108, no. 1, pp. 149–155, 2005.

[60] S. W. Pacala and S. A. Levin, "Biological generated spatial pattern and the coexistence of competing species," in *Spatial Ecology: The Role of Space in Population Dynamics and Interspecific Interactions*, D. Tilman and P. Kareiva, Eds., pp. 204–232, Princeton University Press, Princeton, NJ, USA, 1997.

[61] D. S. Matteucci and A. Colma, *Metodologia para el Estúdio de la Vegetación*, Secretaria General de la Organizacion de los Estados Americanos, Washington, DC, USA, 1982.

[62] P. Binggeli, *An Overview of Invasive Woody Plants in the Tropics*, vol. 13 of *School of Agricultural and Forest Sciences Publication Number*, Unversity of Wales, Bangor, UK, 1998.

[63] T. Goodland and J. R. Healey, *The invasion of Jamaican montane rainforests by the Australian tree* Pittosporum undulatum, School of Agricultural and Forest Sciences, University of Wales, Bangor, UK, 1996.

[64] R. B. Elias and E. Dias, "Gap dynamics and regeneration strategies in *Juniperus-Laurus* forests of the Azores Islands," *Plant Ecology*, vol. 200, no. 2, pp. 179–189, 2009.

[65] P. J. Bellingham, E. V. J. Tanner, and J. R. Healey, "Hurricane disturbance accelerates invasion by the alien tree *Pittosporum*

undulatum in Jamaican montane rain forests," *Journal of Vegetation Science*, vol. 16, no. 6, pp. 675–684, 2005.

[66] A. F. Souza and F. R. Martins, "Spatial distribution of an undergrowth palm in fragments of the Brazilian Atlantic Forest," *Plant Ecology*, vol. 164, no. 2, pp. 141–155, 2003.

[67] P. J. van der Meer, F. J. Sterck, and F. Bongers, "Tree seedling performance in canopy gaps in a tropical rain forest at Nouragues, French Guiana," *Journal of Tropical Ecology*, vol. 14, no. 2, pp. 119–137, 1998.

[68] B. Rumeu, M. Nogales, R. B. Elias et al., "Contrasting phenology and female cone characteristics of the two Macaronesian island endemic cedars (*Juniperus cedrus* and *J. brevifolia*)," *European Journal of Forest Research*, vol. 128, no. 6, pp. 567–574, 2009.

[69] P. J. Diggle, *Statistical analysis of spatial and spatio-temporal point patterns*, vol. 128 of *Monographs on Statistics and Applied Probability*, CRC Press, Boca Raton, Fla, USA, Third edition, 2014.

17

Economic Valuation of Nontimber Forest Products under the Changing Climate in Kilombero District, Tanzania

Chelestino Balama,[1,2] Suzana Augustino,[1] Danford Mwaiteleke,[3]
Leopord P. Lusambo,[3] and Fortunatus B. S. Makonda[1]

[1]Department of Wood Utilization, Sokoine University of Agriculture, P.O. Box 3014, Chuo Kikuu, Morogoro, Tanzania
[2]Directorate of Forest Utilisation Research, Tanzania Forestry Research Institute, P.O. Box 1854, Morogoro, Tanzania
[3]Department of Forest Economics, Sokoine University of Agriculture, P.O. Box 3011, Chuo Kikuu, Morogoro, Tanzania

Correspondence should be addressed to Chelestino Balama; balamapc@gmail.com

Academic Editor: Kihachiro Kikuzawa

Sustainable collection of Nontimber Forest Products (NTFPs) for trade is an appropriate measure to increase people's adaptive capacity against adverse effects of climate change. However, information on the economic value for NTFPs for subsistence use and trade under the changing climate is inadequate, particularly in households around Iyondo Forest Reserve (IFR), in Kilombero District, Tanzania. The study identified and quantified NTFPs used for subsistence and trade, estimated its economic value, and examined factors influencing supply of NTFPs at household level. Data were collected through Focus Group Discussions, key informant interviews, questionnaire survey of 208 sample households, and spot market analysis to randomly selected NTFPs collectors, sellers, and buyers. The study identified 12 NTFPs used for subsistence and trade, which was evaluated in terms of the mean annual value per household. The mean annual value of the identified NTFPs ranged from TZS 4700 to 886 600. The estimated economic value of the studied NTFPs was TZS 51.4 billion (USD 36 million). The supply of NTFPs at household level was influenced by distance to the forest, change in forest management regime, seasonality, and change in rainfall pattern. NTFPs around IFR have high economic value which portrays the potential of developing them to enhance households' adaptive capacity against climate change adverse effects.

1. Introduction

Nontimber Forest Products (NTFPs) play a significant and critical role in improving livelihoods to a large part of the world's population [1–3] particularly at the current change in climate. Climate related hazards and more subtle trends interact with nonclimatic stressors such as multiple deprivations, market shifts, conflict and insecurity, and loss of access to resources to exacerbate the vulnerability of agricultural systems [4, 5]. While not all the poor are equally affected by such stressors, nonpoor groups can also be vulnerable. Rural populations living in poverty often suffer more than others when extreme climate change events like heavy rains and floods, prolonged dry spells, and extreme heat occur [5]. These climate extreme events have been noted to affect much

natural asset, particularly agriculture which is the main livelihood of the rural people in Africa [6–8].

Agricultural production and food security in many parts of Africa are likely to be severely compromised by climate change, in particular by damaging high temperatures and the greater incidence of drought [9]. Agriculture is a significant household activity in many countries of Sub-Saharan Africa because the majority of the people who live in rural areas depend on both substance and local trade in agricultural produce [8]. Much of the cropland is rain-fed and is therefore vulnerable to climate change that is characterised by drought and floods that frequently cause crop failure. Households with low adaptive capacity are thought to be more vulnerable to the adverse effects of climate change, which contribute to the loss of their natural resources [10, 11]. According to URT

[12], the change in temperature and precipitation patterns has led to increased risk of recurrent droughts and devastating floods and has been threatened with biodiversity loss, an expansion of plant and animal diseases, and a number of potential challenges for public health. Such recurrent risks have affected the main conventional livelihoods of the households, thus increasing reliance on NTFPs for both subsistence use and cash income [13]. Increased reliance on NTFPs calls for economic valuation for rational planning of the resource. The economic value concept entailed the value of the goods that would be missing if a forest suddenly disappeared [14]. This can also be referred to as the total cost to households in the study area if the forest is completely removed or total access is denied.

The need to identify the economic value of NTFPs in rural households in the developing countries is gaining importance in both the conservation and development phenomena [15–18]. NTFPs are reported to significantly contribute to economic benefits of rural households in developing countries in three major ways: first, providing domestic subsistence and consumption requirements [19] for increased disposable income to the household; second, serving as an immediate safety net against experienced climate change adverse effects, constituting an important part of adaptive capacity [11, 20, 21]; and, third, contributing to direct monetary benefits through trade [17, 22]. Studies show that most attempts to valuate NTFPs have been undertaken at the local level [22]. To estimate the contribution of NTFPs to social and economic well-being, their valuation at broader scales is acclaimed as pertinent [23, 24], due to being among the effective climate change adaptation strategies in Africa [13]. Heubach et al. [18] in Northern Benin argued that species with high economic value can buffer possible cash shortfalls; this especially holds true for women since they are the main collectors and traders of NTFPs.

Sustainable collection of NTFPs for trade is expected to increase the adaptive capacity of households in rural Tanzania to climate change by majority of people including those surrounding IFR. Schaafsma et al. [25] revealed that local people bear economic loss if they are denied of collection of NTFPs. Similarly, Msalilwa et al. [26] revealed that the demand for some NTFPs was increasing by forest dependent households around New Dabaga Ulongambi Forest Reserve in Kilolo District more than it used to be 30 years ago, due to climate change effects. This is also the case for the households around IFR in Kilombero District where demand of NTFPs by households to increase resilience to adverse effects of changing climatic conditions is significantly eminent [27]. However there is increasing demand of NTFPs, and information on its economic value for subsistence use and trade in terms of benefits and its importance to livelihoods of households under the changing climate is inadequate in Tanzania including IFR and its surroundings. Also, most of the NTFPs are collected, traded, and consumed outside the cash economy and therefore are not adequately captured in national economy statistics [8]. Lack of economic contribution of NTFPs to the GDP has led to insufficient recognition in national planning for local livelihoods adaptation measures. There is a growing need at national and international policy

levels for projections at large spatial scales of the economic values that households derive from forests, including the collection of NTFPs [28]. It is against this background that the study to estimate economic value of NTFPs in IFR and its implication on adaptation to climate change in Kilombero District, Tanzania, was designed. Specifically the study identified and quantified NTFPs that are used for subsistence and trade, estimated their economic value, and examined factors influencing their supply in the study area. Estimating the economic value of NTFPs has advantage since it helps to ascertain the true value of the standing forest, thus leading to more rational decisions about the alternative uses of the forest [29, 30]. The results from this study are also useful to various stakeholders in the overall implementation of the National REDD+ strategy in Tanzania to enhance the adaptive capacity of households through promotion and development of NTFPs as carbon cobenefits for sustainable management of natural resources. Ferreira et al. [31] argued that sustainable management of NTFPs plays two potential roles in REDD+. Firstly, it helps to reduce NTFPs extraction that contributes to forest degradation and associated emission. Secondly, it enables sustainable exploitation of NTFPs that can contribute to reducing degradation and deforestation caused by other factors, through (a) increasing value of the forests and thereby reducing pressure on them and (b) providing alternative sources of income to activities that deplete forest carbon stock. However, the potential for increasing the sustainability of NTFPs use depends on the products extracted and the characteristics of the species and the forests [31]. For example, harvesting of dead wood, fruits, seeds, and mushrooms has been pointed out to have the high potential for sustainability.

2. Methodology

2.1. Study Area. Kilombero District is among the six districts of Morogoro region [32]. A large part of Kilombero District is located in a floodplain area, which is important for main economic activities of the people including agriculture, livestock keeping, fishing, and wild game hunting [33]. However, these livelihood activities have been affected by the recurrent stresses due to changing climatic conditions including floods, dry spells during rainy season, extreme heat, and some other stresses, which have adverse effects on the livelihoods of the people [34–36]. Specifically the study was conducted in four villages of Kilombero District, namely, Mpofu, Igima, Njage, and Mngeta (Figure 1). The choice of the study area was based on villages that surround Iyondo Forest Reserve (IFR) which is part of Kilombero Nature Reserve (KNR) at the moment with resources important for increasing resilience of the households against adverse effects of climate change. Other villages that surround the reserve are Mbingu and Mchombe.

The climate in the study area is marked by wet and dry seasons which are further distinguished into four subseasons, namely, hot wet season from December to March, cool wet season April–June, cool dry season July-August, and hot dry season September–November. The area receives between 1200 and 1800 mm of rainfall per year and temperatures range from 26°C to 32°C [38, 39]. However, the rains are currently unpredictable, whereby the onset and cessation are

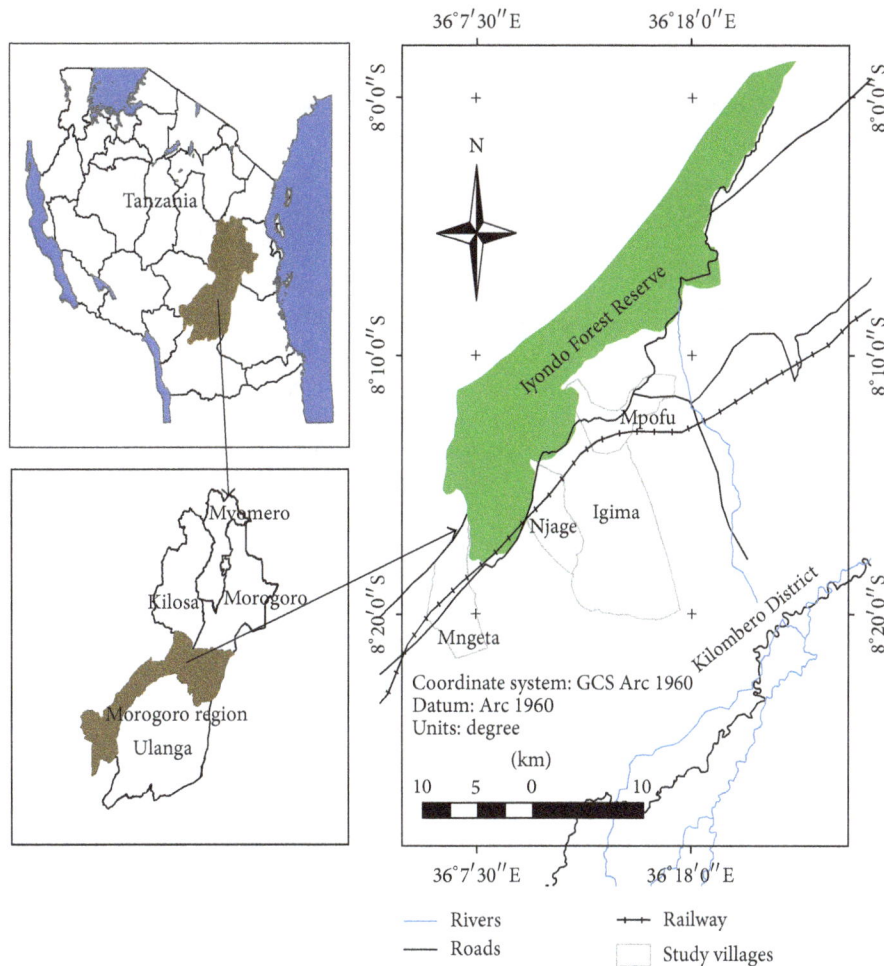

FIGURE 1: Map showing location of study villages in Kilombero District, Tanzania. Source: NBS [37].

inconstant [27]. The general topography in the study villages is a flat land with mean elevation ranging from 274 to 358 m a.s.l. Soils are mainly loamy and sandy while some cotton black soils in flooded areas are found. In hilly areas, the soils are sandy loam over crystalline rocks [40]. The main vegetation found in the study area is Miombo woodlands [40], with some open grassland areas in the floodplains [33].

Access to the study villages is through an earth road running from Ifakara Town to Mlimba as well as the railway line of Tanzania-Zambia Railway Authority (TAZARA) that extends from Dar es Salaam (Tanzania) to Kapiri Mposhi (Zambia). Ethnic groups (Table 1) are seemingly diverse but still share similar livelihoods and sociocultural norms [32]. The main economic activities in the study area are agriculture (crops grown are in Table 1) and livestock keeping.

2.2. Data Collection. Data were collected using Focus Group Discussions (FGDs), key informant interviews, household questionnaires survey, and spot market analysis. FGDs are aimed at capturing information on available NTFPs in the area, places where they are collected, unit value of each NTFP, main sources of household income, and historical use of

NTFPs. FGDs comprised 10–12 people in each village, aged 18 years and above, representing various livelihoods and sex. Key informant interviews were conducted to village leaders, village elders, NTFPs vendors, herbalists (i.e., traditional healers and medicinal plant sellers), village environmental committee members in the respective villages to supplement data collected through FGDs, and household questionnaire survey. Key informants were purposely selected based on their experience on the subject matter and special level of information they provided. Moreover, household survey was conducted in four villages that were purposely selected on the basis of their relative distance from the forest for questionnaire interviews. Moreover, household survey was conducted in four villages, whereby a total of 208 households were randomly selected based on sampling intensity of 5% (Table 2).

Pilot testing of the survey instruments was conducted among 30 randomly selected households prior to implementation of the survey and then the questions were adjusted accordingly. Pilot testing was carried out in order to improve validity of the survey tools [41]. The questionnaires were then administered to randomly selected households in each village to capture information on quantity of NTFPs collected, pattern of use (subsistence or trade), places where NTFPs are

TABLE 1: Some physical and demographic characteristics of the study area.

Characteristics	Mpofu	Igima	Njage	Mngeta
Geographic position	08°12′57″S; 36°14′33″E	08°15′59″S; 36°13′27″E	08°15′26″S; 36°10′08″E	08°19′24″S; 36°06′28″E
Mean altitude (m)	295	274	312	387
Population of people*	3 123	5 146	3 402	5 116
Average household size**	4.72	4.82	4.78	4.93
Crops grown	Main (banana, maize, rice); others (sesame, cocoa, sunflower, cassava)	Main (rice, banana); others (maize, cassava)	Main (rice, banana, maize); others (sesame, cassava)	Main (rice, maize); others (cassava, banana)
Ethnic groups	Main (Hehe, Nyakyusa, Bena, Ndali); others (Sukuma, Makua, Kerewe, Gita, Ndamba, Pogolo, Nyamwezi, Haya, Safwa, and Kinga)	Main (Ndamba, Hehe, Nyakyusa); others (Bena, Sukuma, Pogolo, Kerewe, Chaga, and Nyamwezi)	Main (Hehe, Bena, Nyakyusa, Ndamba); others (Sukuma, Pogolo, Kerewe, Gita, Chaga, and Matumbi)	Main (Ndamba, Hehe, Bena, and Sukuma); others (Pogolo, Nyakyusa, Ndali, Luguru, Kinga, and Matumbi)

*Housing and population census of 2012 [32]; **number of households in each village is indicated in Table 2.

TABLE 2: Sample size distribution in the study area.

Village	Relative distance	Number of households	Sampling intensity	Sample size (n)
Mpofu	Very close*	714	0.05	36
Igima	Far**	1 282	0.05	64
Njage	Very close	868	0.05	43
Mngeta	Far	1 293	0.05	65
Total		4 157		208

*These are villages whose households were very close to IFR with a walking distance of about 0.2 km.
**These are villages whose households were relatively far from the IFR with a walking distance of about 5 and 10 km for Igima and Mngeta, respectively.

collected, unit value of each NTFP, and factors influencing supply of NTFPs at household level.

Spot market analysis was carried out using a checklist which was randomly administered to collectors, sellers, and buyers in the local markets and around households where NTFPs were being displayed for sale. Information collected included unit value/market price and average amount of NTFPs collected per week and their potential in enhancing adaptive capacity to climate change at household level.

Secondary data on climate, mainly rainfall from the Tanzania Meteorological Agency (TMA) for a span of 30 (1980–2010) years, were used to show trend of rainfall in the study area.

2.3. Data Analysis. Qualitative data collected through FGDs and key informant interview were analysed using content analysis. The data were categorized into meaningful units and themes in keeping with the research questions. The summaries of the narrations were used in the discussion. Quantitative data including amount of NTFPs collected and unit value of NTFPs were coded, processed, and descriptively analysed using Statistical Package for Social Sciences (SPSS) computer software tools. Descriptive statistics generated percentage and means on demographic and socioeconomic characteristics; quantities and values of direct use benefit from NTFPs collected factors influencing supply of NTFPs. The results were summarized in tables to simplify interpretation. The market price method was used to value NTFPs.

Market price is an approach that is commonly used to value environmental goods and services that have established markets [17, 30]. Quantification of identified NTFPs was calculated in terms of mean annual values as indicated in the following:

$$a_m = Q_m * V_u, \tag{1}$$

where a_m is mean annual value in TZS, Q_m is mean annual quantity of NTFPs collected per household (Kg), and V_u is unit value in TZS per unit measure. This is an annual average price of a particular NTFP from different dealers in study area. The calculated unit value is assumed to be applied throughout the year. It also makes consideration of seasonal price fluctuation.

The following discounting formula was used to estimate the economic value of NTFPs that was expressed in terms of annual present value (PV):

$$PV = \frac{a\left[(1 + r)^n - 1\right]}{\left[r(1 + r)^n\right]}, \tag{2}$$

where PV is present value of NTFPs in TZS, a is the estimated annual actual value of NTFPs in TZS, and r is social discount rate. The discount rate was useful in this study because the study dealt with benefits of the forests and woodlands which are public property and attached to community values that count more than individual preference. Therefore, the social discount rate chosen was 10% which is recommended by the World Bank [42]. n is time horizon (infinity for the context of this study).

Assuming continuous flow of benefits to the community from the forest and woodlands for infinity annual series, then "n" in (2) approaches to infinity and a useful formula becomes

$$PV = \left[\frac{a}{r} \right]. \tag{3}$$

From (3), the annual actual value "a" was calculated as given in

$$a = Q_m * V_u * P_r * H_t, \tag{4}$$

where P_r are proportions of respondents using the product in percentage and H_t is total number of households in the study area. In this case a total of 8 308 households from all villages (six) surrounding the IFR were used.

In calculating the economic values of NTFPs, the following key assumptions and considerations were adhered to:

(i) Assume that each NTFP collected had the same market value.

(ii) Economic value of NTFPs was in terms of gross values because estimation of cost of harvesting of each NTFP separately seemed very involving. This is because of the nature of trips into forest and woodlands, which were multipurpose, like doing it along with farming activities and some of the NTFPs were collected jointly, for example, extraction of withies and ropes, so estimating them could lead to double counting. According to Morse-Jones et al. [43] double counting may occur where competing ecosystem services are valued separately and the values aggregated or where an intermediate service is first valued separately but also subsequently through its contribution to a final service benefit.

3. Results and Discussions

3.1. Characteristics of the Respondents. The characteristics of respondents included were age, sex, marital status, education level, and occupation (Table 3). These characteristics were reported to influence behaviour of respondents in collection and trading of NTFPs. Results showed that 65% of the respondents were aged between 31 and 60 years (Table 3). This showed that the majority of the respondents were physically and economically active to engage in various production activities including collection of NTFPs. Studies, for example, Dolisca et al. [44] and Tazeze et al. [45], show that age is significantly related to farmer's decisions during adoption to climate change adaptation strategies. The same authors also reported that as age of the household head increases the person is expected to acquire more experience in weather forecasting that helps in increasing the likelihood of practicing different adaptation strategies to climate change including collection and trade of NTFPs. This causal relationship between age and ability to decide and practice various adaptation options including collection of NTFPs was also

TABLE 3: Socioeconomic characteristics of the respondents in the study area.

Socioeconomic attribute	Response (%) $n = 208$
Age (years)	
18–30	28
31–60	65
Above 60	7
Sex	
Male	39
Female	61
Marital status	
Single	2
Married	91
Widow/widowed	4
Separated/divorced	3
Education level	
No formal education	13
Primary level	83
Secondary level	3
Postsecondary education	1
Occupation	
Peasant farming	95
Petty trade	3
Salaried employment	2

revealed during FGDs whereby people aged 60 years and above declared to have fewer adaptation strategies than youths. This is because most of the livelihood activities were claimed to be carried out manually/physically in such a way that they did not afford.

The results further showed that 61% of respondents were women and 39% were men. However, it should be noted that all women interviewed were not necessarily heads of households because in some instance heads of households (men) were not available at homes during interview for various reasons, so they were represented by their wives. During FGDs, it was revealed that collection of NTFPs like mushrooms and firewood as among the coping strategies for climate change involves both men and women. Participants in the FGDs further argued that, in the past, collection of such products was only done by women, whereby men were mainly involved in game meat hunting, collection of building poles, and timber harvesting. This implies that current collection of NTFPs aims not only for subsistence use but also for cash income generation which is used for various households needs. Regarding marital status, about 91% of the respondents were married (Table 3), implying that households with married couples have ability to carry out different livelihoods activities including collection and trade of NTFPs.

Eighty-three per cent of the respondents had primary education, whereas few of them had secondary and postsecondary education (Table 3). As most of the respondents were

TABLE 4: Mean annual value of identified NTFPs in the study area.

Products	Units	Mean annual quantity per household		Mean annual quantity per household	Unit value (TZS)	Mean annual value (TZS) per household
		Subsistence	Trade			
Firewood	Head load	115 (83)	328.3 (12)	443.3	2 000	886 600
Bush meat	Kgs	45.3 (36)	110 (38)	155.3	3 000	465 900
Wild mushroom	Kgs	68 (36)	224 (21)	292	1 000	292 000
Medicinal plants	Kgs	2.6 (36)	26.3 (2)	28.9	8 000	231 200
Honey	Litres	4.6 (10)	21.1 (22)	25.7	6 000	154 200
Poles	Pieces	27.3 (30)	52 (5)	79.3	1 000	79 300
Thatch grass	Head load	26 (45)	25 (28)	51	1 000	51 000
Ropes	Bundles	2 (23)	0	2	8 000	16 000
Wild vegetables	Kgs	28.65 (70)	0	28.65	500	14 325
Withies	Bundles	2.5 (25)	0	2.5	5 000	12 500
Wild fruits	Kgs	19.7 (51)	0	19.7	400	7 880
Tool handles	Pieces	3.7 (51)	10 (11)	4.7	1 000	4 700

Number in parenthesis is a proportion of respondents (%) of the households.
A unit of head load for firewood and thatch grass was equivalent to 16.55 ± 3.33 and 14.12 ± 3.19 Kg, respectively.

literate, it is shown that they are aware of various livelihoods related to climate change adaptation including collection and trade of NTFPs. A study by Tazeze et al. [45] found that literate farmers are more likely to respond to climate change by making best adaptation options based on preferences and influences individual decision making. On the other hand, the majority of respondents (95%) were peasant farmers followed by few respondents with other occupations. This implied that farming was the main economic activity in the study area which is claimed not stable because of climate change adverse effects. This further implied that households were probably engaged in other livelihoods like collection and trading of NTFPs in order to gain income.

3.2. Identified and Quantified NTFPs.
Results showed that 12 NTFPs ranging between food products, firewood, and construction materials were collected and traded by households living adjacent to IFR. The identified and quantified NTFPs in the study area were firewood, bush meat, wild mushrooms, medicinal plants, honey, poles, thatch grass, ropes, wild vegetables, withies, wild fruits, and tool handles (Table 4). The mean annual value of the identified NTFPs ranged from TZS 4700 to 886 600. Firewood had the highest mean annual value followed by others (Table 4). The NTFPs were collected mainly from farmlands and village forests and illegally from IFR for some NTFPs like dried firewood, mushrooms, wild vegetable, and wild fruit. The most dominant NTFPs in terms of mean annual value per household were firewood, bush meat, wild mushroom, medicinal plants, and honey.

The majority (83%) of respondents collected firewood for subsistence and trade, respectively (Table 3). A mean of 443.3 head loads per household was collected, which had a mean annual value of TZS 881 200 (Table 4). The annual mean head loads of firewood recorded in this study were higher than 98 and 99 head loads recorded by Msemwa [46] in Kilosa District and Kilonzo [47] in households around Nyanganje

Forest Reserve in Kilombero Districts, respectively. The observed difference could be attributed by the use pattern of the product, as this study recorded quantities of firewood used for subsistence and trade, while past studies recorded subsistence use only. Also higher mean head loads for trade could probably be associated with reduced income of the households from conventional sources, mainly agriculture, due to the current adverse effects of climate change.

On the other hand, bush meat hunted from IFR was for subsistence (36%) and trade (38%). A mean annual value of TZS 465 900 per household was obtained from 155.3 kg of bush meat per year. Only few respondents reported to collect bush meat because this activity was done illegally in the IFR, and thus only few people dared to go for hunting. Even though there were few hunters of bush meat, they collected bush meat in high quantities because it fetched relatively low unit value of TZS 3 000 per kg (Table 4) compared to TZS 6 000 for meat from cattle. Larger proportion of the collected bush meat was traded (110 kg per year) entailing that hunting for the bush meat was aimed for cash income. During FGDs it was noted that trading for bush meat started in 2000 for income generation due to unpromising yields from conventional livelihoods including agriculture. This could probably be influenced by adverse effects of climate change in the study area. The mean annual amount of bush meat collected in the study area was relatively higher than that reported by Masam [48] who found about 11 Kg of bush meat being consumed by 73% of the respondents living adjacent to New Dabaga Ulongambi Forest Reserve, in Kilolo District. Hamza et al. [49] reported a mean of 60.1 Kg was consumed per household annually to villages surrounding Mgori Forest Reserve, Singida Rural District. Differences in consumption of wild meat could be caused by availability, abundance, accessibility to the resources, reluctance of household probably in giving true figures, and sometimes religious beliefs. The study has revealed large amount of bush meat being illegally hunted from IFR. The IFR is now part of the Kilombero Nature Reserve (KNR) where its management policy does not allow anthropogenic activities in the reserve. Continued illegal hunting of bush meat indicates high demand of animal protein by the households; this could also be due to insufficient alternative sources. Policies governing conservation of the reserve could be reviewed in order to regulate bush meat hunting. Also it is important for the households to diversify sources of protein through livestock keeping.

On the other hand, wild mushroom was collected for both subsistence and trade by 36% and 21% of the respondents, respectively. The mean annual value of TZS 292 000 per household was obtained from 155.3 kg of wild mushroom per household (Table 3). The current amount of mushrooms used for subsistence and traded in this study was much higher than that reported by Paulo [50] in Kilwa District, whereby annual average of about 29 kg was collected. Other results within Tanzania have reported different findings; for instance, Msemwa [46] reported that 29% of households in Kilosa District collected wild mushrooms from general land at an average quantity of 7.5 kg per household per year. Similarly, Kilonzo [47] reported 71 kg of wild mushrooms collected annually per household (66%). Higher quantities regarding mushrooms collection in this study as compared to other studies in Tanzania could probably be due to the fact that the current study included mushrooms that were collected for both subsistence and trade. On the other hand, about 36% of the respondents declared to collect medicinal plants for subsistence use, and only few (2%) collected them for trade. A mean annual value of TZS 231 200 per household was obtained from 28.9 kg of medicinal plants per household. Results from other studies in Tanzania including the study by Masam [48] in Kilolo District reported an average of 38 kg being collected per year per household, while Paulo [50] in Kilwa District reported an estimate of 13 kg per year per household. Similarly Msemwa [46] recorded a mean annual value of TZS 22 088 of medicinal plants per household in Kilosa District. Variation in collection could be due to existing local knowledge on the available medical plants, accessibility to the products, and religious beliefs, since some respondents associate use of medicinal plants with witchcraft.

Honey was also collected for both subsistence and trade by 10% and 22% of the respondents. The mean annual value of TZS 254 000 per household was obtained from 25.7 kg of honey per household (Table 3). The amount of honey collected per year in this study was much higher than those reported by Msemwa [46] and Paulo [50] which were 11.76 and 15 kg with mean annual values of TZS 14 739 and 10 500 per household in Kilosa and Kilwa Districts, respectively. The differences in mean annual value between the current and previous studies could be due to unit value of the honey. In the current study, unit value was TZS 6000 per kg while that by Msemwa [46] was TZS 1755 per kg. The high unit value of honey observed in the study is a reflection of scarcity and high marketing potential of bee products in Kilombero District which could also be linked to current adverse effects of climate change on households' livelihoods.

3.3. Economic Value of NTFPs. Economic value of the studied NTFPs for both subsistence and trade at a discounting rate of 10% was TZS 51.4 billion, equivalent to USD 36 million (Table 5). The economic value from the NTFPs indicated how households can increase their adaptive capacity through sustainable use of the resources for improved livelihoods. The economic value from the current study was relatively low compared to the one obtained by Schaafsma et al. [28] working around Eastern Arc Mountains which was TZS 59 billion (USD 42 million) for firewood, charcoal, thatch grass, and poles. However, it was high compared to URT [14] for Uluguru Catchment Forest reserves, in Morogoro region where TZS 39.6 billion (USD 39.6 million) for firewood, poles, and withies was obtained. The differences between the three studies could probably be due to the differences in market prices of the products as attributed by time differences, number of NTFPs dealt, and population size of the people in the study areas. The economic value of NTFPs from this study was mainly contributed by firewood (46%) followed by bush meet (28%), mushrooms (11%), thatch grasses (4%), medicinal plants (3%), and others (8%). Other NTFPs studied that had economic value included honey, poles, wild vegetable, tool handles, wild fruits, withies, and ropes. However, the present study discussed the most dominant five NTFPs.

In this study, the annual present value of firewood for both subsistence and trade, at discount rate of 10%, was TZS 23.5 billion which was equivalent to USD 11.3 million (Table 5). High contribution (72%) of the annual present value was from firewood used for subsistence and the remaining proportion was for trade. This indicated that firewood use pattern is under transition from subsistence to trade for generating income. The annual present value from this study was lower than those reported by URT [14] and Schaafsma et al. [28] which were TZS 32.6 billion (USD 32.6 million) and 36 billion (USD 25.33 million) from households around Uluguru Catchment Forest reserves, in Morogoro region and the Eastern Arc Mountains, respectively. Low value obtained from this study compared to the two past studies could probably be due to large number of populations involved in their study as well as availability and access to the firewood. The findings further implied that firewood as part of the natural capital asset had economic value which is significant. As the economic value of firewood was significantly high, this calls for households to be assisted in establishment of woodlots of fast growing tree species for firewood production. In future such woodlots can be used as collateral to various financial institutions and access development funds, therefore increasing adaptive capacity of households to adverse climate change effects as well as extreme events.

On the other hand, the present study recorded high annual present value compared to TZS 4.6 billion (USD 4.2 million) reported by Msemwa [46] from forests on general land in Kilosa District. The current findings were also higher than that reported by Kilonzo [47] from Nyanganje Forest Reserve in Kilombero District. The differences observed could be attributed by the average amount of head loads collected per year per household. Another plausible reason could probably be change in use pattern from subsistence to trade of firewood for income generation in order to carter for

TABLE 5: Economic value of recorded NTFPs for subsistence use and trade around Iyondo Forest Reserve, Kilombero District, Tanzania.

NTFPs	Unit	Average quantity	Unit value	Proportion	Actual value TZS	Present value TZS	Present value USD
Subsistence							
Firewood	Head load	115	2000	0.88	1 681 539 200.00	16 815 392 000	8 095 805.11
Bush meat	Kg	45.3	3000	0.36	409 847 763.60	4 098 477 636	1 973 220.50
Mushroom	Kg	68	1000	0.36	201 120 064.00	2 011 200 640	968 296.69
Thatch grass	Head load	26	1000	0.65	140 405 200.00	1 404 052 000	675 983.73
Medicinal plants	Kg	3	8000	0.56	110 861 952.00	1 108 619 520	533 747.15
Poles	Pieces	27.3	1000	0.29	67 135 286.40	671 352 864	323 224.22
Honey	Kg	4	6000	0.10	20 736 768.00	207 367 680	99 837.60
Wild vegetable	Kg	18.65	500	0.70	54 230 470.00	542 304 700	261 093.71
Tool handles	Pieces	3.7	1000	0.51	15 769 414.80	157 694 148	75 922.17
Withies	Bundle	2.5	5000	0.25	25 962 500.00	259 625 000	124 996.99
Wild fruits	Kg	19.7	400	0.51	33 087 042.02	330 870 420.16	159 298.25
Ropes	Bundle	2	800	0.24	3 137 100.80	31 371 008	15 103.64
Subtotal 1					2 763 832 761.62	27 638 327 616.16	13 306 529.75
Trade							
Firewood	Head load	328.3	2000	0.12	670 969 034.40	6 709 690 344	3 230 394.23
Bush meat	Kg	110	3000	0.38	1 052 789 760.00	10 527 897 600	5 068 677.98
Mushroom	Kg	224	1000	0.21	390 808 320.00	3 908 083 200	1 881 554.70
Thatch grass	Head load	25	1000	0.28	58 156 000.00	581 560 000	279 993.26
Medicinal plants	Kg	26.3	8000	0.02	41 952 076.80	419 520 768	201 979.14
Poles	Pieces	52	1000	0.05	21 600 800.00	216 008 000	103 997.50
Honey	Kg	21	6000	0.12	128 757 384.00	1 287 573 840	619 905.08
Tool handles	Pieces	10	1000	0.11	9 138 800.00	91 388 000	43 998.94
Subtotal 2					2 374 172 175.20	23 741 721 752.00	11 430 500.83
Total (subsistence + trade)					5 149 435 437.65	51 391 479 868.99	36 025 294.84

The economic values are expressed in terms of gross benefits to NTFP producing households, based on a mean July 2015 exchange rate of USD 1 = TZS 2 077.05 [55]; a unit of head load for firewood and thatch grass was equivalent to 16.55 ± 3.33 and 14.12 ± 3.19 Kg, respectively.

various household needs in the face of changing climate with adverse effects on conversional livelihoods including agriculture. According to Nkem et al. [13], NTFPs are essentially the niche for the poor population, which make them relevant for addressing poverty, health problems, and adaptation to external shocks and stresses due to climate change effects. The findings implied that firewood as part of the natural capital asset had economic value which is significant.

Bush meat was mentioned to be hunted and consumed in the study area. Wild animals that were mostly hunted were puku antelope (Sheshe), bush pigs (Nguruwe pori), and giant pouched rat (Ndezi). The annual present value of bush meat at discounted rate of 10% was TZS 14.6 billion equivalent to USD 7 million (Table 5). The annual present value recorded in this study was higher than TZS 434 450 520 reported by Msemwa [46] in Kilosa District. The difference between the two studies could probably be attributed by insufficiency of other sources of meat especially from domestic animals. It was also noted that the unit value of available meat from domesticated animals like cattle was high (TZS 6 000 per kg), therefore increasing demand of bush meat. Households seem to have adopted various sources of income apart from agriculture to carter for daily human needs.

The annual present value for wild mushroom for subsistence and trade at a discount rate of 10% was TZS 5.9 billion, which was equivalent to USD 2.8 million (Table 5). The annual present value recorded in this study was higher than TZS 317.7 million (USD 288 841.8) reported by Msemwa [46] in Kilosa District. Similarly, Kilonzo [47] reported annual present value of about 86.3 million (USD 66 412), which was low compared to the current study. High annual present values recorded in the current study compared to some past studies could probably be due to amount consumed, availability, and use knowledge of the products. High amount of wild mushroom collected was also related to the current changing climate as most of the households earned income that was used for other human needs. Another plausible reason on the difference in amount of wild mushroom consumed can be due to difference in rainfall intensity or pattern and condition of the forests as well as indigenous knowledge. Most of the forests in general land are not well protected and hence highly disturbed for growth of wild mushroom. It has also been reported by Edouard et al. [51] that harvesting of wild mushrooms is likely to be sustainable as long as the habitat of the fungi is not unduly disturbed.

The mean annual present value for thatch grasses for both subsistence use and trade for 8 308 households, at a discounting rate of 10%, was estimated at TZS 1.9 billion, which was equivalent to USD 813 740.41 (Table 5). The findings from this study are higher than those reported by Schaafsma et al. [28] study in the Eastern Arc Mountains which was about TZS 220 million (USD 0.16 million), probably due to differences in study population size. Another reason could be due to differences in market price, indigenous knowledge, and purpose of collection of the thatch grasses in a particular location. Another plausible reason for the high value of the thatch grasses in the current study could be due to scarcity of the resources especially in Mpofu Villages whereby people collected thatch grasses informally in open

areas within the IFR and formally at the Roman Catholic Mission farm in Mbingu Village. High amount of thatch grasses collected were also related to current adverse effects of climate change in the area. Increased temperatures were related to increase in termite activity on thatch grasses in the houses. Also during flooding, houses were demolished, making thatch grasses the immediate roofing materials. The results from the current study were however lower than TZS 7.5 billion (USD 6.8 million) reported by Msemwa [46] in Kilosa District. The difference observed could be due to population size, availability, and the accessibility to the products. The latter study was carried out in general lands of Kilosa District, therefore depicting full access to the products unlike that to the IFR and surrounding environment. According to Schaafsma et al. [25], better enforcement of conservation policies is expected to increase costs of NTFPs collection and therefore reduce extraction levels, either directly through costs of licensing, fines, or bribery or indirectly through the risk premium on illegal collection when avoiding fines. As the study has recognised economic value of the thatch grasses, it is important for households to retain some of the open grasslands that are found within their farms for this purpose. The government should prepare land use management plan that will accommodate some areas thatch grass production.

The annual present value for medicinal plants used for subsistence and trade at a discounting rate of 10% was TZS 1.5 billion equivalent to USD 735 726.29 (Table 5). The annual present value recorded in this study was lower than TZS 3.7 billion (USD 3.3 million) recorded by Msemwa [46] in Kilosa District. The reason behind the differences could be due to population size, accessibility, availability of the products, and religious beliefs, since some respondents associate use of medicinal plants with witchcraft. Free access to forest resources in Kilosa District might have increased the quantity collected and hence there was high annual present value compared to the current study area where there was no free access at to products. On the other hand Kilonzo [47] study in Nyanganje Forest Reserve in Kilombero District reported an annual present value from medicinal plants of TZS 3.3 million (USD 2 585). The present study has recorded higher values of annual present values than that by Kilonzo [47]. This could be attributed by the difference in price of the medicinal plants which was TZS 8 000 per Kg for present study compared to TZS 500 per Kg in 2009. The findings imply that medicinal plants have significant economic value that can be contributed to the natural capital asset and its value is increasing daily. Therefore, households should be encouraged to retain on farms some of the tree and shrub species with medicinal values. This will increase availability and therefore serve households against diseases that currently prevail due to adverse effects of climate change effects.

3.4. Factors Influencing Supply of NTFPs at Household Level. Supply of NTFPs was influenced by distance to the forest, change in forest management regime, seasonality, and change in rainfall pattern (Table 6).

About 66% of respondents agreed that increasing distance to the forest hinders rate of collection of NTFPs (Table 6). The amount of NTFPs collected decreased as distances increased.

TABLE 6: Factors influencing supply of NTFPs at household level.

Factors	*Response (%) n = 208
Distance to the forest	66
Change in forest regime	43
Seasonality	41
Change in rainfall pattern	16

*Applied multiple response analysis.

This means that opportunity cost of labour time spent for collections increases with distance, implying that people living closer to the forest are more dependent on NTFPs despite the restrictions imposed. Similarly, for households surrounding Nyanganje Forest Reserve, collection of firewood was significantly decreasing with increasing distance [47]. This implied that households residing near IFR still depend more on NTFPs such as firewood and mushroom than those located at distance, in order to increase their adaptive capacity to adverse climate change and vice versa.

In 2007, IFR changed management regime to nature reserve. A nature reserve is the highest category of protected areas which do not allow human consumptive activities [36]. About 43% of the respondents (Table 6) claimed that limiting access to the forest has decreased the supply of NTFPs at their households, hence increasing their vulnerability to adverse effects of climate change. However in discussion with the forest officer in charge, it was revealed that currently there is informal arrangement that allows households to enter into the forests to collect some nondestructive NTFPs like dried firewood, mushrooms, fruits, and wild vegetables. Therefore, restriction on forest access to promote conservation requires additional policies to prevent a consequent increase in poverty and tradeoff between conservation and extraction of NTFPs [25]. It could be appropriate if surrounding households could be allowed to use the nature reserve for activities which do not pose threat but enhance conservation such as beekeeping. According to Chidumayo [8], activities like beekeeping, collection of firewood, wild fruits, and medicinal plants can be conducted in a reserved forest if carefully planned.

Seasonality was another constraint which hindered collection of NTFPs by 41% of the respondents (Table 6). Wunder et al. [52] found that seasonality was a substantial constraint during collection of some NTFPs particularly during wet season as physical access was limited. According to Chidumayo [8], increasing temperatures in southern Africa are predicted to either extend the growing season in some ecosystems or shorten it in others. During FGDs it was reported that mushrooms are normally available from February to April. The current growing season has changed because of shift for both short and long rains, therefore affecting yields. Households reported to dry and store mushrooms for some period of time but processing and storage conditions were claimed to be poor. Therefore, mushrooms availability can be improved if adoption of simple but improved technologies of processing and storage could be promoted to households to enhance anticipatory adaptation.

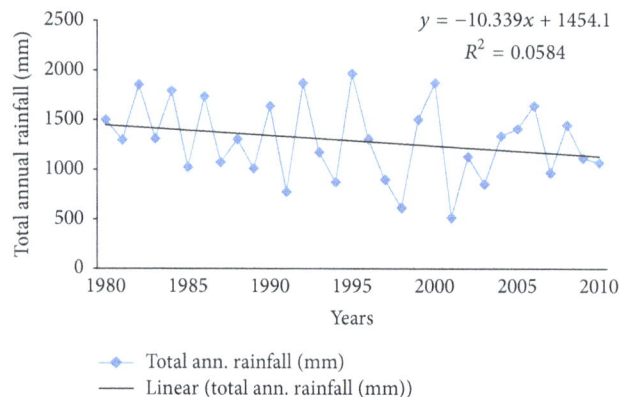

$$y = -10.339x + 1454.1$$
$$R^2 = 0.0584$$

-◆- Total ann. rainfall (mm)
—— Linear (total ann. rainfall (mm))

FIGURE 2: Total annual rainfall (mm) recorded between 1980 and 2010 around the study area.

Change of rainfall pattern and intensity has led to lower supply of some NTFPs like mushrooms. Normally in December varieties of mushrooms collected are *Amanita zambiana* (wilelema) and *Termitomyces le-testui* (wikulwe), but due to changing of onset and cessation of rainfall in recent years, they are less available. The empirical data showed that rainfall in the study area has been unpredictable (Figure 2).

The rainfall pattern from 1980 to 2010 showed a trend of decrease in total rainfall received for the past three decades. This implied that the study area has been receiving high rains for short period, as was perceived by the majority. Also, from the indigenous knowledge perspective of some respondents, it was reported that prolonged drought has increased the chance of occurrence of wildfire which burns all litters in the forest that form substrate for mushrooms to grow, hence decreasing the supply of mushrooms. Similarly, Chidumayo [8] reported effect of climate change on availability of forest product and services through forest fires which are associated with drought, resulting in biodiversity loss.

During FGDs it was reported that wild mushrooms, thatch grasses, wild vegetables, and bush meet were more sensitive to adverse effects of changing climatic conditions particularly dry spells and floods. Other NTFPs are directly attached to trees found in the natural forests which are more resilient to climate change impacts than monoculture plantations or any artificial forest [53, 54]. Similarly, Nkem et al. [20] argued that natural forests are less sensitive to climate change effects compared to agricultural crops. Heubes et al. [3] revealed that as NTFPs are mostly derived from plants, their supply is related to the species' occurrence probabilities which might differ from one species to another depending on the sensitivity to adverse effects of changing climatic conditions when other productions circumstances are kept constant.

4. Conclusion and Recommendations

4.1. Conclusion. Various NTFPs ranging between food products, firewood, and construction materials were collected for subsistence and traded by households around IFR and village forests and on farms during adverse effects of climate change.

Collection of NTFPs from the IFR was revealed to be done illegally because of change in management regime from forest reserve to nature reserve. The study has recorded high mean annual value and the economic value of the studied NTFPs from IFR and village forests and on farms. This amount shows the magnitude of the economic loss that households would bear if NTFPs collection was fully and effectively banned in the IFR and the surrounding village forests. The economic value of NTFPs was mainly contributed by firewood, followed by other NTFPs. High contribution of firewood to the economic value is associated with high collection of the product due to daily demand, unlike for other NTFPs. Due to this value the majority of the households seem to rely on firewood as immediate sources of income when conventional livelihood activities were adversely affected by climatic stresses. The main body of understanding has so far focused on the value of prominent forest products like timber and the impacts of climate change on such resources. This empirical documentation of the economic value of priority NTFPs adds to scientific understanding of forest ecosystem based adaptation by showing the potential of NTFPs in increasing adaptive capacity of the households. Last, the supply of NTFPs at household level was influenced by distance to the forests, change in forest management regime, seasonality, and change in rainfall pattern.

4.2. Recommendations. First, change in policies governing forest resources has affected access to the NTFPs in IFR. In this study, this has been regarded as another stress affecting households. Study findings suggest that conservations rules and regulations should be adjusted in order to facilitate access by households to forest resources. Access to the forests by vulnerable populations is as important as the conservation of biodiversity. Access to forests as well as conservation of biodiversity is important to adaptive capacity. The potential synergies and conflicts between these two aspects of NTFPs are managed by rules and regulations, indicating the importance of transforming structures.

Second, the study has documented that households in Kilombero are collecting NTFPs illegally from IFR because the type of management regime does not allow human activities. Together with this, it was as well noted that, within the villages' land, there were remnants of forests that could be managed through formal processes into village forest reserves. Establishment of village forest reserves in which households will have full access is important for sustainable utilization of NTFPs, thus enhancing their adaptive capacity. The study proposes that the established village forest reserves could be managed through community based forest management (CBFM), a management regime that will allow total participation of households in decision making about sustainable forest management.

Third, the study found that households lack skills and knowledge about value addition and marketing of the NTFPs collected. This situation has led to low market value of the products. Capacity building to households on NTFPs value addition and strengthening its markets are pertinent for improved income and eventually increase adaptive capacity to adverse effects of the changing climatic conditions. Also marketing of the products was revealed to be poor as most of the products were sold in isolation at scattered homesteads. This situation might have contributed to low value of the products as most of the traders were not aware of the real market value of the products. From this point of view, capacity building on marketing of the products could include establishment of collection/trading centres in order that it becomes easy for price monitoring as well as control of the quality of the products.

Last, as the NTFPs had high annual economic value, more research could be undertaken to compare economic value of some NTFPs in rainy and dry seasons. Study findings have shown that availability of most of the NTFPs depended on the season. Some of them were reported to be available during rainy season like mushrooms and wild vegetable. Some of the fruits were only available during either rain or dry season. For example, peak season of Tamarind fruits occurred during the dry season when other fruits for making beverages were not available. Households were reported to dry wild fruits for some period of time in order to be used during time of scarcity. Thatch grasses were harvested during dry season, therefore making availability during rainy season be low. The same applies to firewood, as collection was mainly done during dry season rather than in the rainy season. In the rainy season, most households were preoccupied with agricultural activities rather than firewood collection. Season availability of NTFPs was observed to affect market value of the products based on the supply and demand theory. For example, the market value of thatch grasses and firewood was recorded to be high during rainy season compared to dry season. The case is different to mushroom, as it fetched high market value during dry season.

Competing Interests

The authors declare that they have no competing interests.

Acknowledgments

The authors thank the Climate Change Impacts Adaptation and Mitigation (CCIAM) Programme, Tanzania, for financial support. They are also highly indebted to the people in Mpofu, Igima, Njage, and Mngeta Villages and other stakeholders in Kilombero District for knowledge sharing and their cooperation during data collection.

References

[1] B. M. Belcher, "Forest product markets, forests and poverty reduction," *International Forestry Review*, vol. 7, no. 2, pp. 82–89, 2005.

[2] B. Belcher and K. Schreckenberg, "Commercialisation of non-timber forest products: a reality check," *Development Policy Review*, vol. 25, no. 3, pp. 355–377, 2007.

[3] J. Heubes, K. Heubach, M. Schmidt et al., "Impact of future climate and land use change on non-timber forest product provision in Benin, West Africa: linking niche-based modeling with ecosystem service values," *Economic Botany*, vol. 66, no. 4, pp. 383–397, 2012.

[4] I. Niang, O. C. Ruppel, M. A. Abdrabo et al., "Africa," in *Climate Change 2014: Impacts, Adaptation, and Vulnerability. Part B: Regional Aspects. Contribution of Working Group II to the Fifth Assessment Report of the Intergovernmental Panel on Climate Change*, V. R. Barros, C. B. Field, D. J. Dokken et al., Eds., pp. 1199–1265, Cambridge University Press, Cambridge, UK, 2014.

[5] L. Olsson, M. Opondo, P. Tschakert et al., "Livelihoods and poverty," in *Climate Change 2014: Impacts, Adaptation, and Vulnerability. Part A: Global and Sectoral Aspects. Contribution of Working Group II to the Fifth Assessment Report of the Intergovernmental Panel on Climate Change*, C. B. Field, V. R. Barros, D. J. Dokken et al., Eds., pp. 793–832, Cambridge University Press, New York, NY, USA, 2014.

[6] J. N. Nkem, O. A. Somorin, C. Jum, M. E. Idinoba, Y. M. Bele, and D. J. Sonwa, "Profiling climate change vulnerability of forest indigenous communities in the Congo Basin," *Mitigation and Adaptation Strategies for Global Change*, vol. 18, no. 5, pp. 513–533, 2013.

[7] P. Rowhani, D. B. Lobell, M. Linderman, and N. Ramankutty, "Climate variability and crop production in Tanzania," *Agricultural and Forest Meteorology*, vol. 151, no. 4, pp. 449–460, 2011.

[8] E. Chidumayo, "Climate change and the woodlands of Africa," in *Climate Change and African Forest and Wildlife Resources*, E. Chidumayo, D. Okali, G. Kowero, and M. Larwanou, Eds., pp. 85–101, African Forest Forum, Nairobi, Kenya, 2011.

[9] C. Gordon, *The Science of Climate Change in Africa: Impacts and Adaptation*, Department for International Development, London, UK, 2008.

[10] S. H. Eriksen, K. Brown, and P. M. Kelly, "The dynamics of vulnerability: locating coping strategies in Kenya and Tanzania," *Geographical Journal*, vol. 171, no. 4, pp. 287–305, 2005.

[11] J. Paavola, "Livelihoods, vulnerability and adaptation to climate change in Morogoro, Tanzania," *Environmental Science and Policy*, vol. 11, no. 7, pp. 642–654, 2008.

[12] URT, *National Climate Change Strategy*, United Republic of Tanzania, Vice President's Office, Division of Environment, Dar es Salaam, Tanzania, 2012.

[13] J. Nkem, F. B. Kalame, M. Idinoba, O. A. Somorin, O. Ndoye, and A. Awono, "Shaping forest safety nets with markets: adaptation to climate change under changing roles of tropical forests in Congo Basin," *Environmental Science and Policy*, vol. 13, no. 6, pp. 498–508, 2010.

[14] URT, *Resource Economic Analysis of Catchment Forest Reserves in Tanzania*, Ministry of Natural Resources and Tourism, Forest and Beekeeping Division, Dar es Salaam, Tanzania, 2003.

[15] W. Cavendish, "Empirical regularities in the poverty-environment relationship of rural households: evidence from Zimbabwe," *World Development*, vol. 28, no. 11, pp. 1979–2003, 2000.

[16] S. K. Pattanayak and E. O. Sills, "Do tropical forests provide natural insurance? The microeconomics of non-timber forest product collection in the Brazilian Amazon," *Land Economics*, vol. 77, no. 4, pp. 595–612, 2001.

[17] V. Ingram and G. Bongers, *Valuation of Non-Timber Forest Product Chains in the Congo Basin: A Methodology for Valuation*, FAO-CIFOR-SNV-World Agroforestry Centre-COMIFAC, CIFOR, Yaounde, Cameroon, 2009.

[18] K. Heubach, R. Wittig, E.-A. Nuppenau, and K. Hahn, "Local values, social differentiation and conservation efforts: the impact of ethnic affiliation on the valuation of NTFP-species in Northern Benin, West Africa," *Human Ecology*, vol. 41, no. 4, pp. 513–533, 2013.

[19] P. Vedeld, A. Angelsen, J. Bojö, E. Sjaastad, and G. Kobugabe Berg, "Forest environmental incomes and the rural poor," *Forest Policy and Economics*, vol. 9, no. 7, pp. 869–879, 2007.

[20] J. Nkem, H. Santoso, D. Murdiyarso, M. Brockhaus, and M. Kanninen, "Using tropical forest ecosystem goods and services for planning climate change adaptation with implications for food security and poverty reduction," *Semi-Arid Tropical Agricultural Research*, vol. 4, pp. 1–23, 2007.

[21] J. Sumukwo, A. Wario, M. Kiptui, G. Cheserek, and A. K. Kipkoech, "Valuation of natural insurance demand for non-timber forest products in South Nandi, Kenya," *Journal of Emerging Trends in Economics and Management Sciences*, vol. 4, no. 1, pp. 89–97, 2013.

[22] C. M. Shackleton and S. E. Shackleton, "The importance of non-timber forest products in rural livelihood security and as safety nets: a review of evidence from South Africa," *South African Journal of Science*, vol. 100, no. 11-12, pp. 658–664, 2004.

[23] R. P. Neumann and E. Hirsch, *Commerciallisation of Non-Timber Forest Products: Review and Analysis of the Research*, Centre for International Forestry Research, Bogor, Indonesia, 2000.

[24] L. Croitoru, "Valuing the non-timber forest products in the Mediterranean region," *Ecological Economics*, vol. 63, no. 4, pp. 768–775, 2007.

[25] M. Schaafsma, S. Morse-Jones, P. Posen et al., "The importance of local forest benefits: economic valuation of Non-Timber Forest products in the Eastern Arc Mountains in Tanzania," *Ecological Economics*, vol. 80, pp. 48–62, 2012.

[26] U. Msalilwa, S. Augustino, and P. R. Gillah, "Community perception on climate change and usage patterns of non-timber forest products by communities around Kilolo District, Tanzania," *Ethiopian Journal of Environmental Studies and Management*, vol. 6, no. 5, pp. 507–516, 2013.

[27] C. Balama, S. Augustino, S. Eriksen, and F. B. S. Makonda, "Forest adjacent households' voices on their perceptions and adaptation strategies to climate change in Kilombero District, Tanzania," *SpringerPlus*, vol. 5, article 792, 2016.

[28] M. Schaafsma, S. Morse-Jones, P. Posen et al., "The importance of local forest benefits: economic valuation of non-timber forest products in the eastern Arc mountains in Tanzania," *Global Environmental Change*, vol. 24, no. 1, pp. 295–305, 2014.

[29] D. W. Pearce and D. Moran, *The Economic Value of Biological Diversity*, IUCN the World Conservation Union, Earthscan Publications Ltd, London, UK, 1994.

[30] T. Tietenberg and L. Lewis, *Environmental and Natural Resource Economics*, Peason Education, Upper Saddle River, NJ, USA, 9th edition, 2012.

[31] J. Ferreira, M. Guariguata, L. P. Koh et al., "Impacts of forests and land management on biodiversity and carbon," in *Understanding Relationships between Biodiversity, Carbon, Forests and People. The Key to Achieving REDD+ Objectives*, J. A. Parrotta and R. L. Trosper, Eds., pp. 53–80, International Union of Forest Research Organization, Vienna, Austria, 2012.

[32] URT (United Republic of Tanzania), *Tanzania in Figures 2012*, National Bureau of Statistics, Ministry of Finance, Dar es Salaam, Tanzania, 2013.

[33] F. Kato, "Development of a major rice cultivation area in the Kilombero Valley, Tanzania," *African Studies Monograph*, vol. 36, pp. 3–18, 2007.

[34] A. Chamwali, *Survival and Accumulation Strategies at the Rural-Urban Interface: A Study of Ifakara Town, Tanzania*, Research on Poverty Alleviation, Dar es Salaam, Tanzania, 2000.

[35] M. Starkey, N. Birnie, A. Cameron et al., *The Kilombero Valley Wildlife Project: An Ecological and Social Survey in the Kilombero Valley, Tanzania*, Kilombero Valley Wildlife Project, Edinburgh, UK, 2002.

[36] P. Harrison, *Socio-Economic Study of Forest-Adjacent Communities from Nyanganje Forest to Udzungwa Scarp: A Potential Wildlife Corridor. Incorporating Livelihood Assessments and Options for Future Management of Udzungwa Forests*, World Wide Fund (WWF) for Nature, Dar es Salaam, Tanzania, 2006.

[37] NBS, "Shapefiles-Level one and two," 2012, http://www.nbs.go.tz/.

[38] T. E. Erlanger, A. A. Enayati, J. Hemingway, H. Mshinda, A. Tami, and C. Lengeler, "Field issues related to effectiveness of insecticide-treated nets in Tanzania," *Medical and Veterinary Entomology*, vol. 18, no. 2, pp. 153–160, 2004.

[39] M. A. W. Hetzel, S. Alba, M. Fankhauser et al., "Malaria risk and access to prevention and treatment in the paddies of the Kilombero Valley, Tanzania," *Malaria Journal*, vol. 7, article 7, 2008.

[40] J. C. Lovett and T. Lovett and Pocs, *Assessment of the Condition of the Catchment Forest Reserves, a Botanical Appraisal*, Morogoro Region, 1993.

[41] P. Barribeau, B. Butler, J. Corney et al., "Survey Research," 2015, http://writing.colostate.edu/guides/pdfs/guide68.pdf.

[42] G. C. Monela, S. A. O. Chamshama, R. Mwaipopo, and D. M. Gamasa, *A Study on the Social, Economic and Environmental Impacts of Forest Landscape Restoration in Shinyanga Region, Tanzania*, Forestry and Beekeeping Division of the Ministry of Natural Resources and Tourism, of Tanzania, and IUCN—The World Conservation Union Eastern Africa Regional Office, Dar es Salaam, Tanzania, 2005.

[43] S. Morse-Jones, T. Luisetti, R. K. Turner, and B. Fisher, "Ecosystem valuation: some principles and a partial application," *Environmetrics*, vol. 22, no. 5, pp. 675–685, 2011.

[44] F. Dolisca, D. R. Carter, J. M. McDaniel, D. A. Shannon, and C. M. Jolly, "Factors influencing farmers' participation in forestry management programs: A case study from Haiti," *Forest Ecology and Management*, vol. 236, no. 2-3, pp. 324–331, 2006.

[45] A. Tazeze, J. Haji, and M. Ketema, "Climate change adaptation strategies of smallholder farmers: the case of Babilie District, East Harerghe Zone of Oromia Regional State of Ethiopia," *Journal of Economics and Sustainable Development*, vol. 3, no. 14, pp. 1–13, 2012.

[46] S. C. Msemwa, *Economic valuation of forests on general land in Kilosa District, Tanzania [M.S. dissertation]*, Sokoine University of Agriculture, Morogoro, Tanzania, 2007.

[47] M. Kilonzo, *Valuation of non-timber forest products used by communities around Nyanganje Forest Reserve, Morogoro, Tanzania [M.S. dissertation]*, Sokoine University of Agriculture, Morogoro, Tanzania, 2009.

[48] I. S. Masam, *Contribution of non-timber forest products to the livelihoods of communities living adjacent to New Dabaga Ulongambi Forest Reserve, Kilolo District [M.S. thesis]*, Sokoine University of Agriculture, Morogoro, Tanzania, 2009.

[49] K. F. S. Hamza, U. L. Msalilwa, and R. J. L. Mwamakimbullah, "Contribution of some Non-Timber Forest Products (NTFPs) to household food security and income generation: a case study of villages around Mgori Forest Reserve in Singida, Tanzania," *Journal of Tanzania Association of Foresters*, vol. 11, pp. 11–22, 2007.

[50] T. Paulo, *The contribution of non-timber forest products in improving livelihood of rural community in Kilwa District,* *Tanzania [Dissertation for Award of MSc. Degree]*, Sokoine University of Agriculture, Morogoro, Tanzania, 2007.

[51] F. Edouard, Q. Raday, and E. Marshall, "Fresh, dried and exported mushrooms: community business and entrepreneurs," in *Commercialization of Non-timber Forest Products: Factors Influencing Success. Lessons Learned from Mexico and Bolivia and Policy Implications for Decision Makers*, E. Marshal, K. Schrecken Berg, and A. C. Newton, Eds., pp. 49–52, UNEP World Conservation Monitoring Centre, Cambridge, UK, 2006.

[52] S. Wunder, J. Börner, G. Shively, and M. Wyman, "Safety nets, gap filling and forests: a global-comparative perspective," *World Development*, vol. 64, no. 1, pp. S29–S42, 2014.

[53] P. Smith, M. Bustamante, H. Ahammad et al., "Agriculture, Forestry and Other Land Use (AFOLU)," in *Climate Change 2014: Mitigation of Climate Change. Contribution of Working Group III to the Fifth Assessment Report of the Intergovernmental Panel on Climate Change*, O. Edenhofer, R. Pichs-Madruga, Y. Sokona et al., Eds., pp. 811–922, Cambridge University Press, Cambridge, UK, 2014.

[54] N. H. Ravindranath, "Mitigation and adaptation synergy in forest sector," *Mitigation and Adaptation Strategies for Global Change*, vol. 12, no. 5, pp. 843–853, 2007.

[55] Bank of Tanzania, "Bank of Tanzania (BoT)," 2015, https://www.bot-tz.org/.

18

Detecting the Early Genetic Effects of Habitat Degradation in Small Size Remnant Populations of *Machilus thunbergii* Sieb. et Zucc. (Lauraceae)

Shuntaro Watanabe,[1,2] **Yuko Kaneko,**[3] **Yuri Maesako,**[4] **and Naohiko Noma**[2]

[1] *Field Science Education and Research Center, Kyoto University, Oiwake-cho, Kitashirakawa, Sakyo-ku, Kyoto 606-8502, Japan*
[2] *The University of Shiga Prefecture, No. 2500, Hassaka-cho, Hikone, Shiga 522-8533, Japan*
[3] *Toyo University, 5-28-20 Hakusan, Bunkyo-ku, Tokyo 112-8606, Japan*
[4] *Osaka Sangyo University, Nakagaito, Daito, Osaka 574-8530, Japan*

Correspondence should be addressed to Yuko Kaneko; kaneko065@toyo.jp

Academic Editor: Ignacio García-González

Habitat degradation caused by human activities has reduced the sizes of many plant populations worldwide, generally with negative genetic impacts. However, detecting such impacts in tree species is not easy because trees have long life spans. *Machilus thunbergii* Sieb. et Zucc. (Lauraceae) is a dominant tree species of broad-leaved evergreen forests distributed primarily along the Japanese coast. Inland habitats for this species have become degraded by human activities. To investigate the effects of habitat degradation on genetic structure, we compared the genetic diversities of mature and juvenile trees of five *M. thunbergii* populations around Lake Biwa in Japan. Allelic diversity was influenced by past lineage admixture events, but the effects of forest size were not clear. On the other hand, the inbreeding coefficient of the juvenile stage was higher in small populations, whereas large populations maintained panmictic breeding. Also, the extent of genetic differentiation was greater in juveniles than in mature trees. We detected the early genetic effects of habitat degradation in small, isolated *M. thunbergii* populations, indicating that habitat degradation increases inbreeding and genetic differentiation between populations.

1. Introduction

Warm-temperate evergreen forests in East Asia occur primarily at low elevations and feature the dominant genera *Machilus (Persea), Castanopsis, Quercus, Lithocarpus, Cinnamomum,* and *Neolitsea* [1]. However, in the main island of Japan, such forests are highly degraded because of human disturbance. Generally, habitat degradation caused by humans has reduced the sizes and increased the spatial isolation of many plant populations worldwide. Small population size and increased isolation can cause genetic erosion, increased genetic divergence via random drift, increased inbreeding, reduced gene flow, disrupted pollination, and increased probability of local extinction [2–4]. Genetic erosion caused by habitat fragmentation can be of immediate concern if genetic changes directly influence individual fitness and the

short-term viability of remnant populations. One long-term evolutionary consequence of genetic erosion is limitation of the ability to respond to environmental changes, which is expected to increase the probability of extinction [2, 4]. Recent meta-analyses suggested that plant species generally exhibit negative genetic responses to habitat fragmentation [5, 6]. However, detecting the impact of habitat fragmentation on the genetic structures of tree species is not easy [7]. Many tree species may be buffered from the effects of such fragmentation by individual longevity, high intrapopulation genetic diversity, and the potential for long-distance pollen flow. Thus, several earlier studies of trees found no evidence that fragmentation influenced genetic parameters (e.g., [8, 9]). In addition, any reduction in genetic diversity or increase in the level of inbreeding may not be immediately evident after forest fragmentation because the older trees remaining after habitat isolation often have the same extent of genetic diversity as that observed prior to fragmentation [7]. Therefore, assessment of only mature members of a population may yield no information on the contemporary genetic effects of isolation. One possible approach to study of the impact of forest fragmentation on the genetic diversity of tree species is to compare the genetic diversity and structure between predisturbance adult populations and postdisturbance generations cohorts [10]. Such comparisons can reveal whether significant, potentially deleterious genetic changes are occurring in the present generation (e.g., [10–13]).

Lake Biwa is the largest freshwater lake in Japan and is located in the west-central region of the country. The lakeside hosts many coastal plants, including *Calystegia soldanella* (L.) R. Br. (Convolvulaceae), *Vitex trifolia* subsp. *litoralis* Steenis, *Lathyrus japonicus* Willd. (Leguminosae), *Arabidopsis kamchatica* subsp. *kawasakiana* (Makino) K. Shimizu & Kudoh (Brassicaceae), *Raphanus raphanistrum* subsp. *sativus* (L.) Domin, *Dianthus japonicus* Thunb. (Caryophyllaceae), and *Pinus thunbergii* Parl. (Pinaceae) [14]. *Machilus thunbergii* Sieb. et Zucc. (Lauraceae) is a broad-leaved evergreen tree species in warm-temperate forests of Japan and also distributed around Lake Biwa. These populations are important because this species is rarely found inland and regarded as a flagship species around Lake Biwa [15]. However, these *M. thunbergii* populations are heavily degraded because their distribution overlaps areas of human activity [16]. As a result of such disturbance, *M. thunbergii* forests around Lake Biwa now occur almost exclusively around Shinto shrines or on islands in the lake. Therefore, some *M. thunbergii* populations around the lake are expected to exhibit low levels of genetic diversity, high levels of inbreeding, and high extents of genetic differentiation.

The objectives of the present study were to analyze and to assess the effects of habitat degradation on local genetic diversity and differentiation of *M. thunbergii* populations that persist around Lake Biwa. We compared the genetic diversities of mature and juvenile *M. thunbergii* in each population, using seven microsatellite markers and evaluated the extent of genetic divergence among populations.

2. Materials and Methods

2.1. Study Species. *Machilus thunbergii* Sieb. et Zucc. is a representative dominant tree species of warm-temperate evergreen forests in Japan. It is distributed mainly around the sea coast, and over 90% of all trees are distributed along the shore [17, 18]. The tree grows to approximately 15 m and blooms from April to June. Bees, flies, and beetles visit flowers and disperse pollen [19]. *Machilus thunbergii* is a heterodichogamous species, consisting of two types of protogynous and bisexual flowers at the individual-tree level: a morning female-afternoon male morph and a morning male-afternoon female morph [20]. Seeds generally mature from June to August [21, 22] and are dispersed by birds, mainly the brown-eared bulbul (*Hypsipetes amaurotis* Temminck) and the white-cheeked starling (*Sturnus cineraceus* Temminck).

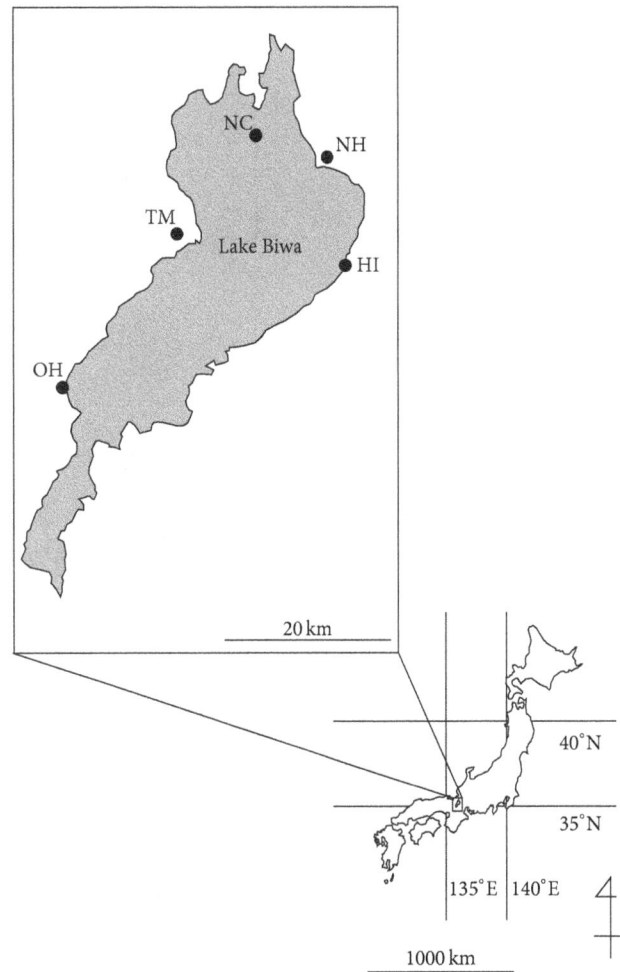

FIGURE 1: Location of the study sites around Lake Biwa, Japan.

2.2. Study Site. We worked in five warm-temperate evergreen forests around Lake Biwa (Figure 1; Table 1). Lake Biwa is the largest freshwater lake in Japan and is located in the west-central region of the country.

Chikubushima Island (NC; Nagahama, Shiga, Japan), which is 2 km away from the shoreline, is uninhabited, and the level of conservation is high. The forest canopy of the island is dominated by evergreen broad-leaved species including *M. thunbergii*, *Quercus myrsinifolia* Blume, and *Ilex integra* Thunb. This forest has maintained almost the same primeval state since 15th century. This forest contained more than 100 individuals of *M. thunbergii*. The Hikitari shrine site (NH; Nagahama, Shiga, Japan) is a shrine forest dominated by *M. thunbergii*, *Q. myrsinifolia*, and *Zelkova serrata* (Thunb.) Makino. This forest is surrounded by agricultural area and highly fragmented. This forest contained 15 individuals of *M. thunbergii*. The Inukami River site (HI; Hikone, Shiga, Japan) is a riverside forest located at the outlet of the Inukami River and has a canopy dominated by *M. thunbergii*, *Aphananthe aspera* (Thunb.) Plach., *Quercus aliena* Blume, and *Celtis sinensis* Pers. Area of this forest is decreasing since 1960s owing to river improvement but relatively large size forest

TABLE 1: Location and size of each *M. thunbergii* study site.

Study site	Locality	Abbreviation	Latitude	Longitude	Forest area (ha)
Hassho—shrine	Otsu, Shiga	OH	35°10′N	135°54′E	1.46
Myounji—temple	Takashima, Shiga	TM	35°19′N	136°04′E	0.13
Chikubushima—island	Nagahama, Shiga	NC	35°25′N	136°08′E	14.17
Hikitari—shrine	Nagahama, Shiga	NH	35°26′N	136°12′E	0.88
Inukami—river	Hikone, Shiga	HI	35°15′N	136°13′E	9.41

still maintained. This forest contained about 60 individuals of *M. thunbergii*. The Hassho shrine site (OH; Otsu, Shiga, Japan) is another shrine forest, with a canopy dominated by *M. thunbergii*, *Castanopsis cuspidata* (Thunb.) Schottky, and *Cryptomeria japonica* (Thunb. ex L.f.) D. Don. This forest is surrounded by urban area and highly fragmented. This forest contained about 24 individuals of *M. thunbergii*.

The Myounji temple site (TM; Takashima, Shiga, Japan) is a temple forest with a canopy dominated by *M. thunbergii*, *C. cuspidata*, *Q. myrsinifolia*, and *Z. serrata*. This forest is surrounded by paddy fields and highly fragmented. This forest contained 9 individuals of *M. thunbergii*. In this paper, we used areas of forest (ha) as an indicator of population size because we could not evaluate exactly the number of individuals in HI and NC population.

2.3. Sample Collection.

We collected leaf of mature *M. thunbergii* trees. The average sample size per population was 29 (range: 9–55), and 147 individual trees were sampled in total. In this study, a mature tree was defined as an individual >130 cm in height and >5 cm in diameter at breast height. Samples from mature trees were randomly collected at HI and NC, whereas all individuals within a population were sampled at OH, TM, and NH. In the TM population, we sampled just nine mature trees because there were only nine mature trees.

We also collected leaf samples from juvenile trees, including current-year seedlings and small saplings (>1 year old) to eliminate year-specific effects such as differences in flower or seed mass. The average sample size per population was 34 (range: 18–55), and 172 individuals were sampled in total. Juvenile samples were randomly collected from HI in 2006, OH in 2010, TM and NH in 2014, and NC in 2015. Leaf samples were stored at 4°C with silica gel prior to DNA extraction using a modified cetyltrimethylammonium bromide method [23].

2.4. Microsatellite Typing.

Seven loci, corresponding to the microsatellite markers Mt03, Mt04, Mt05, Mt13, Mt14, Mt16, and Mt20 identified by Kaneko et al. [24], were used in the analysis. We scored the genotypes of these microsatellite loci in each DNA sample. We performed PCR amplification using a DNA thermal cycler (Eppendorf, Mastercycler ep gradient S) under the following conditions: initial denaturation at 94°C for 9 min, followed by 40 cycles of denaturation at 94°C for 30 s and 1 min annealing at 72°C with a final extension step of 72°C for 5 min. The reaction mixtures (5 μl) contained 2 μl PCR Master Mix (Qiagen Multiplex PCR

Buffer, pH 8.7, consisting of dNTPs, Qiagen HotStarTaq DNA Polymerase, and $MgCl_2$ to a final concentration of 3 mM) and 10 ng template DNA. The sizes of the PCR products were determined by automated fluorescence scanning using a 3130xl Genetic Analyzer (Applied Biosystems) running the GeneMapper software (Applied Biosystems).

2.5. Genetic Diversity and Genetic Differentiation.

For each population, the genetic diversity was evaluated by calculating allelic richness (A_R; [25]), expected heterozygosity (H_E), and observed heterozygosity (H_O) values and the fixation index (F). All parameters were calculated using version 2.9.3 of the FSTAT software [26]. Deviations from Hardy–Weinberg equilibrium (HWE) were assessed using FSTAT. The null alleles affect population parameters which are estimated based on the proportion of heterozygotes. So, we also estimated frequency of null allele for mature and juvenile population of each site and posterior distribution of $F_{(null)}$ values. $F_{(null)}$ is inbreeding coefficient which is estimated in consideration of the presence of null alleles. Estimations of $F_{(null)}$ and the average null allele frequency were conducted under an individual inbreeding model (IIM) using INEst 2.0 (Table 2) [27].

To investigate the relationships between genetic diversity (A_R, H_E, F) and forest area (ha), we applied a generalized linear model with an identity link function and Gaussian distribution using the R 2.11.1 software [28]. Nei's estimates [29] of heterozygosity (H_S, H_T) were calculated for multiple loci using FSTAT. To estimate the population size reduction at the mature stage, recent population bottlenecks were evaluated using version 1.2.02 of the BOTTLENECK software [30]. We simulated equilibrium conditions (10000 replications) using both an infinite alleles model (IAM) and a two-phase model (TPM). We used the Wilcoxon signed-rank test to determine if significant excess heterozygosity existed. The TM population was excluded from this analysis due to its small sample size.

To estimate genetic differentiation among populations at both the mature and the juvenile stages, F_{ST} [31], R_{ST} [32], G'_{ST} [33] values were calculated. F_{ST}, R_{ST}, H_S, H_T were computed using FSTAT, and G'_{ST} values were computed using GenAlEx 6.5 [34].

3. Results

3.1. Within-Population Genetic Diversity.

A_R ranged from 2.67 to 4.11 for mature trees and from 2.58 to 4.20 for juveniles (Table 2). A_R values of mature trees were highest in NC and

TABLE 2: Genetic diversity of five *M. thunbergii* populations around Lake Biwa.

Site	Population	N	A_R	H_O	H_E	F	$F_{(null)}$	Null allele
OH	Mature	24	2.67	0.60	0.51	−0.17	0.02	0.02
	Juvenile	39	2.58	0.41	0.50	0.19*	0.05	0.12
TM	Mature	9	3.39	0.41	0.49	0.23	0.10	0.19
	Juvenile	20	3.05	0.41	0.51	0.23*	0.16	0.17
NC	Mature	55	4.11	0.58	0.62	0.08	0.05	0.05
	Juvenile	40	4.20	0.58	0.64	0.11	0.08	0.07
NH	Mature	15	3.84	0.61	0.62	0.06	0.08	0.04
	Juvenile	18	3.79	0.51	0.62	0.2*	0.16	0.09
HI	Mature	44	3.58	0.66	0.64	−0.009	0.02	0.01
	Juvenile	55	3.43	0.52	0.58	0.12	0.07	0.06

N, number of samples; A_O, allelic richness (calculated from 14 gene copies); H_O, observed heterozygosity; H_E, expected heterozygosity; F, inbreeding coefficient for juvenile and mature life stages of *M. thunbergii*; * population significantly deviated from Hardy–Weinberg equilibrium ($P \leq 0.05$); $F_{(null)}$, Bayesian estimated posterior mean value of F in consideration of presence of null alleles; *null allele*, the average null allele frequency.

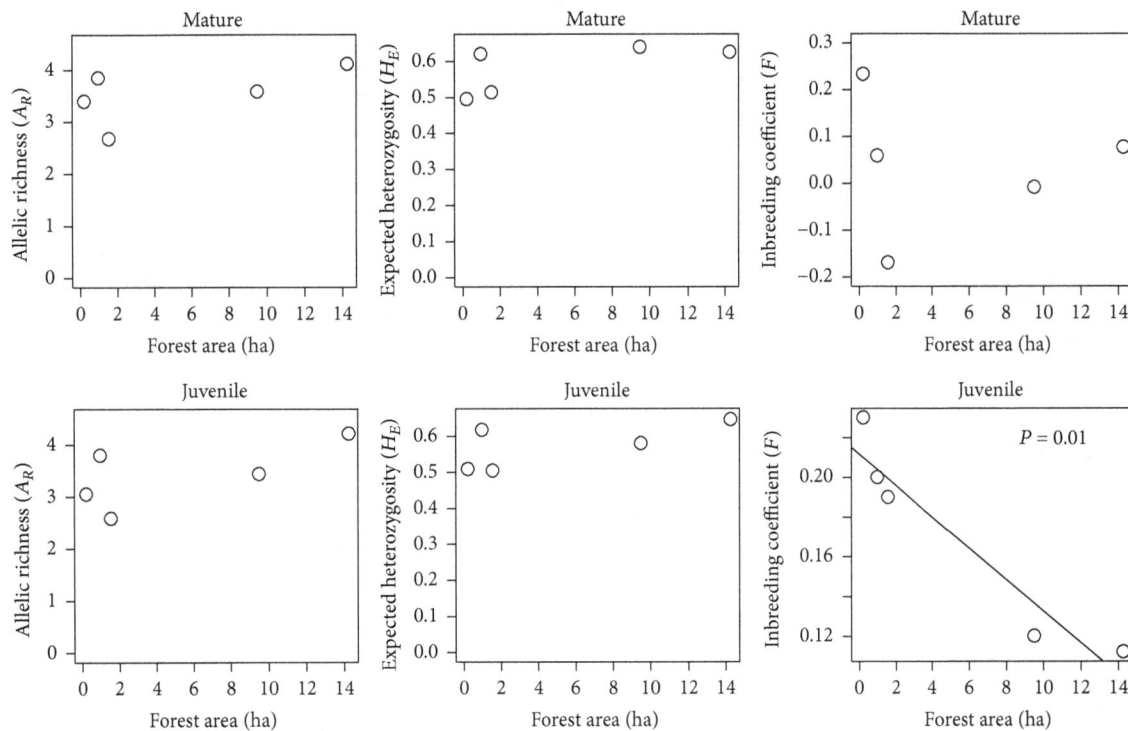

FIGURE 2: Relationships between the genetic diversity parameters (A_R, H_E, F) and forest area in mature and juvenile *Machilus thunbergii* trees.

lowest in OH. H_O and H_E values of mature trees ranged from 0.41 to 0.66 and 0.49 to 0.64, respectively, and those of juveniles ranged from 0.41 to 0.58 and 0.5 to 0.64. F values ranged from −0.17 to 0.23 for mature trees and from 0.11 to 0.23 for juveniles. F values of mature trees were highest at TM and lowest at OH, whereas those of juvenile trees were highest at TM and lowest at NC. No mature tree population deviated significantly from HWE, but deviation from HWE was observed in juvenile populations at OH, TM, and NH ($P < 0.05$). The posterior mean values of $F_{(null)}$ ranged from 0.02 to 0.10 for mature trees and from 0.05 to 0.16 for juveniles. Frequency of null allele ranged from 0.01 to 0.19 for mature trees and from 0.04 to 0.17 for juveniles.

Genetic variation within populations, as revealed by H_S, was higher for mature trees (0.60) than for juveniles (0.58) (Table 3). Values of H_T were the same in mature trees and juveniles (0.67). Genetic variation among populations was higher among juvenile compared to adult trees ($F_{ST} = 0.17$ and 0.13; $R_{ST} = 0.27$ and 0.21; $G_{ST}' = 0.35$ and 0.30, resp., Table 3).

3.2. Relationships between Forest Area and Genetic Diversity. Statistical analyses revealed that F values of juvenile trees were significantly negatively correlated with forest area ($P = 0.01$, Figure 2). F values of mature trees and A_R and H_E values of both mature and juvenile trees were not significantly correlated with forest area.

TABLE 3: Genetic variation within and among populations of *M. thunbergii*.

Populations	H_S	H_T	G'_{ST}	F_{ST}	R_{ST}
Among mature trees	0.60	0.67	0.30	0.13	0.21
Among juveniles	0.58	0.67	0.35	0.17	0.27

H_S, expected genetic diversity within subpopulations; H_T, total genetic diversity of overall population; G'_{ST}, genetic differentiation index of Hedrick (2005); F_{ST}, genetic differentiation index of Weir and Cockerham (1984); R_{ST}, genetic differentiation index of Rousset (1996).

3.3. Bottleneck Test. The BOTTLENECK analysis (using Wilcoxon's signed-rank test) indicated that recent population bottlenecks had occurred. When the Wilcoxon test was run under the IAM and TPM conditions, significant excess heterozygosity was detected in the OH, NH, and HI populations (Table 4). When the Sign test was run under the IAM conditions, significant excess heterozygosity was detected in the NH and HI populations (Table 4).

4. Discussion

4.1. Genetic Diversity. Previous meta-analysis indicated that the number of alleles is generally correlated with population size and rapidly declines with population size reduction [35]. However, in our result, the relationship between A_R value and forest area was not significant (Figure 2). We suppose that population history may play an important role in A_R value. In the prior study, we have revealed that populations on the western and eastern sides of Lake Biwa have different lineages, and these two lineages were admixed on the northern side of the lake from the phylogenetic analysis of *M. thunbergii* in the Kinki region [36]. Admixture of genetically differentiated populations increases the genetic heterogeneity of the newly established population [37]. In our results, NH and TM that locate at north side of Lake Biwa maintained relatively high levels of A_R value (Table 3) although these population sizes are small. These facts suggest that the effect of past lineage admixture events is grater than that of current population size on A_R value for *M. thunbergii* populations around Lake Biwa.

Values of H_E were not clearly different among these populations (Table 2). Generally, allelic diversity declines more rapidly than heterozygosity when the population size decreases, and a change in heterozygosity may take several generations to become apparent [2, 7]. Aguilar et al. [5] reported significant correlations between the number of generations and the extent of reduction of H_E. We suspect that because the reduction in population size of *M. thunbergii* around Lake Biwa occurred recently, values of H_E reflect the situation before the population size declined. Bottleneck testing supported this speculation and demonstrated significant excess heterozygosity in the OH, NH, and HI populations (Table 4, TM population was not analyzed). Bottleneck testing assumes that the reduction of allele number is faster than that of H_E [38], and it is effective for assessing recently experienced population size reduction [30]. Thus, our results indicate that the NC population has not experienced recent

population size reduction and maintained current population size for a long time, whereas the OH, NH, and HI populations experienced a population reduction within the past few generations.

4.2. Inbreeding Coefficient. Our analyses indicated that HWE is maintained in all populations at the mature stage (Table 2), although the statistical significance of the trend in the TM population needs further consideration owing to the small population size. Selection against homozygotes during recruitment could have affected the maintenance of HWE at the mature stage, as tree species generally experience high levels of inbreeding depression early in the life cycle [39]. *Machilus thunbergii* is a heterodichogamous tree, and this breeding system promotes outcrossing and regulates selfing [20]. The magnitude of inbreeding depression is strong in naturally outbreeding species [40], and the effects of inbreeding depression occur successively at different developmental stages, suggesting reduced levels of inbreeding at later stages due to enhanced mortality caused by inbreeding depression. Thus, selective death of homozygote individuals during growth from the juvenile to the mature stage may have reduced F values of mature trees.

We detected significant deviation from HWE at the juvenile stage in small and disturbed populations (OH, TM, and NH) and maintenance of HWE in relatively large populations (HI and NC; Table 2). In addition, F values of the juvenile stage were significantly correlated with forest area (Figure 2). The most likely explanation for these results is that human exploitation and habitat degradation have reduced the effective population size of adult trees in remnant populations, thus decreasing pollen availability and increasing the level of either self-fertilization or biparental inbreeding in the small, disturbed populations. Some previous researches indicate habitat degradation increases mating with relatives and genetic differentiation of future generations [7, 41–43]. On the other hand, result of $F_{(null)}$ and frequency of null allele indicate some populations contain substantial proportion of null alleles and this also contributed to the excessive homozygosity. Thus, the values of F of TM and OH population need further consideration.

Pollen flow among *M. thunbergii* trees occurs primarily within 200 m (Watanabe et al., unpublished data). In addition, insect pollinators often stay within a habitat fragment, rather than moving among fragments (e.g., [44, 45]). Moreover, habitat degradation has a strong influence on animal pollinators because it affects their population size and foraging behavior [46].

In this study, we were not able to evaluate contemporary gene flow, but results of inbreeding coefficient suggest that the chance of outside gene flow of *M. thunbergii* is possibly diminished by forest degradation in the OH, TM, and NH populations, whereas panmictic breeding is maintained in relatively large populations such as NC.

The genetic differentiation index values may also reflect an increase of inbreeding at the juvenile stage in these populations. The values were higher among juveniles than among mature trees (Table 3). When contemporary gene flow is restricted by habitat degradation, we would expect to find

TABLE 4: Probability of a bottleneck estimated using the program BOTTLENECK.

Population	Sign		Wilcoxon	
	IAM	TPM	IAM	TPM
OH	0.069	0.089	0.008*	0.027*
TM	NA	NA	NA	NA
NC	0.579	0.105	0.188	0.766
NH	0.018*	0.403	0.004*	0.039*
HI	0.015*	0.122	0.004*	0.008*

The probabilities of significant heterozygosity excess (evidence of bottleneck) for two-tailed Sign and Wilcoxon tests under the IAM and the TPM are marked with an asterisk ($^*P < 0.05$). NA, population that was not analyzed.

greater differentiation between seedlings (which are experiencing contemporary, limited gene flow) compared to adults (which experienced historic gene flow).

4.3. Conclusions. Our data suggest that the patterns of allelic diversity were more influenced by the past lineage admixture events than by the current population sizes for *M. thunbergii* populations around Lake Biwa. The extent of inbreeding at the juvenile stage was influenced by population size, and panmictic breeding was maintained in a large population. Our findings suggest that a recent population size decline will modify the mating pattern and increase inbreeding in these *M. thunbergii* populations.

Competing Interests

The authors declare that there is no conflict of interests regarding the publication of this paper.

Acknowledgments

This research was supported by a Grant-in-Aid for Scientific Research (nos. 25340115, 15H04418) from the Ministry of Education, Culture, Sports, Science and Technology, Japan.

References

[1] T. Kira, "Forest ecosystems of east and southeast Asia in a global perspective," *Ecological Research*, vol. 6, no. 2, pp. 185–200, 1991.

[2] A. Young, T. Boyle, and T. Brown, "The population genetic consequences of habitat fragmentation for plants," *Trends in Ecology and Evolution*, vol. 11, no. 10, pp. 413–418, 1996.

[3] R. Lande, "Anthropogenic, ecological and genetic factors in extinction and conservation," *Researches on Population Ecology*, vol. 40, no. 3, pp. 259–269, 1998.

[4] J. G. B. Oostermeijer, S. H. Luijten, and J. C. M. den Nijs, "Integrating demographic and genetic approaches in plant conservation," *Biological Conservation*, vol. 113, no. 3, pp. 389–398, 2003.

[5] R. Aguilar, M. Quesada, L. Ashworth, Y. Herrerias-Diego, and J. Lobo, "Genetic consequences of habitat fragmentation in plant populations: susceptible signals in plant traits and methodological approaches," *Molecular Ecology*, vol. 17, no. 24, pp. 5177–5188, 2008.

[6] G. Vranckx, H. Jacquemyn, B. Muys, and O. Honnay, "Meta-Analysis of Susceptibility of Woody Plants to Loss of Genetic Diversity through Habitat Fragmentation," *Conservation Biology*, vol. 26, no. 2, pp. 228–237, 2012.

[7] A. J. Lowe, D. Boshier, M. Ward, C. F. E. Bacles, and C. Navarro, "Genetic resource impacts of habitat loss and degradation; reconciling empirical evidence and predicted theory for neotropical trees," *Heredity*, vol. 95, no. 4, pp. 255–273, 2005.

[8] R. G. Collevatti, D. Grattapaglia, and J. D. Hay, "Population genetic structure of the endangered tropical tree species *Caryocar brasiliense*, based on variability at microsatellite loci," *Molecular Ecology*, vol. 10, no. 2, pp. 349–356, 2001.

[9] M. R. Lemes, R. Gribel, J. Proctor, and D. Grattapaglia, "Population genetic structure of mahogany (*Swietenia macrophylla* King, Meliaceae) across the Brazilian Amazon, based on variation at microsatellite loci: implications for conservation," *Molecular Ecology*, vol. 12, no. 11, pp. 2875–2883, 2003.

[10] C. J. Kettle, P. M. Hollingsworth, T. Jaffré, B. Moran, and R. A. Ennos, "Identifying the early genetic consequences of habitat degradation in a highly threatened tropical conifer, *Araucaria nemorosa* Laubenfels," *Molecular Ecology*, vol. 16, no. 17, pp. 3581–3591, 2007.

[11] S. Dayanandan, J. Dole, K. Bawa, and R. Kesseli, "Population structure delineated with microsatellite markers in fragmented populations of a tropical tree, *Carapa guianensis* (Meliaceae)," *Molecular Ecology*, vol. 8, no. 10, pp. 1585–1592, 1999.

[12] J. L. Hamrick, "Response of forest trees to global environmental changes," *Forest Ecology and Management*, vol. 197, no. 1–3, pp. 323–335, 2004.

[13] J. V. M. Bittencourt and A. M. Sebbenn, "Genetic effects of forest fragmentation in high-density *Araucaria angustifolia* populations in Southern Brazil," *Tree Genetics and Genomes*, vol. 5, no. 4, pp. 573–582, 2009.

[14] S. Kitamura, "Phytogeography of Shiga Prefecture," in *Flora of Shiga Prefecture*, S. Kitamura, Ed., p. Hoikusha, 1968 (Japanese).

[15] Y. Maesako, N. Noma, Y. Kaneko, M. Yokogawa, S. Watanabe, and Y. Azuma, "Necessity of biodiversity conservation of the *Persea thunbergii* forest in the Inukami River basin, Shiga Prefecture," *Bulletin of Kansai Organization for Nature Conservation*, vol. 34, no. 2, pp. 165–179, 2012 (Japanese).

[16] Y. Maesako, T. Fujiwaki, and Y. Kaneko, "Present distribution and regional vegetation of the *Persea thunbergii* population in western river basins of Lake Biwa, Shiga Prefecture, central Japan," *Osaka Sangyo University Journal of Human Environment Study*, no. 8, pp. 39–55, 2009 (Japanese).

[17] T. Hattori and S. Nakanishi, "On the distributional limits of the lucidophyllous forest in the Japanese Archipelago," *The Botanical Magazine Tokyo*, vol. 98, no. 4, pp. 317–333, 1985.

[18] T. Hattori, "Synecological study on *Persea thunbergii* type forest: geographical distribution and habitat conditions of *Persea*

thunbergii forest," *Japanese Journal of Ecology*, vol. 42, no. 3, pp. 215–230, 1992.

[19] T. Yumoto, "Pollination systems in a warm temperate evergreen broad-leaved forest on Yaku Island," *Ecological Research*, vol. 2, no. 2, pp. 133–145, 1987.

[20] S. Watanabe, N. Noma, and T. Nishida, "Flowering phenology and mating success of the heterodichogamous tree *Machilus thunbergii* Sieb. et Zucc. (Lauraceae)," *Plant Species Biology*, vol. 31, no. 1, pp. 29–37, 2016.

[21] H. Tagawa, "An investigation of initial regeneration in an evergreen broadleaved forest of MINAMATA special research area of IBP. I. Juvenile production and the distribution of two dominant species," *Report from EBINO Biology Lab Kyushu University*, vol. 1, no. 1, pp. 73–80, 1973.

[22] N. Noma and T. Yumoto, "Fruiting phenology of animal-dispersed plants in response to winter migration of frugivores in a warm temperate forest on Yakushima Island, Japan," *Ecological Research*, vol. 12, no. 2, pp. 119–129, 1997.

[23] B. Milligan, "Plant DNA isolation," in *Molecular Genetic Analysis of Populations: A Practical Approach*, IRL Press, 1992.

[24] Y. Kaneko, C. Lian, S. Watanabe, K.-I. Shimatani, H. Sakio, and N. Noma, "Development of microsatellites in *Machilus thunbergii* (Lauraceae), a warm-temperate coastal tree species in Japan," *American Journal of Botany*, vol. 99, no. 7, pp. e265–e267, 2012.

[25] A. El Mousadik and R. J. Petit, "High level of genetic differentiation for allelic richness among populations of the argan tree [Argania spinosa (L.) Skeels] endemic to Morocco," *Theoretical and Applied Genetics*, vol. 92, no. 7, pp. 832–839, 1996.

[26] J. Goudet, "FSTAT (version 1.2): a computer program to calculate F-statistics," *Journal of Heredity*, vol. 6, no. 6, pp. 485–486, 1995.

[27] I. J. Chybicki and J. Burczyk, "Simultaneous estimation of null alleles and inbreeding coefficients," *Journal of Heredity*, vol. 100, no. 1, pp. 106–113, 2009.

[28] R Development Core Team, *R: A Language and Environment for Statistical Computing*, R Foundation for Statistical Computing, Vienna, Austria, 2010.

[29] M. Nei, *Molecular Evolutionary Genetics*, Columbia University Press, 1987.

[30] S. Piry, G. Luikart, and J.-M. Cornuet, "BOTTLENECK: a computer program for detecting recent reductions in the effective population size using allele frequency data," *Journal of Heredity*, vol. 90, no. 4, pp. 502–503, 1999.

[31] B. S. Weir and C. C. Cockerham, "Estimating F-statistics for the analysis of population structure," *Evolution*, vol. 38, no. 6, pp. 1358–1370, 1984.

[32] F. Rousset, "Equilibrium values of measures of population subdivision for stepwise mutation processes," *Genetics*, vol. 142, no. 4, pp. 1357–1362, 1996.

[33] P. W. Hedrick, "A standardized genetic differentiation measure," *Evolution*, vol. 59, no. 8, pp. 1633–1638, 2005.

[34] R. Peakall and P. E. Smouse, "GenALEx 6.5: genetic analysis in Excel. Population genetic software for teaching and research-an update," *Bioinformatics*, vol. 28, no. 19, pp. 2537–2539, 2012.

[35] R. Leimu, P. Mutikainen, J. Koricheva, and M. Fischer, "How general are positive relationships between plant population size, fitness and genetic variation?" *Journal of Ecology*, vol. 94, no. 5, pp. 942–952, 2006.

[36] S. Watanabe, Y. Kaneko, Y. Maesako, and N. Noma, "Range expansion and lineage admixture of the Japanese evergreen tree *Machilus thunbergii* in central Japan," *Journal of Plant Research*, vol. 127, no. 6, pp. 709–720, 2014.

[37] B. Comps, D. Gömöry, J. Letouzey, B. Thiébaut, and R. J. Petit, "Diverging trends between heterozygosity and allelic richness during postglacial colonization in the European beech," *Genetics*, vol. 157, no. 1, pp. 389–397, 2001.

[38] T. Maruyama and P. A. Fuerst, "Population bottlenecks and nonequilibrium models in population genetics. II. Number of alleles in a small population that was formed by a recent bottleneck," *Genetics*, vol. 111, no. 3, pp. 675–689, 1985.

[39] K. M. Hufford and J. L. Hamrick, "Viability selection at three early life stages of the tropical tree, *Platypodium elegans* (Fabaceae, Papilionoideae)," *Evolution*, vol. 57, no. 3, pp. 518–526, 2003.

[40] B. C. Husband and D. W. Schemske, "Evolution of the magnitude and timing of inbreeding depression in plants," *Evolution*, vol. 50, no. 1, pp. 54–70, 1996.

[41] P. D. Rymer, M. Sandiford, S. A. Harris, M. R. Billingham, and D. H. Boshier, "Remnant *Pachira quinata* pasture trees have greater opportunities to self and suffer reduced reproductive success due to inbreeding depression," *Heredity*, vol. 115, no. 2, pp. 115–124, 2015.

[42] R. O. Manoel, P. F. Alves, C. L. Dourado et al., "Contemporary pollen flow, mating patterns and effective population size inferred from paternity analysis in a small fragmented population of the Neotropical tree *Copaifera langsdorffii* Desf. (Leguminosae-Caesalpinioideae)," *Conservation Genetics*, vol. 13, no. 3, pp. 613–623, 2012.

[43] E. V. Tambarussi, D. Boshier, R. Vencovsky, M. L. M. Freitas, and A. M. Sebbenn, "Paternity analysis reveals significant isolation and near neighbor pollen dispersal in small *Cariniana legalis* Mart. Kuntze populations in the Brazilian Atlantic Forest," *Ecology and Evolution*, vol. 5, no. 23, pp. 5588–5600, 2015.

[44] R. K. Didham, J. Ghazoul, N. E. Stork, and A. J. Davis, "Insects in fragmented forests: a functional approach," *Trends in Ecology & Evolution*, vol. 11, no. 6, pp. 255–260, 1996.

[45] M. Goverde, K. Schweizer, B. Baur, and A. Erhardt, "Small-scale habitat fragmentation effects on pollinator behaviour: experimental evidence from the bumblebee *Bombus veteranus* on calcareous grasslands," *Biological Conservation*, vol. 104, no. 3, pp. 293–299, 2002.

[46] L. Ashworth, R. Aguilar, L. Galetto, and M. A. Aizen, "Why do pollination generalist and specialist plant species show similar reproductive susceptibility to habitat fragmentation?" *Journal of Ecology*, vol. 92, no. 4, pp. 717–719, 2004.

19

Developmental Trends of Black Spruce Fibre Attributes in Maturing Plantations

Peter F. Newton

Canadian Wood Fibre Centre, Canadian Forest Service, Natural Resources Canada, 1219 Queen Street East, Sault Ste. Marie, ON, Canada P6A 2E5

Correspondence should be addressed to Peter F. Newton; peter.newton@canada.ca

Academic Editor: Kurt Johnsen

This study assessed the temporal developmental patterns of commercially relevant fibre attributes (tracheid length and diameters, wall thickness, specific surface area, wood density, microfibril angle, fibre coarseness, and modulus of elasticity) and their interrelationships within maturing black spruce (*Picea mariana* (Mill.) B.S.P.) plantations. A size-based stratified random sample procedure within 5 semimature plantations located in the Canadian Boreal Forest Region was used to select 50 trees from which radial cross-sectional xylem sequences at breast-height (1.3 m) were cut and analyzed. Statistically, the graphical and linear correlation analyses indicated that the attributes exhibited significant ($p \leq 0.05$) relationships among themselves and with morphological tree characteristics. Relative variation of each annually measured attribute declined with increasing size class (basal area quintile). The transitional shifts in temporal correlation patterns occurring at the time of approximate crown closure where suggestive of intrinsic differences in juvenile and mature wood formation processes. The temporal cumulative development patterns of all 8 of the annually measured attributes varied systematically with tree size and exhibited the most rapid rates of change before the trees reached a cambial age of 20 years. At approximately 50 years after establishment, plantation mean attribute values were not dissimilar from those reported for more mature natural-origin stands.

1. Introduction

Black spruce (*Picea mariana* (Mill.) B.S.P.) is a preferred reforestation species used in the establishment of plantations throughout the Canadian Boreal Forest Region [1]. Its desirable fibre characteristics and its ability to grow on a wide range of sites have combined to make black spruce an important industrial feedstock for the production of a broad array of end-products (e.g., dimensional lumber, pulp and paper, and composite wood products [2]). Black spruce plantations are an important component of the current and future industrial wood supply and are expected to provide a broad array of ecosystem services over their rotations (sensu [3]). The expectation is that these plantations will be well managed in accordance with an intensive silvicultural regime and hence will receive the required silvicultural treatments that promote end-product quality over their rotations [4]. For example, when plantations become overstocked and experience the resultant consequences of lower growth and increased mortality, treatments such as commercial thinning (CT) are expected to be carried out.

Operationally, a number of underlying factors that include the establishment of narrowly spaced plantations and high levels of natural ingress have resulted in an increased occurrence of overstocked plantations across the boreal landscape. Summary statistics based on 285 permanent growth and yield sample plots established in upland black spruce plantations across boreal Ontario revealed a relatively high level of density-stress within maturing plantations (i.e., mean absolute density of 2991 stems/ha with an associated relative density index of 0.5 at an average age of 28 years [6]). Given that a relative density index of 0.5 is the approximate threshold value at which imminent competition mortality or self-thinning is likely to occur [6], many plantations are likely to experience the adverse effects of increased competition which could include the loss of merchantable-sized trees. In response, forest managers are actively revising their initial crop plans in terms of adding density control treatments, such

TABLE 1: Product-based performance measures and their relationship with fibre attributes (sensu [5]).

Product category	Performance measure	Functional relationship with fibre attribute
	Tensile strength	$\propto$ fibre length, (wall thickness)$^{-1}$
	Tear strength	$\propto$ fibre length, coarseness
	Stretch	$\propto$ microfibril angle
Pulp and paper	Bulk	$\propto$ wall thickness, (fibre width)$^{-1}$
	Light scattering	$\propto$ (wall thickness)$^{-1}$
	Sheet formation	$\propto$ (fibre length)$^{-1}$
	Collapsibility	$\propto$ wall thickness
	Yield	$\propto$ density
Solid wood and composites	Strength	$\propto$ density, (microfibril angle)$^{-1}$
	Stiffness	$\propto$ density, modulus of elasticity, (microfibril angle)$^{-1}$

as CT [7]. Additionally, apart from ameliorating the effects of density-stress, thinning yields (residues) arising from CT treatments potentially represent an important interim supplemental source of fibre at midrotation (e.g., [8]).

Among the principal determinates governing the end-product potential of harvested trees are their internal fibre characteristics. For example, fibre attributes, such as tracheid dimensions (length and radial and tangential diameters (f_l, d_r, and d_t, resp.)), tracheid wall thickness (w_t), specific surface area (s_a), density (w_d), microfibril angle (m_a or MFA), fibre coarseness (c_o), and modulus of elasticity (m_e or MOE), directly influence end-product quality and quantity (Table 1 (sensu [5])). Consequently, attaining an understanding of the status of these commercially relevant fibre attributes as plantations develop and identifying factors that may influence their development could provide the prerequisite knowledge for forecasting end-product potential, in addition to helping guide forest managers in designing optimal silvicultural treatment schedules.

Thus, the objectives of this study were to investigate static and dynamic correlative patterns of association and the temporal developmental trends of commercially relevant fibre attributes within maturing black spruce plantations. The overall goals of this study were to provide inferences on the nature of the formulation pattern of these attributes, determine if developmental trends varied by tree size (i.e., hierarchical position within the stand as measured by basal area quintile), and infer potential end-products of conceptual CT yields based on the observed attribute values at midrotation relative to those observed within older natural-origin stands.

2. Materials and Methods

2.1. Data Acquisition and Processing. Breast-height (1.3 m) cross-sectional disks were obtained from 50 plantation-grown sample trees during the autumn of 2006 and 2007. These trees were a sample subset taken from a larger investigation studying tree taper and diameter growth patterns of boreal conifers (e.g., see [9–11] for specifics). The 50 trees were sampled from 5 even-aged monospecific black spruce plantations that were situated on upland sites throughout Forest Sections B9 (Superior) and B11 (Upper English River) of the Canadian Boreal Forest Region [1].

Geographically, the plantations were located within a rectangular region bounded by latitudes of 48° N on the south and 50° N on the north and longitudes of 88° W on the east and 92° W on the west (i.e., specific latitude/longitudes of the plantations were 49°05′12″/88°09′58″ (Plantation code W15); 48°56′06″/90°35′51″ (W18); 49°57′52″/92°29′39″ (W19); 50°11″47′/91°42′16″ (W20); and 48°42′11″/90°08′06″ (W23)). All five plantations shared a common silvicultural history: planted at approximately 2500 stems/ha (2 × 2 m intertree spacing) in the late 1960s or early 1970s on sites that were mechanically scarified.

The plantations varied slightly in terms of their productivity (medium-low to medium-high site qualities as inferred from their site index values (mean/min/max values of 16.4/14.8/18.3 m at 50 years, resp. [12]), ages (mean/min/max breast-height ages of 36/34/43, resp.), absolute stand densities (mean/min/max densities (stems/ha) of 3793/2076/5579, resp.) and site occupancy (fully stocked to overstocked as measured by relative density index (mean/min/max values of 0.76/0.51/1.02, resp. [13]). Variation in densities among the plantations was largely attributed to differential rates in natural ingress which occurred immediately following establishment.

Three variable-sized circular temporary sample plots were established within each plantation. The initial size of the sample plots was set at 400 m^2 which was, if necessary, increased by 100 m^2 increments until a minimum of 80 trees per plot were included. All living black spruce trees were sequentially numbered and measured for diameter at breast-height (1.3 m) outside-bark (D; cm) within each plot. The trees were then stratified into basal area quintiles from which 1 tree, which was classified as planted and did not exhibit any visible deformities, such as forks, major stem injuries, and dead or broken tops, was selected for destructive stem analysis. This stratified random sampling protocol ensured that all size and crown classes were represented and indirectly provided an analytical framework for assessing fibre attribute variation and developmental differences throughout the stand structure or size hierarchy. Subsequent to felling each sample tree, total height (H; m), breast-height age (A; yr), live crown ratio (C_r (%), calculated as the length of the live crown divided by total tree height and expressed as a percentage), and diameters of the largest dead (cm) and living

TABLE 2: Descriptive statistical summary of the mensurational characteristics of the 5 plantations and 47 sample trees.

Variable	Symbol	Unit	Mean	Standard error	Minimum	Maximum
Plantation level						
Age at breast-height of dominant trees		yr	36	2	34	43
Quadratic mean diameter		cm	11.0	1.45	6.4	15.8
Mean dominant height		m	12.9	0.90	10.5	16.5
Site index		m	16.4	0.70	14.8	18.3
Density		stems/ha	3793	449	2076	5579
Basal area		m^2/ha	30.0	4.78	17.8	45.4
Total volume		m^3/ha	189.1	43.27	87.1	348.3
Merchantable volume		m^3/ha	125.9	44.97	18.6	292.8
Mean volume per tree		dm^3	60.2	21.30	15.6	146.4
Relative density index		%/100	0.76	0.075	0.51	1.02
Tree level						
Breast-height age	A	yr	33	4	25	43
Height	H	m	10.9	1.03	6.7	17.9
Diameter at breast-height	D	cm	12.2	1.43	5.4	24.8
Maximum dead branch diameter		cm	1.3	0.09	0.5	2.2
Maximum live branch diameter		cm	1.5	0.10	0.1	2.5
Live crown ratio	C_r	%	59.4	4.91	37.2	96.7

(cm) branches were measured. Table 2 provides a summary of the principal mensurational characteristics of the sampled plantations and the individual sample trees selected.

The destructive stem analysis protocol consisted of felling each tree at stump height (approximately 30 cm above the ground surface) and obtaining 13 cross-sectional samples along the stem, specifically at 0.30, 0.5, 0.9, and 1.3 m, and thereafter at every 10% height interval based on remaining distance between breast-height and the stem tip [9]. These cross-sectional samples were then placed in burlap bags, transported to the research laboratory, and stored at − 10°C until 24 hours prior to processing. For the purposes of this study, 2 breast-height cross-sectional samples were randomly selected from each of the basal area quintiles within each plantation. A bark-to-pith-to-bark 2 cm square transverse sample along the geometric mean diameter was sawn from each disk, yielding a total of 50 cross-sectional transverse samples (1 transverse breast-height sample/tree × 2 trees/quintile × 5 quintiles/plantation × 5 plantations). The annual ring-width sequence and age of each sample were assessed for age inconsistencies and the presence of missing or partial rings using a binocular microscope. Age inconsistencies were found in 3 of the samples for which the observed age exceeded the age of the plantation. This finding would suggest that these trees originated from nonplanted residual saplings that were present in the previous stand. Consequently, these 3 trees were removed from the analysis. One pith-to-bark radial sequence was then sawn from each of the transverse samples. This resulted in a total of 47 annual-ring-width radial sequences available for fibre attribute determination.

The anatomical characterization of the fibre attributes along each annual-ring-width sequence was carried out using the SilviScan-3 system [14] and the Fibre Quality Analyzer [15]. Briefly, SilviScan-3 is the latest iteration of the integrated wood analysis system originally developed by Dr. Robert Evans and his research colleagues at the CSIRO's (Commonwealth Scientific and Industrial Research Organisation, Australia) Forestry and Forest Products Division. The system combines automatic image acquisition and analysis (cell scanner), X-ray densitometry, and X-ray diffractometry, to determine a multitude of fibre characteristics. For the purposes of this study, the following attributes were used from the Silviscan-3 output: (1) fibre dimensions inclusive of radial and tangential diameters, wall thickness, and specific surface area as determined via X-ray densitometry [16]; (2) fibre coarseness and density as determined via X-ray densitometry [16]; (3) microfibril angle as determined via X-ray diffraction [17]; and (4) modulus of elasticity as determined from a combination of X-ray densitometry and diffraction measurements [18]. Note that the samples were extracted with acetone in order to remove resins which may influence the density estimates. Specifically, the samples were soaked in acetone for 12 hours and then extracted for 8 hours at 70°C using a modified Soxhlet extraction system. After extraction, the samples were air-dried for approximately 12 hours and then conditioned to a 40% relative humidity at 20°C.

Separately, mean fibre length was determined for the entire sequence using a high resolution Fibre Quality Analyzer (OpTest Equipment Inc.) [15, 19] employing the following methodology (Source: Dr. Tong, FPInnovations Inc., Vancouver, British Columbia, Canada). The transverse radial sequence was sectioned into 2 age class segments for samples less than 30 years of age and into 3 age class segments for samples greater than or equal to 30 years. The high resolution FQA, which is comprised of hydraulic, optical, and image processing systems, was used to derive a mean fibre length estimate from a composite (bulk) macerated sample

TABLE 3: Statistical summary of the area-weighted fibre attribute values derived from the 47 breast-height cross-sectional sample disks at the time of sampling.

Variable	Symbol (nominal equivalent)	Unit	Mean	Standard error	Minimum	Maximum	CV[a] (%)
Fibre length (weight-weighted)	f_l	mm	2.4	0.2	1.8	2.8	8.8
Wood density	w_d	kg/m^3	487.3	42.5	399.7	574.3	8.7
Microfibril angle	m_a (MFA)	°	15.0	3.6	9.4	23.6	23.9
Modulus of elasticity	m_e (MOE)	GPa	13.0	2.7	8.5	18.0	20.4
Coarseness	c_o	μg/m	332.4	31.2	249.7	379.4	9.4
Wall thickness	w_t	μm	2.3	0.2	1.9	2.6	7.9
Tracheid radial diameter	d_r	μm	27.6	2.0	22.7	33.5	7.1
Tracheid tangential diameter	d_t	μm	25.9	1.4	22.6	29.0	5.4
Specific surface area	s_a	m^2/kg	336.1	25.0	305.0	400.8	7.4

[a]Coefficient of variation.

consisting of fibres from all the segments combined. More precisely, a dilute suspension of pulp fibres from each segment was obtained and processed through optical and imaging systems by a flow cell that orients fibres hydrodynamically into a 2-dimensional plane. The digital imaging system acquired and analyzed the images of the oriented fibres from which their length and other morphological properties were determined. To obtain the pulp sample, each segment was boiled in deionized water for approximately 4 hours after which the temperature was reduced to 70°C. Subsequently, the samples were soaked for an additional 12 hours. They were then macerated in a solution of hydrogen peroxide (35%) and glacial acetic acid (1:1 v/v) for 48 hours at a temperature of 70°C. The obtained pulp was washed, dispersed using a mixer, and filtered using a 150 mesh screen. Fibre length was then determined by sampling and weighing a predetermined amount of the dry pulp from the resulting handsheet, diluting it to a predetermined consistency, and analyzing the sample (in duplicate) using the FQA. Based on the number of fibres in each length class and mean fibre length within each class, an overall mean weight-weighted fibre length was calculated. Table 3 provides a statistical summary of the fibre attributes of the selected sample trees.

2.2. Computations and Analyses

2.2.1. Correlative Relationships.
Graphical analyses were used to determine the most appropriate linear expression for investigating (1) bivariate correlative relationships among attributes at the time of sampling (i.e., pairwise linear association among $f_l, w_d, m_a, m_e, c_o, w_t, d_r, d_t$, and s_a); (2) bivariate correlative relationships between attributes ($f_l, w_d, m_a, m_e, c_o, w_t, d_r, d_t$, and s_a) at the time of sampling and individual-tree morphological metrics (D, H, A, and C_r); and (3) temporal annual patterns of bivariate correlation between attributes (i.e., pairwise linear association among ring width area (a_r (mm^2)), $w_d, m_a, m_e, c_o, w_t, d_r, d_t$, and s_a by cambial age). Pairwise linearity was verified graphically using bivariate scatterplots and then statistically quantified via the Pearson moment correlation coefficient (r).

Attribute variation among the sample trees at the time of sampling was measured using the coefficient of variation.

Specifically, the ring area-weighted mean and associated standard derivation were calculated for $w_d, m_a, m_e, c_o, w_t, d_r, d_t$, and s_a along each pith-to-bark radial sequence, from which the coefficient of variation was computed:

$$cv_k = \frac{s_k}{m_k} = \frac{\left(s_k^2 \sum_{j=1}^{J} a_j^2 / \left(\sum_{j=1}^{J} a_j \right)^2 \right)^{1/2}}{\sum_{j=1}^{J} v_{k(j)} a_j / \sum_{j=1}^{J} a_j}, \tag{1}$$

where cv_k, s_k, s_k^2, and m_k are, respectively, the area-weighted sequence-specific coefficient of variation, standard deviation, variance, and mean value of the kth fibre attribute, a_j is the area of the annual ring (mm^2) corresponding to the jth cambial age ($j = 1, \ldots, J$), and $v_{k(j)}$ is the mean annual ring-width-area weighed value specific to the kth attribute and jth cambial age.

2.2.2. Accumulative Computations and Resultant Indices.
For each sequence, the cumulative annual-ring-area-weighted moving average was calculated for each attribute ($w_d, m_a, m_e, c_o, w_t, d_r, d_t$, and s_a) in the pith-to-bark direction:

$$V_{k(i)} = \frac{\sum_{j=1}^{i} v_{k(j)} a_j}{\sum_{j=1}^{i} a_j}, \tag{2}$$

where $V_{k(i)}$ is the cumulative area-weighted moving average value specific to the kth attribute calculated to the ith cambial age. Note that $W_d, M_a, M_e, C_o, W_t, D_r, D_t$, and S_a are used to denote the cumulative moving average value (i.e., $V_{k(i)}$ in (2)) for attributes $w_d, m_a, m_e, c_o, w_t, d_r, d_t$, and s_a, respectively. The cumulative moving average value reflects the accumulated status of a given attribute up to the ith cambial age and hence is indicative of the overall product potential of a tree if harvested at that specific age. Three-dimensional graphics were utilized in evaluating and summarizing the pattern of cumulative development of each variable over time by tree size class (basal area quintile). Note that Fortran-based analytical programs were written in order to calculate the cumulative values, evaluate and select the most appropriate bivariate correlative relationships, and generate the descriptive statistics. The graphical analysis and the statistical computations and testing were carried out employing Statistica (V12; Dell Inc.).

TABLE 4: Bivariate linear association among fibre attribute values at midrotation as measured by the Pearson product moment correlation coefficient.

Attribute[a]	Attribute[a]							
	f_l	w_d	m_a	m_e	c_o	w_t	d_r	d_t
f_l	1.0							
w_d	0.0400	1.0						
m_a	−0.3343*	−0.4840*	1.0					
m_e	0.2995*	0.7765*	−0.8981*	1.0				
c_o	0.6992*	0.1969*	−0.1569	0.2044	1.0			
w_t	0.4476*	0.8201*	−0.4429*	0.6765*	0.7201*	1.0		
d_r	0.4799*	−0.5952*	0.2820	−0.4477*	0.5913*	−0.0769	1.0	
d_t	0.5542*	−0.4677*	0.1585	−0.3037*	0.6700*	0.0609	0.7072*	1.0
s_a	−0.5090*	−0.7164*	0.3791*	−0.5812*	−0.8103*	−0.9743*	0.0560	−0.2047

Note: *significant correlation at the 0.05 probability level.
[a]Denotations are defined in Table 3.

TABLE 5: Bivariate linear association between tree level variables and fibre attribute values at midrotation as measured by the Pearson product moment correlation coefficient.

Attribute[a]	Tree level variables[b]			
	D	H	A	C_r
f_l	0.3636*	0.4017*	0.4117*	0.1808
w_d	−0.6089*	−0.4977*	−0.2559	0.2540
m_a	0.3757*	0.3337*	0.2293	−0.4099*
m_e	−0.5144*	−0.4068*	−0.2499	0.3752*
c_o	0.2493	0.1898	0.2946*	0.1878
w_t	−0.2832	−0.2330	0.0025	0.2820
d_r	0.7186*	0.5728*	0.4377*	−0.0671
d_t	0.4764*	0.3795*	0.3477*	−0.0096
s_a	0.1785	0.1704	−0.0296	−0.3185*

Note: *significant correlation at the 0.05 probability level.
[a]Denotations are defined in Table 3.
[b]Denotations are defined in Table 2.

3. Results and Discussion

3.1. Static Correlative Relationships. The correlation patterns between the fibre attributes largely followed expectation (Table 4). Increasing fibre lengths were associated with enhanced lumber related attributes such as lower m_a and increased m_e (stiffness). Decreases in tracheid diameters were associated with increasing wood density and stiffness as evident by the inverse d_r–w_d^{-1}, d_r–m_e^{-1}, d_t–w_d^{-1}, and d_t–m_e^{-1} correlative relationships. Increasing wall thickness which is associated with increased wood density and wood stiffness was also reflected in the results: that is, the positive w_t–w_d and m_e–w_t correlations. Strong correlations ($r \geq 0.70$) were observed among (1) wood density and the modulus of elasticity which are both related to solid wood products and lumber grades (w_d–m_e) and (2) fibre dimensional attributes associated with pulp yields and paper end-products (f_l–c_o, c_o–w_t, c_o–s_a^{-1}, w_t–s_a^{-1}, and d_r–d_t). Essentially, these results suggest that increasing wood density is accompanied by a suite of other favourable attributes which could translate into greater solid wood product yields and grades (Table 1). Similarly, long fibre lengths are accompanied with a suite of attributes which collectively are associated with enhanced pulp and paper yields and quality (Table 1).

Linear bivariate graphical patterns and significant ($p \leq 0.05$) correlations existed between some of the fibre attributes and tree level variables (Table 5). Fibre length was positively correlated with individual-tree diameter, height, and age. Tracheid length is associated with the degree of fibre-bonding and hence is directly related to the tensile and tear strength of derived paper products [20]. Consequently, the observed correlations for fibre length, suggest that the suitability of these plantation black spruce trees for pulp and paper production is enhanced with increasing tree size and age. Similar patterns of correlation between fibre attributes and diameter were observed for height (f_l, w_d^{-1}, m_a, m_e^{-1}, d_r, and d_t–H). Positive correlations between age and c_o, d_r, and d_t were also detected. Attributes m_a and s_a were negatively correlated with live crown ratio whereas m_e was positively correlated. Overall, the attributes were most strongly and frequently correlated the individual-tree size metrics (diameter and height).

3.2. Dynamic Correlative Relationships. Figure 1 graphically illustrates the pairwise correlation among the attributes on

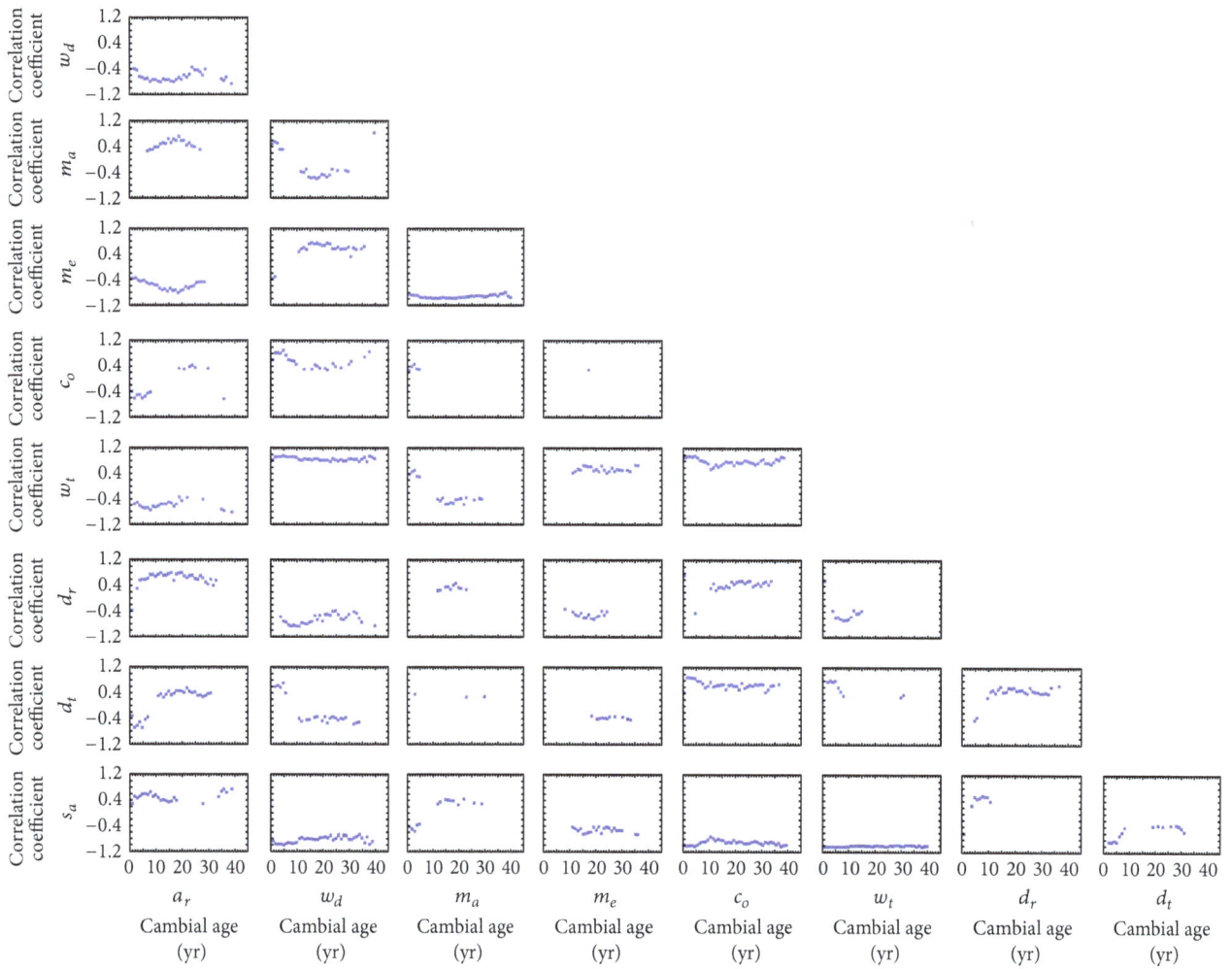

FIGURE 1: Matrix of temporal annual correlation patterns for bivariate relationships between individual attributes by cambial age. Note: (1) only Pearson moment correlation coefficients which were significant ($p \leq 0.05$) are displayed, and (2) a_r, w_d, m_a, m_e, c_o, w_t, d_r, d_t, and s_a denote annual-ring-width area, wood density, microfibril angle, modulus of elasticity, fibre coarseness, tracheid wall thickness, radial tracheid diameter, tangential tracheid diameter, and specific surface area, respectively.

an annual basis (cambial age) as the plantations developed. Six general trends emerged: (1) noncontinuous inverse correlative patterns for the a_r–w_d, a_r–m_e, a_r–w_t, w_d–d_r, m_e–d_r, m_e–d_t, m_e–s_a, and d_t–s_a relationships; (2) noncontinuous directly proportional patterns for the a_r–m_a, a_r–s_a, w_d–c_o, m_a–d_r, m_e–w_t, c_o–d_r, and c_o–d_t relationships; (3) approximately constant and continuous positive correlations across all cambial ages for the w_d–w_t and c_o–w_t relationships; (4) approximately constant and continuous negative correlations across all cambial ages for the w_d–s_a, m_a–m_e, c_o–s_a, and w_t–s_a relationships; (5) shifting negative-to-positive correlations for the a_r–c_o, a_r–d_r, a_r–d_t, w_d–m_e, m_a–s_a, and d_r–d_t relationships; and (6) shifting positive-to-negative correlations for the w_d–m_a, w_d–d_t, and m_a–w_t relationships.

These results suggest the presence of a set of developmental (age independent) invariant bivariate associations which included the w_d–w_t, w_d–s_a, m_a–m_e, c_o–w_t, c_o–s_a, and w_t–s_a relationships. The underlying morphological (functional) relationships that exist among the attributes may be

partially responsible for these results (e.g., $w_d = f(w_t)$). Furthermore, these temporal invariant relationships provide an opportunity to employ surrogate variables for estimating the values of other attributes, through a sequential set of nested bivariate relationships: $w_t = f(w_d) \rightarrow s_a = f(w_t) \rightarrow c_o = f(s_a)$.

The shifting positive-to-negative and negative-to-positive correlative patterns are plausible transitions points as the plantations reached the crown closure stage of development. This is the stage of stand development that is commonly associated with the cessation of juvenile or crown-based wood formation. These transition points coincided with cambial ages ranging from approximately 8 to 13 years as inferred graphically from Figure 1. Assuming a 5-year developmental period is required for the sample trees to reach breast-height (1.3 m), this cambial age range corresponds to a total stand age of approximately 13–18 years. This total age range is similar to the juvenile-mature wood transition age that has previously been reported in the literature for black spruce (i.e., 11–21 year range [21, 22]) and is also similar to that reported for white spruce (*Picea glauca* (Moench) Voss) [23]. Based on predictions derived from a structural stand density management diagram for these sampled plantations [6], the observed transition age range corresponds to the approximate time of crown closure.

Furthermore, the temporal positive-to-negative and negative-to-positive shifts in the correlations may indicate intrinsic differences in the underlying wood formation processes during the juvenile and postjuvenile growth stages. The majority of the attributes which exhibited transitions in their correlative patterns were directly or indirectly associated with cell geometry; for example, increasing radial diameters were associated with decreasing tangential diameters during the juvenile phase whereas they were directly proportional to each other during the maturing phase. Results derived from other studies are in accordance with this inference. For example, Lenz et al. [23] suggested that the accelerated growth which is likely the outcome of frequent anticline divisions of cambial initials partially underlies the unique geometrical characteristics of cells produced during the juvenile phase.

Collectively, these results provide confirmatory support for previous research in which cambial age has been found to be associated with unique fibre formation patterns [24]. For example, microfibril angle has been used as an investigative tool to differentiate between juvenile and mature wood types [25]. The higher growth rate and production of juvenile wood during the precrown closure stage may also elicit a temporary decline in wood quality in terms of end-product potential (e.g., lower w_d and m_e and higher m_a (Table 1)). Overall, the results and supporting references suggest that physiologically induced changes in allometric relationships were partially responsible for the observed shift in correlation patterns as the black spruce plantations transitioned from the precrown closure stage of development to the postcrown closure stage of development.

3.3. Temporal Developmental Patterns. The temporal development trends of the cumulative area-weighted moving average values by tree size (basal area quintile) for wood density, microfibril angle, modulus of elasticity, coarseness, tracheid wall thickness, tracheid radial diameter, tracheid tangential diameter, and specific surface area are graphically presented in Figure 2. Irrespective of attribute, the patterns revealed that the most rapid period of change was observed within the first decade of growth and that the rate of change varied inversely with tree size (e.g., rate of change declined with increased tree size). Among the attributes, W_d, C_o, W_t, D_r, D_t, and S_a exhibited the greatest rate of change during this period whereas M_a and M_e exhibited the least. Following this initial period of rapid change, (1) W_d, C_o, W_t, D_r, D_t, and S_a exhibited a period of relative stability that was characterized by a marginal but continuous increase (W_d, C_o, W_t, D_r, and D_t) or decrease (S_a) with increasing cambial age, and (2) M_a and M_e exhibited a continuous declining (M_a) or increasing (M_e) trend with increasing cambial age. For a given cambial age, W_d, M_e, C_o, and W_t exhibited size-dependency in which values declined with increasing basal area quintile. After

approximately 40 years, the rate of change within a given basal area quintile was minimal.

The results indicated that the most rapid change within the trajectories occurred before the trees reached a cambial age of 10 years at breast-height which corresponds to a plantation age of less than 20 years. This is approximately the termination point of the juvenile corewood or crown-based wood formation process as the plantations achieved crown closure status. These trends and inferences are in general accordance with the observed temporal correlation patterns (e.g., juvenile-mature wood transition occurring at the point of crown closure) and consistent with the developmental patterns reported for other softwood species [24–26].

Simultaneously controlling for age-dependent (ontogenetic) and growth-dependent (environmental) influences on fibre attributes is an analytical challenge when quantifying their temporal developmental trends. Attempts to include both ring area and cambial age as predictor variables when specifying fibre attribute prediction models have resulted in mixed success. As shown in Figure 1, annual ring area (growth rate) was significantly ($p \leq 0.05$) correlated to all 8 attributes for the majority of the years assessed. Consequently, accounting for fluctuations in annual growing conditions on fibre attributes by analyzing temporal developmental patterns by including the chronological calendar year of formation may help in minimizing growth-dependent effects underlying attribute variation. The effect of intraspecific tree competition on growth rate and by extension on fibre attribute variability is partially reduced by stratifying trees into hierarchical size-based classes (e.g., basal area quintiles). Given these inferences, composite multivariate allometric (e.g., [6, 27]) or hierarchical mixed models [28] are plausible analytical frameworks for consideration when attempting to quantify attribute developmental patterns.

3.4. Patterns of Attribute Variation and Modeling Implications. As forest management objectives migrate from a volumetric maximization to a value-added product-based paradigm, initiatives to develop fibre attribute prediction models have increased. In regard to coniferous boreal species, the fibre prediction equations developed for jack pine (*Pinus banksiana* Lamb.) [29], Scot pine (*Pinus sylvestris* L.) [30]; and black spruce [31] are representative examples of this renewed research focus. Analytically, arriving at the correct specification is largely an iterative process in which observed patterns of attribute development and associated variation across potential covariates are used to select a set of candidate fibre attribute prediction models. In this study, for planted black spruce trees, the variation in fibre attributes was found to decline with increased tree size (Figure 3). Specifically, the coefficient of variation for all 8 attributes decreased from a mean value of approximately 11-12% within the 1st quintile to a mean of 8-9% within the 5th quintile. The range of variation also declined with increasing basal area quintile (tree size). These patterns are suggestive of a size-dependent ordering of attribute variation throughout the stand structure. Combined with the observed size-dependent temporal development trends of the cumulative moving average values, plausible

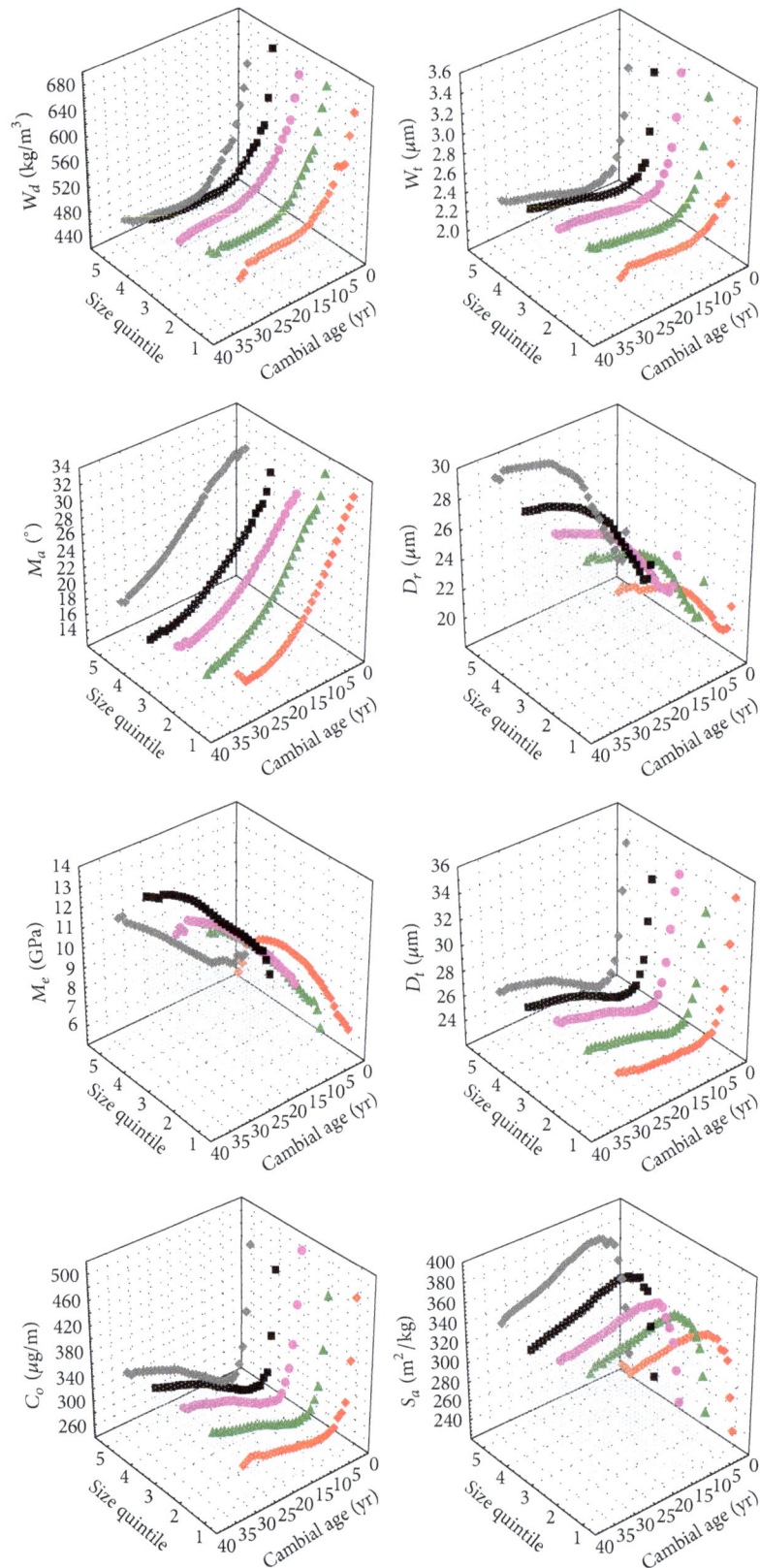

FIGURE 2: Three-dimensional visualization of mean temporal developmental trends illustrating patterns of change in attribute-specific cumulative area-weighted moving average values with increasing cambial age and size quintile. Note: W_d, M_a, M_e, C_o, W_t, D_r, D_t, and S_a denote the mean value up to the ith cambial age for wood density, microfibril angle, modulus of elasticity, fibre coarseness, tracheid wall thickness, radial tracheid diameter, tangential tracheid diameter, and specific surface area, respectively.

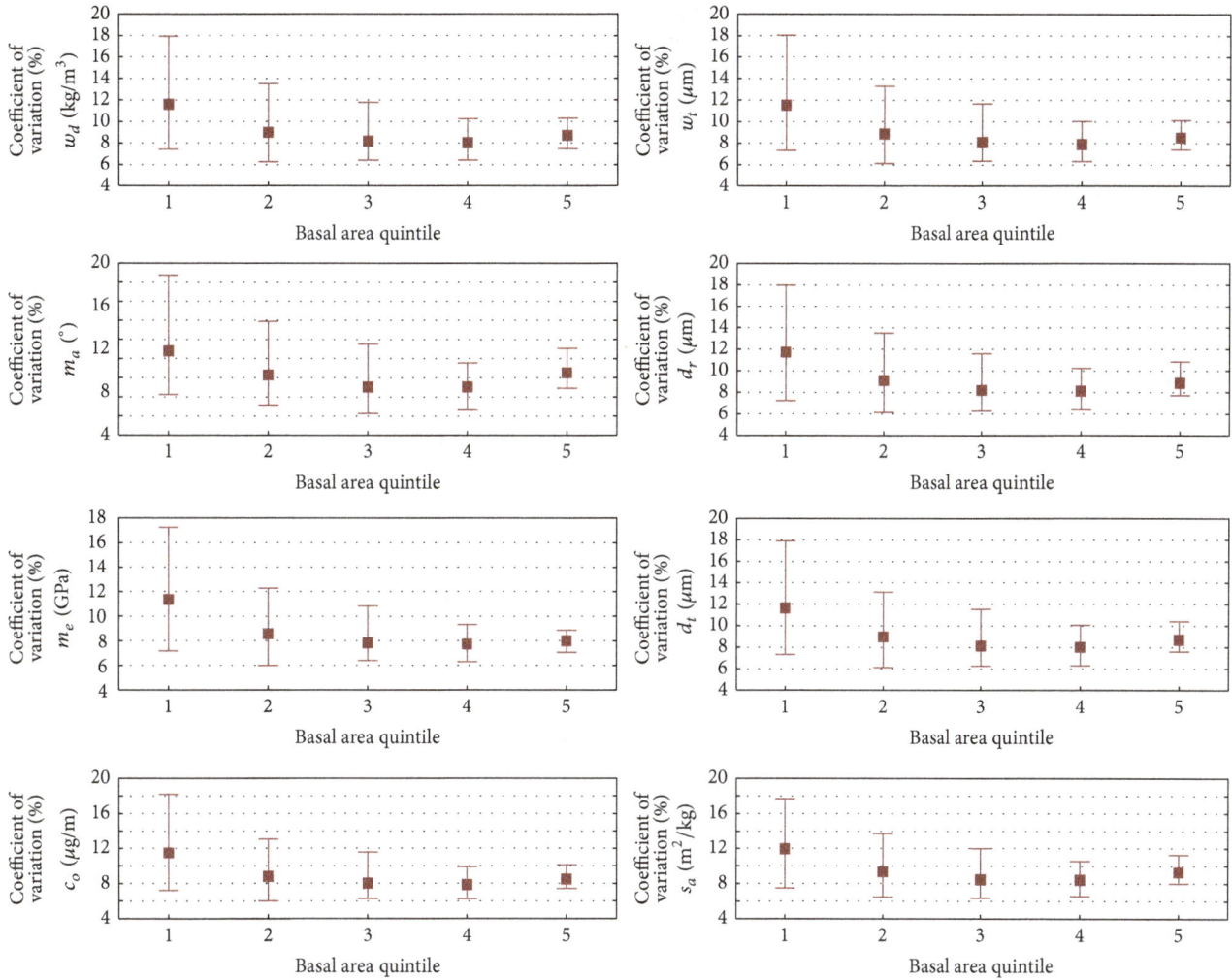

FIGURE 3: Fibre attribute variation as measured by the coefficient of variation by size quintile. Note: (1) the solid square denotes the mean value whereas the lower and upper whiskers denote minimum and maximum values, respectively, and w_d, m_a, m_e, c_o, w_t, d_r, d_t, and s_a are abbreviative notation for wood density, microfibril angle, modulus of elasticity, fibre coarseness, tracheid wall thickness, radial tracheid diameter, tangential tracheid diameter, and specific surface area, respectively.

candidate model specifications are those which include size-based explanatory variables. Furthermore, as shown in this study (Figure 1), all 8 attributes were significantly ($p \leq 0.05$) correlated with annual ring area (basal area growth rate). Consequently, accounting for fluctuations in annual growing conditions and their corresponding effect on fibre attribute formulation through the inclusion of a growth rate predictor variable may also increase the explanatory power of potential fibre attribute prediction models (e.g., [28]).

In summary, the results of this study suggest the fibre attribute models for plantation black spruce should consider the inclusion of predictor variables that reflect age-dependence (intrinsic ontogenetic development (e.g., cambial age)), growth-dependence (climatic-based environmental effects (e.g., annual-ring-width area)), and size-dependence (intraspecific competition (e.g., diameter)) sources of variation.

3.5. Relative Quality of Potential Commercial Thinning (CT) Residuals. Given the lack of definitive design specifications for the attributes considered in this study in terms of end-product potential, it was necessary to implement a supplementary comparative analysis in order to attain an appreciation of the quality of the fibre attributes of the black spruce plantations at midrotation. The conceptual rationale for this comparison is that the quality of attributes from more mature and natural-origin black spruce stands would be indicative of the black spruce fibre resource that has historically met or exceeded end-product quality expectations (e.g., [32]). Thus, if attributes from the plantations at midrotation were not substantially different from those observed within natural-origin stands at rotation, it would be reasonable to infer that the end-product potential of residuals arising from plausible commercial thinning treatments would be not substantially dissimilar.

Consequently, results from 2 regionally diverse and independent Silviscan-based studies on black spruce attributes were used in this comparison. The first external set was comprised of Silviscan results from 111 breast-height cores obtained from black spruce trees within 75 plots established in 45 stands that were situated on principally upland site-types within Forest Section B4 (Northern Clay) of the Canadian Boreal Forest Region [1] (Source: Dr. B. Pokharel, Nipissing University, North Bay, Ontario, Canada). Although the sample plots were established in stands ranging in age from 25 to 163 years, the majority were within developmentally mature and older stands, as evident from their structural characteristics at time of sampling: mean dominant height of 17.8 m (standard error/minimum/maximum: 3.6/9.0/25.9); mean basal area of 27.4 m^2/ha (10.3/17.0/51.4); mean quadratic diameter (cm) of 16.7 (3.8/11.0/28.3); and mean density of 1324 stems/ha (577/1750/2601). The second external data set consisted of results derived from the Silvican analysis of 160 breast-height cores sampled from merchantable-sized (breast-height diameter $\geq$ 9 cm) black spruce trees located within 16 black spruce stands situated throughout Forest Sections B28a (Grand Falls), B28b (Corner Brook), and B29 (Northern Peninsula) of the Canadian Boreal Forest Region [1] (Source: Table 1 in [33]). Representative mensurational characteristics of the merchantable-sized tree population within these sampled stands suggested that most were at a mature stage of development: mean dominant height of 11.9 m (standard error/minimum/maximum: 2.8/5.9/17.1); mean basal area of 25.2 m^2/ha (12.6/1.9/46.3); mean quadratic diameter (cm) of 13.4 (2.5/9.2/18.3); and mean density of 1725 stems/ha (712/250/3200).

Denoting these Silviscan data sets from Ontario and Newfoundland as SS-ON and SS-NF, respectively, the results from this comparison revealed slight to moderate differences between the plantations and the mature natural-origin stands in terms of mean lower wood density (487 versus 537 (SS-ON) and 562 (SS-NF)), higher microfibril angle (15.0 versus 12.8 (SS-ON) and 13.9 (SS-NF)), lower modulus of elasticity (13.0 versus 14.9 (SS-ON) and 15.2 (SS-NF)), lower fibre coarseness (332 versus 375 (SS-ON) and 395 (SS-NF)), lower wall thickness (2.3 versus 2.6 (SS-ON) and 2.8 (SS-NF)), higher radial diameter (27.6 versus 27.5 (SS-ON) and 25.9 (SS-NF)), lower tangential diameter (25.9 versus 26.4 (SS-ON) and 26.5 (SS-NF)), and higher specific surface area (336 versus 301 (SS-ON) and 285 (SS-NF)). Similarly, differences among the attributes in terms of their relative variation were not substantially dissimilar: coefficient of variation (%) values of 8.7 versus 9.3 (SS-ON) and 6.8 (SS-NF) for wood density, 24.0 versus 24.3 (SS-ON) and 15.8 (SS-NF) for microfibril angle, 20.8 versus 16.8 (SS-ON) and 11.2 (SS-NF) for modulus of elasticity, 9.4 versus 10.3 (SS-ON) and 7.1 (SS-NF) for fibre coarseness, 8.7 versus 11.5 (SS-ON) and 7.6 (SS-NF) for wall thickness, 7.2 versus 5.1 (SS-ON) and 2.2 (SS-NF) for radial diameter, 5.4 versus 5.3 (SS-ON) and 2.3 (SS-NF) for tangential diameter, and 7.4 versus 9.2 (SS-ON) and 6.7 (SS-NF) for specific surface area.

Thus, based on these descriptive comparisons, plantation attributes at midrotation were not substantially different from their mature natural-origin stand counterparts with the possible exception of microfibril angle and fibre coarseness. However, given the importance of these two attributes in terms of their effect on pulp quality (sensu Table 1), these differences may be consequential in cases were the thinning residuals are directed towards pulp mills for the production of paper-related end-products. More generally, caution must also be exercised when interpreting such comparative results, in that the size-dependent effect on the developmental patterns of the attributes which was clearly evident for the trees analyzed in this study, is not reflected when employing simple measures of central tendency (sample means).

3.6. Potential Management Implications. Although the development of fibre attributes is under strong ontogenetic control and hence largely predetermined, phenotypic plasticity arising from mechanical (support) or hydraulic (physiology) requirements enables trees to modify their developmental patterns while under external stress. This intrinsic plasticity capability helps trees in maximizing their survival probabilities (sensu [34]). As evident by the size-dependent attribute developmental patterns and variability trends observed in this study, black spruce exhibits a considerable range of phenotypic plasticity. Thus, the potential to indirectly influence fibre attribute developmental variation through the regulation of size structures via density control treatments, offers forest managers the opportunity to optimally manipulate not only the quantity but also the quality of the fibre yields. Furthermore, natural pruning, branch diameter growth, juvenile wood production, and the rotational end-products are directly influenced by site occupancy regulation. Specifically, the establishment and maintenance of adequate site occupancies that facilitate rapid crown closure and encourage intraspecific competition have been shown to improve end-product quality at rotation (e.g., black spruce [25], Norway spruce (*Picea abies* (L.) Karst) [35], Jack pine [36], and Douglas-fir (*Pseudotsuga menziesii* (Mirb.) Franco) [37]).

The size-dependent ordering of fibre attribute development observed by midrotation in this study partially arose from intense intraspecific competition among neighbouring trees once the plantations achieved crown closure status. Asymmetrical competition, principally for above-ground resources (light), has been found to underlie the hierarchical patterns of size-dependent growth within black spruce stands [38]. This competitive effect eventually governs structural development patterns and self-thinning processes. The results of this study suggest that fibre attributes were also adhering to a similar pattern of differentiation as a consequence of competition. Hence, the ability to manipulate competition processes offers forest managers a pathway to indirectly influence fibre attribute developmental patterns and ultimately end-product potentials.

4. Conclusions

The objectives of this study were to analyze the temporal developmental patterns of commercially relevant fibre attributes and their interrelationships within maturing black spruce plantations. The results from the analyses indicated that fibre attributes were correlated among themselves and with morphological tree characteristics and exhibited unique size-dependent temporal developmental patterns. After approximately 40 years of development, the rate of change for the majority of the attributes was minimal. The period of most rapid change was during the precrown closure period of stand development which encompassed the first 20 years of growth. A pattern of size-dependent ordering of attribute variation within the black spruce plantations was evident from the analyses. Although the results of this study suggested that commercial row thinning of similar density-stressed plantations at midrotation may potentially yield satisfactory fibre quality for specific end-products, the lack of product-based design specifications negated a conclusive determination. Identification of some of the sources of variation underlying the development of the studied attributes enabled inferences to be derived for potential use in future modeling studies.

Competing Interests

The author declares no conflict of interests.

Acknowledgments

The author expresses his appreciation to (1) Dr. Sharma of the Ontario Forest Research Institute, Ontario Ministry of Natural Resources and Forestry, for jointly facilitating the field sampling and overseeing the storage logistics of the cross-sectional disk samples, (2) Mike Laporte of the Canadian Wood Fibre Centre, Canadian Forest Service, for cutting out the transverse samples during the laboratory preparation phase, (3) staff at FPInnovations Inc., for completing the SilviScan-3 and the fibre quality analyses, and (4) the Canadian Wood Fibre Centre for fiscal support.

References

[1] J. S. Rowe, "Forest regions of Canada," Publication 1300, Government of Canada, Department of Environment, Canadian Forestry Service, Ottawa, Canada, 1972.

[2] S. Y. Zhang and A. Koubaa, *Softwoods of Eastern Canada: Their Silvics, Characteristics, Manufacturing and End-Uses*, Special Publication SP-526E, FPInnovations, Quebec City, Canada, 2008.

[3] C. Malouin, G. Larocque, M. Doyle, F. Bell, J. Dacosta, and K. Liss, "Considerations of ecosystem services in ecological forest management," in *Ecological Forest Management Handbook*, G. R. Larocque, Ed., Applied Ecology and Environmental Management, pp. 107–138, CRC Press, 2015.

[4] F. W. Bell, J. Parton, D. Joyce et al., "Developing a silvicultural framework and definitions for use in forest management planning and practice," *Forestry Chronicle*, vol. 84, no. 5, pp. 678–693, 2008.

[5] M. Defo, "SilviScan-3: a revolutionary technology for high-speed wood microstructure and properties analysis," 2008, http://chaireafd.uqat.ca/midiForesterie/pdf/20080422PresentationMauriceDefo.pdf.

[6] P. F. Newton, "A decision-support system for forest density management within upland black spruce stand-types," *Environmental Modelling & Software*, vol. 35, pp. 171–187, 2012.

[7] L. M. McKinnon, G. J. Kayahara, and R. G. White, "Biological framework for commercial thinning evenaged single-species stands of jack pine, white spruce, and black spruce in Ontario," 2006, http://www.forestresearch.ca/Projects/fibre/Framework-CTofPjSwSb.pdf.

[8] J. V. Hatton and S. S. Johal, "Mechanical pulping of commercial thinnings of six softwoods from New Brunswick: TMP and CTMP from spruce thinnings are a viable source for printing and writing grades," *Pulp and Paper Canada*, vol. 97, no. 12, pp. 93–97, 1996.

[9] P. F. Newton and M. Sharma, "Evaluation of sampling design on taper equation performance in plantation-grown *Pinus banksiana*," *Scandinavian Journal of Forest Research*, vol. 23, no. 4, pp. 358–370, 2008.

[10] M. Sharma and J. Parton, "Modeling stand density effects on taper for jack pine and black spruce plantations using dimensional analysis," *Forest Science*, vol. 55, no. 3, pp. 268–282, 2009.

[11] N. Subedi and M. Sharma, "Individual-tree diameter growth models for black spruce and jack pine plantations in northern Ontario," *Forest Ecology and Management*, vol. 261, no. 11, pp. 2140–2148, 2011.

[12] W. H. Carmean, G. Hazenberg, and K. C. Deschamps, "Polymorphic site index curves for black spruce and trembling aspen in northwest Ontario," *Forestry Chronicle*, vol. 82, no. 2, pp. 231–242, 2006.

[13] P. F. Newton, "Asymptotic size-density relationships within self-thinning black spruce and jack pine stand-types: parameter estimation and model reformulations," *Forest Ecology and Management*, vol. 226, no. 1–3, pp. 49–59, 2006.

[14] M. Defo, A. Goodison, and N. Uy, "A method to map within-tree distribution of fibre properties using SilviScan-3 data," *Forestry Chronicle*, vol. 85, no. 3, pp. 409–414, 2009.

[15] G. Robertson, J. Olson, P. Allen, B. Chan, and R. Seth, "Measurement of fiber length, coarseness, and shape with the fiber quality analyzer," *Tappi Journal*, vol. 82, no. 10, pp. 93–98, 1999.

[16] R. Evans, "Rapid measurement of the transverse dimensions of tracheids in radial wood sections from *Pinus radiata*," *Holzforschung*, vol. 48, no. 2, pp. 168–172, 1994.

[17] R. Evans, S. A. Stuart, and J. Van Der Tou, "Microfibril angle scanning of increment cores by X-ray diffractometry," *Appita Journal*, vol. 49, no. 6, pp. 411–414, 1996.

[18] R. Evans, "Wood stiffness by X-ray diffractometry," in *Characterization of the Cellulosic Cell Wall*, D. D. Stokke and L. H. Groom, Eds., pp. 138–146, Blackwell Publishing, Ames, Iowa, USA, 1st edition, 2006.

[19] R. J. Trepanier, "Automatic fiber length and shape measurement by image analysis," *Tappi Journal*, vol. 81, no. 6, pp. 152–154, 1998.

[20] D. D. S. Perez and T. Fauchon, "Wood quality for pulp and paper," in *Wood Quality and Its Biological Basis*, J. R. Barnett and G. Jeronimidis, Eds., pp. 157–186, Blackwell Publishing, Ames, Iowa, USA, 1st edition, 2003.

[21] K. C. Yang and G. Hazenberg, "Impact of spacing on tracheid length, relative density, and growth rate of juvenile wood and mature wood in *Picea mariana*," *Canadian Journal of Forest Research*, vol. 24, no. 5, pp. 996–1007, 1994.

[22] J. Alteyrac, A. Cloutier, C.-H. Ung, and S. Y. Zhang, "Mechanical properties in relation to selected wood characteristics of black

spruce," *Wood and Fiber Science*, vol. 38, no. 2, pp. 229–237, 2006.

[23] P. Lenz, A. Cloutier, J. MacKay, and J. Beaulieu, "Genetic control of wood properties in *Picea glauca*—an analysis of trends with cambial age," *Canadian Journal of Forest Research*, vol. 40, no. 4, pp. 703–715, 2010.

[24] S. D. Mansfield, R. Parish, C. M. Di Lucca, J. Goudie, K.-Y. Kang, and P. Ott, "Revisiting the transition between juvenile and mature wood: a comparison of fibre length, microfibril angle and relative wood density in lodgepole pine," *Holzforschung*, vol. 63, no. 4, pp. 449–456, 2009.

[25] M. Wang and J. D. Stewart, "Determining the transition from juvenile to mature wood microfibril angle in lodgepole pine: a comparison of six different two-segment models," *Annals of Forest Science*, vol. 69, no. 8, pp. 927–937, 2012.

[26] I. D. Cave and J. C. F. Walker, "Stiffness of wood in fast-grown softwoods: the influence of microfibril angle," *Forest Products Journal*, vol. 44, no. 5, pp. 43–48, 1994.

[27] P. F. Newton, "Forest production model for upland black spruce stands—optimal site occupancy levels for maximizing net production," *Ecological Modelling*, vol. 190, no. 1-2, pp. 190–204, 2006.

[28] W. Xiang, M. Leitch, D. Auty, E. Duchateau, and A. Achim, "Radial trends in black spruce wood density can show an age- and growth-related decline," *Annals of Forest Science*, vol. 71, no. 5, pp. 603–615, 2014.

[29] P. F. Newton, "Development of an integrated decision-support model for density management within jack pine stand-types," *Ecological Modelling*, vol. 220, no. 23, pp. 3301–3324, 2009.

[30] H. Mäkinen and J. Hynynen, "Predicting wood and tracheid properties of Scots pine," *Forest Ecology and Management*, vol. 279, pp. 11–20, 2012.

[31] B. Pokhare, J. P. Dech, A. Groot, and D. Pitt, "Ecosite-based predictive modeling of black spruce (*Picea mariana*) wood quality attributes in boreal Ontario," *Canadian Journal of Forest Research*, vol. 44, no. 5, pp. 465–475, 2014.

[32] P. Watson and M. Bradley, "Canadian pulp fibre morphology: superiority and considerations for end use potential," *Forestry Chronicle*, vol. 85, no. 3, pp. 401–408, 2009.

[33] J. E. Luther, R. Skinner, R. A. Fournier et al., "Predicting wood quantity and quality attributes of balsam fir and black spruce using airborne laser scanner data," *Forestry*, vol. 87, no. 2, pp. 313–326, 2014.

[34] B. Lachenbruch, J. R. Moore, and R. Evans, "Radial variation in wood structure and function in woody plants, and hypotheses for its occurrence," in *Size- and Age-Related Changes in Tree Structure and Function*, C. Meinzer, B. Lachenbruch, and T. E. Dawson, Eds., vol. 4 of *Tree Physiology*, pp. 121–164, Springer, Dordrecht, The Netherlands, 2011.

[35] K. Johansson, "Effects of initial spacing on the stem and branch properties and graded quality of *Picea abies* (L.) Karst," *Scandinavian Journal of Forest Research*, vol. 7, no. 1–4, pp. 503–514, 1992.

[36] K.-Y. Kang, S. Y. Zhang, and S. D. Mansfield, "The effects of initial spacing on wood density, fibre and pulp properties in jack pine (*Pinus banksiana* Lamb.)," *Holzforschung*, vol. 58, no. 5, pp. 455–463, 2004.

[37] A. Rais, W. Poschenrieder, H. Pretzsch, and J.-W. G. van de Kuilen, "Influence of initial plant density on sawn timber properties for Douglas-fir (*Pseudotsuga menziesii* (Mirb.) Franco)," *Annals of Forest Science*, vol. 71, no. 5, pp. 617–626, 2014.

[38] P. F. Newton and P. A. Jolliffe, "Assessing processes of intraspecific competition within spatially heterogeneous black spruce stands," *Canadian Journal of Forest Research*, vol. 28, no. 2, pp. 259–275, 1998.

20

Allometric Models for Estimating Tree Volume and Aboveground Biomass in Lowland Forests of Tanzania

Wilson Ancelm Mugasha,[1] **Ezekiel Edward Mwakalukwa,**[2] **Emannuel Luoga,**[3]
Rogers Ernest Malimbwi,[3] **Eliakimu Zahabu,**[3] **Dos Santos Silayo,**[4] **Gael Sola,**[5]
Philippe Crete,[5] **Matieu Henry,**[5] **and Almas Kashindye**[6]

[1]*Tanzania Forestry Research Institute (TAFORI), P.O. Box 1854, Morogoro, Tanzania*

[2]*Department of Forest Biology, Sokoine University of Agriculture, P.O. Box 3010, Morogoro, Tanzania*

[3]*Department of Forest Mensuration and Management, Sokoine University of Agriculture, P.O. Box 3013, Morogoro, Tanzania*

[4]*Department of Forest Engineering, Sokoine University of Agriculture, P.O. Box 3012, Morogoro, Tanzania*

[5]*UN-REDD Programme, Food and Agriculture Organization of the United Nations (FAO), Viale delle Terme di Caracalla,*
00153 Rome, Italy

[6]*Forest Training Institute, Olmotonyi, P.O. Box 943, Arusha, Tanzania*

Correspondence should be addressed to Wilson Ancelm Mugasha; wilmugasha@gmail.com

Academic Editor: Timothy Martin

Models to assist management of lowland forests in Tanzania are in most cases lacking. Using a sample of 60 trees which were destructively harvested from both dry and wet lowland forests of Dindili in Morogoro Region (30 trees) and Rondo in Lindi Region (30 trees), respectively, this study developed site specific and general models for estimating total tree volume and aboveground biomass. Specifically the study developed (i) height-diameter (ht-dbh) models for trees found in the two sites, (ii) total, merchantable, and branches volume models, and (iii) total and sectional aboveground biomass models of trees found in the two study sites. The findings show that site specific ht-dbh model appears to be suitable in estimating tree *height* since the tree allometry was found to differ significantly between studied forests. The developed general volume models yielded unbiased mean prediction error and hence can adequately be applied to estimate tree volume in dry and wet lowland forests in Tanzania. General aboveground biomass model appears to yield biased estimates; hence, it is not suitable when accurate results are required. In this case, site specific biomass allometric models are recommended. Biomass allometric models which include basic wood density are highly recommended for improved estimates accuracy when such information is available.

1. Introduction

In Tanzania, lowland forests are located close to the Indian Ocean, and occasionally further inland up to the base of the Eastern Arc Mountains below 1000 m above sea level, often embedded within larger areas of miombo woodlands and Montane/humid forests [1]. The total area covered by lowland forest in Tanzania is estimated to be about 1.7 mil. ha [2]. Depending on the magnitude of precipitation, lowland forests may be categorised into dry (<1 000 mm) and wet (>1 000 mm) [3]. In the northern part of Tanzania,

the lowland forest strips are very thin but as one moves south, the strips expand further to the inland. Based on the National Forest Resource Monitoring and Assessment (NAFORMA) classification, of the eight land cover types, the lowland forest belongs to *"forest"* cover [2]. Other lands classified in these cover types include humid Montane, Mangrove, and plantations.

Lowland forest supports the livelihood of thousands of people directly (fuel wood, food, medicine, and construction materials) and indirectly by offering environmental services which include biodiversity, catchment values, and carbon sequestration. The latter has recently received global attention due to climate mitigation function they offer [4]. However, there is uncertainty of the quantities of carbon stocks in the lowland forests in Tanzania since no local biomass allometric model is available.

Volume models which are able to quantify merchantable tree volume and total volume are also required when trees are warranted for commercial purposes. Timber licensing and pricing system in Tanzania based on volume estimation [5] requires also that tree-sectional volume models are developed. Such models will aid in obtaining accurate quantitative information on the amount of wood for specific uses, that is, saw timber and fuel wood. To date there are no total or tree sectional volume estimation models for lowland forest of Tanzania. Preferred trees species for timber in the lowland forests include *Pterocarpus angolensis, Afzelia quanzensis,* and *Sterculia quinqueloba* [6]. However, due to diminishing rate of these tree species and large demand of timber, lesser known timber tree species has been exploited [7, 8]. Therefore, this necessitates the need to develop multispecies volume models other than for only known timber tree species [9]. Though not common in the scientific literature, many multispecies volume models have been developed and can be found in the international allometric equations database GlobAllomeTree [10].

The need for quantification of carbon stocks for different forest types is also relevant for the emerging carbon credit market mechanism such as Reducing Emission from Deforestation and Forest Degradation (REDD). This requires that appropriate allometric models specific for a given forest type are in place [11, 12]. Allometric models use the easy to measure individual tree parameters such as diameter at breast height (dbh) and total tree height (ht) from forest inventories to estimate volume and AGB. Another important explanatory variable for biomass estimating allometric model is wood basic density (WD) which is determined from wood samples in laboratory as a ratio of dry mass to the green volume [13]. Literatures list these variables according to their importance as dbh, WD, and ht in explaining tree biomass variations and dbh and ht for tree volume [12, 13]. Among the three explanatory variables, tree dbh and ht have been often used as only explanatory variables to develop biomass allometric models because they are readily available compared to WD which results in overall poor estimation of AGB [14, 15] especially for forests where WD of trees varies considerably [12, 13]. This calls for the need of developing biomass allometric models which integrate WD in estimating tree biomass.

Conventionally, forest inventories measure dbh of all trees in each plot but often few are randomly selected and measured for ht for development of simple and local ht-dbh models that are used to estimate ht of trees not measured in the field [16, 17]. This implies that biomass allometric model, in practice, requires local ht-dbh models for ht estimation. Although Mugasha et al. [18] recently developed ht-dbh models for four main forest types including lowland forest, none of the sites were selected from the lowland forests in the coast. Furthermore, due to large variations in ht from one forest to another as a function of climate and other

environmental factors, it is imperative that a local ht-dbh model is developed to improve the tree volume or biomass estimates [19, 20].

It is against this background that this study entails develop site specific and general models for estimating total tree volume and aboveground biomass. Specifically the study aims to develop and compare (i) height-diameter (ht-dbh) models for trees found in the two sites, (ii) total, merchantable, and branches volume models, and (iii) total and sectional aboveground biomass models of trees found in the two study sites.

2. Material and Methods

2.1. Study Sites Description. The study area covered two forest reserves, namely, Rondo forest reserve located in Lindi Region and Dindili forest reserve located in Morogoro Region in Tanzania (Figure 1). Rondo forest reserve is located along the coast of Indian Ocean (39.08°E, 10.04°S) 46 km from the Indian Ocean shores in Lindi Region (Figure 1). The area of the forest is about 14 060 ha and it is managed by the government. The forest is described as lowland forest (wet) and situated at the top of the plateau in a relatively flat terrain between 465 and 885 m above sea level. The average annual rainfall is 1 215 mm and the mean annual temperature is between 15 and 31°C. Dindili forest reserve is located in the inland (37.87°E, 6.70°S) about 117 km away from the Indian ocean shores. The forest is situated about 50 km east west of Morogoro municipality, the administrative capital of Morogoro Region. The area of the forest is 1 009.9 ha and it is managed by the government as a catchment forest. The forest is described as lowland forest (dry) and situated at the ridge top of a mountain between 465 and 765 m above sea level. The average annual rainfall is 1 000 mm and the mean annual temperature is between 21 and 26°C.

2.2. Field Sampling. This study implemented a nested 1 ha plot design. This was necessary to capture as much as possible the large trees which are normally excluded when a small concentric circular sample plot design is used [22].

For each study sites, the following plot design was implemented:

(i) two 1 ha plots (100 × 100 m) where all trees greater or equal to dbh of 50 cm were measured for dbh and ht,

(ii) one 0.5 ha plot (50 × 100 m) nested in (i) above where all trees with dbh greater or equal to 20 cm and less than dbh of 50 cm were measured for dbh and at least 25% of the trees were selected randomly and measured for ht,

(iii) one 0.1 ha plot (50 × 20 m) nested in (ii) above where all trees with dbh less than 20 cm and greater or equal to dbh of 5 cm were measured for dbh and at least 10% of trees were measured for ht.

The measured trees were marked with paint to ensure that no measurement repetition was made. Total number of sample trees measured for both dbh and ht were 153 and 322 for Dindili and Rondo forests, respectively.

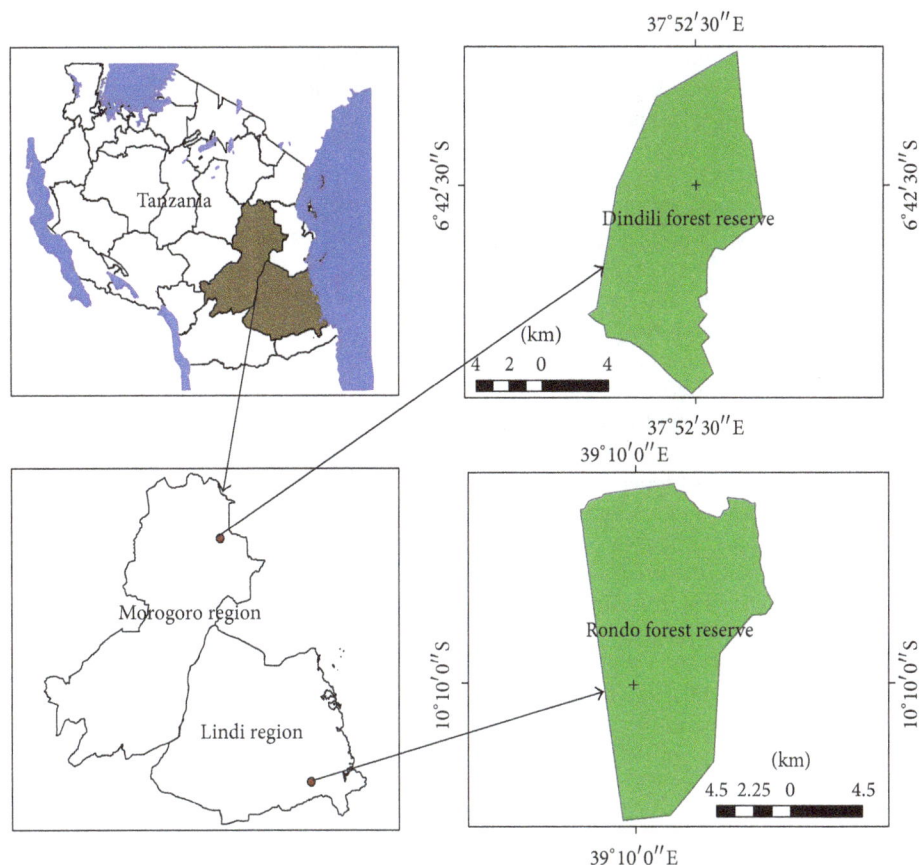

FIGURE 1: Location of Dindili and Rondo forest reserves.

TABLE 1: Sample trees selection corresponding to tree size distribution in each plot.

Diameter class (cm)	Trees ha^{-1}		Trees felled		Total number of trees felled
	Rondo	Dindili	Rondo	Dindili	All
5–15	540	435	3	5	8
15–25	169	128	7	1	8
25–35	74	59	1	6	7
35–45	41	26	2	4	6
45–55	23	17	2	4	6
55–65	12.5	8	0	6	6
65–75	8.5	1	3	2	5
75–85	2.5	0	3	2	5
85–100	2	0	5	0	5
>100	2	0	4	0	4
Total	874.5	674	30	30	60

2.3. Selection and Destructive Sampling of Sample Trees. To secure an appropriate distribution of sample trees with regard to tree sizes and tree species, information collected from sample plot inventories was used. The tree size, species distribution, and dominance/abundance of species from forest inventory information were used for the selection of sample trees to be used in volume and biomass modelling. A total of 30 trees were selected from each site to represent typical size and species distributions for each tree species (Table 1). Prior to the destructive procedure, all sample trees were recorded

for species name, while dbh were measured with calipers or a diameter tape and ht measured using Suunto hypsometer. Trees were further divided into five major sections, namely,

(i) buttress (if any),

(ii) bole stem (merchantable section),

(iii) branches including tops (up to a minimum diameter of 2.5 cm),

(iv) twigs with diameter less than 2.5 cm,

(v) leaves.

For small trees with dbh < 10 cm, no merchantable stem part was considered. For trees with dbh ≥ 10 cm no specific minimum diameters were set to distinguish between merchantable stem biomass and branch biomass. However, the decision between these ranges was based on subjective judgment of the researchers and districts forest department personnel experience on the total length of the stem that can be used to produce timber. All leaves were separated from twigs and weighed separately.

Stems and branches were trimmed and crosscut into manageable billets ranging from 1 to 2.5 m in length and then weighed for green weight. In addition, the length and the mid-diameter of billets were measured for the purpose of estimating tree volume. At least two small wood samples of 2 cm thick from the tree core to the outside excluding bark were extracted from stem sections (depending on the stem length) and three samples from branches and weighed immediately in the field. Twigs were tied into separate bundles and weighed in the field and the green weights of each were recorded. Small wood samples from each bundle were extracted, labelled, and measured for green weight in the field. Leaves were collected in bundles, weighted in the field and small sample (small bunch of leaves), extracted, and weighed. Samples from all components were sent to the laboratory in order to determine dry to green weight ratio and WD.

2.4. Laboratory Measurements. In the laboratory all stems, branches, twigs, and leaves subsamples collected from the field were oven dried at $103 \pm 2°C$ to constant weight. Dried samples were weighed and the biomass ratios for each pile of stems, branches, and twigs components were computed as the ratio of oven-dry weight to green weight. Green volumes of the sample disks/wood samples were obtained after soaking the disks/wood samples in water for at least four days until all disks are saturated. Using the water displacement method, the volume of each disk/wood sample was determined [23]. Wood basic density (WD, $g\,cm^{-3}$) for each disk/wood sample was determined as the ratio of dry weight (g) to green volume (cm^3).

2.5. Data Preparation. Components biomass was estimated as the product of dry to green ratio and total green weight (kg) of the respective tree component. The total biomass for each tree was obtained as the sum of stump, stem, branches, twigs, and leaves component tree biomass. Huber's formula [24] was applied to compute billet volume. Volume of tree merchantable stem and branches was obtained by summing the volumes of the billets of the respective sections for that particular tree. Total tree volume was finally obtained through summation of stem and branches component volume. The resulting dataset was used for developing volume and biomass models.

2.6. Data Analysis

2.6.1. Height-Diameter Models Development. Five nonlinear model forms outlined below were used to model ht for

the sample tree measured for both ht and dbh during the forest inventory exercise. Their characteristics, that is, flexibility and shape, are well documented in the literature [25]:

$$\text{ht} = 1.3 + a \times \left[1 - \exp\left(-b \times \text{dbh}\right)\right]^c \tag{1}$$

(see [26]),

$$\text{ht} = 1.3 + a \times \left[\exp\left(-\frac{b}{(\text{dbh} + c)}\right)\right] \tag{2}$$

(see [27]),

$$\text{ht} = 1.3 + a \times \left[\exp\left(-b \times \exp\left(-c \times \text{dbh}\right)\right)\right] \tag{3}$$

(see [28]),

$$\text{ht} = 1.3 + \left[\frac{\text{dbh}^2}{a + b \times \text{dbh} + c \times \text{dbh}^2}\right] \tag{4}$$

(see [29]),

$$\text{ht} = 1.3 + \left[\frac{a}{\exp\left(-b \times \exp\left(-c \times \text{dbh}\right)\right)}\right] \tag{5}$$

(see [30]).

2.6.2. Volume and Biomass Models Development. Prior to the analysis, dependent variables (volume and biomass) were plotted against each of the explanatory variables to examine the range and shape of the functional relationship and to assess the heterogeneity of the variance. The following general nonlinear model forms for prediction of volume and biomass were fitted:

$$Y = a \times \text{dbh}^b, \tag{6}$$

$$Y = a + b \times \text{dbh}^2, \tag{7}$$

$$Y = a \times \text{dbh}^b \times \text{ht}^c, \tag{8}$$

$$Y = a \times \left(\text{ht} \times \text{dbh}^2\right)^b, \tag{9}$$

$$Y^1 = a \times \left(\text{WD} \times \text{dbh}^2 \times \text{ht}\right)^b, \tag{10}$$

where Y is the volume ($m^3\,tree^{-1}$) or biomass ($kg\,tree^{-1}$); WD is wood basic density ($g\,cm^{-3}$); a, b, and c are model parameters to be estimated. WD (in model (10)) was not used as a predictor in modelling tree volume.

The NLP procedure (Nonlinear Programming) in SAS software [31] was used to fit the models parameters. The procedure fits both model parameters and variance parameters (Variance = $n^2 \times \text{dbh}^{2m}$, where n and m are parameters to be estimated) simultaneously by applying maximum likelihood regression approach. This type of procedure was used due to its flexibility to work with equations forms and its recognized robustness over nonlinear models with additive error and log transformed models [32]. A broad range of initial values for the model and variance parameters were used to ensure an optimal solution to the Root Mean Square Error (RMSE) minimization. Selection of our final models was based on high adjusted R^2, low RMSE, and finally low Akaike Information Criterion (AIC). The selected biomass and volume models were evaluated by computing prediction error and model efficiency [13, 33] as follows:

$$\text{MPE} = \left(\frac{100}{n}\right) \times \sum \left[\frac{\left(\overset{\prime\prime}{Y} - y_i\right)}{y_i}\right] \% \tag{11}$$

$$\text{EF} = 1 - \left[\frac{\sum \left(y_i - \overset{\prime\prime}{Y}\right)^2}{\sum \left(y_i - \overline{Y}\right)^2}\right], \tag{12}$$

where MPE is prediction error, EF is model efficiency, y_i is observed volume or biomass, $\overset{\prime\prime}{Y}$ is predicted volume or biomass, $\overline{Y}$ is the mean of observed volume or biomass, and n is the number of trees.

In addition, the generic biomass model developed by Chave et al. [12] for tropical forest, volume model for miombo woodlands [9], biomass and volume models developed for montane/humid forests [21, 34], and ht-dbh model for lowland forests in Tanzania were also tested to the modelling data.

3. Results

3.1. Height-Diameter Models. Parameter estimates and model performance criteria for ht-dbh models are presented in Table 2. For Dindili, model (4) performed better in terms of R^2 (68%), RMSE (2.65 m), and smaller AIC than other fitted models while for Rondo model (3) performed better with R^2 (61%), RMSE (2.89 m), and smaller AIC than other fitted models (result for other poor performing models not shown). When all data sets were fitted to develop a general model, model (3) performed better than other models. However, general ht-dbh model had larger RMSE (around 3 m) and lower R^2 (0.57) than site specific models. Trees found in Rondo forest were found to be relatively taller than those in Dindili forest at a given dbh (Figure 2). Height-diameter model developed by Mugasha et al. [18] overestimated and underestimated trees smaller and larger than dbh of about 40 cm, respectively, in Rondo while in Dindili, the model overestimated ht of trees larger than dbh of about 14 cm (Figure 2). Equations (13) represent the selected site specific and general ht-dbh models:

$$\text{ht (Dindili)}$$

$$= 1.3$$

$$+ \left[\frac{\text{dbh}^2}{0.4239 + 0.8893 \times \text{dbh} + 0.0398 \times \text{dbh}^2}\right],$$

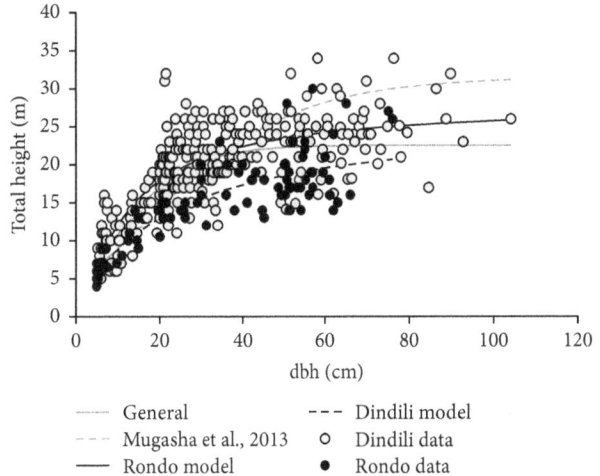

FIGURE 2: Pattern of observed and predicted ht against dbh of the selected models.

$$\text{ht (Rondo)}$$

$$= 1.3 + \left[\frac{22.8525}{\exp\left(-1.9824 \times \exp\left(-0.0888 \times \text{dbh}\right)\right)}\right],$$

$$\text{ht (General)}$$

$$= 1.3 + \left[\frac{21.2679}{\exp\left(-2.1776 \times \exp\left(-0.0993 \times \text{dbh}\right)\right)}\right]. \tag{13}$$

3.2. Total, Stem, and Branch Tree Volume Models. Parameter estimates and model performance criteria for total tree volume and tree sections are presented in Table 3. By fitting the four alternative volume models to entire data set, over 92% and 73% of the variations of total and sectional tree volume, respectively, were explained. Based on AIC, model (7) was the best performing one for models with only dbh and model (8) for models with both dbh and ht. Although model (8) outperformed model (7), the performance differences were quite insignificant. As expected, there was a significant decrease in RMSE from general to site specific volume models.

A comparison between allometric models for total tree volume of miombo woodlands [9] and Montane/humid forests [21] and volume allometric model developed in this study with dbh only as explanatory variable is shown in Figure 3. Miombo woodlands volume model overestimated trees with dbh greater than 65 cm.

The selected general models (models (7) and (8)) were also tested to each study site (Table 4). For total tree volume, the prediction error was found to be not significantly different from zero ($p > 0.05$) and found to be more efficient (model efficiency above 0.87) than sectional models. When general sectional models were tested, except for model (8) for branches volume, mean prediction error was found to be significantly different from zero ($p < 0.05$). Site specific sectional model had low mean prediction error and is more efficient than general sectional volume models.

TABLE 2: Parameter estimates and performance criteria of five ht-dbh models.

Site	Model	Parameter estimates			Error	RMSE	R^2	AIC
		a	b	c				
Dindili	1	19.5933	0.0415	0.9023	$0.314 \times dbh^{0.6304}$	2.65	0.67	402.76
	2	24.0599	18.3594	5.6557	$0.317 \times dbh^{0.6271}$	2.64	0.68	402.32
	3	17.9649	2.0739	0.0797	$0.317 \times dbh^{0.6315}$	2.65	0.66	404.64
	4	**0.4239**	**0.8893**	**0.0398**	$0.318 \times dbh^{0.6264}$	2.64	0.68	402.28
	5	19.8856	0.0554	0.9246	$0.315 \times dbh^{0.6296}$	2.65	0.67	402.7
Rondo	1	23.3447	0.0639	1.0728	$2.239 \times dbh^{0.1469}$	2.90	0.61	1593.02
	2	27.5692	11.2866	2.0427	$2.374 \times dbh^{0.1306}$	2.92	0.61	1595.18
	3	**22.8525**	**1.9824**	**0.0888**	$2.145 \times dbh^{0.1591}$	2.89	0.61	1591.6
	4	1.9292	0.3691	0.0369	$2.399 \times dbh^{0.1273}$	2.92	0.61	1594.98
	5	23.2808	0.0534	1.0454	$2.233 \times dbh^{0.1477}$	2.904	0.61	1592.98
All	**3**	**21.2679**	**2.1776**	**0.0993**	$1.681 \times dbh^{0.2539}$	3.03	0.57	2099.2
	4	2.3468	0.3489	0.0398	$1.835 \times dbh^{0.2088}$	3.06	0.57	2101.98

TABLE 3: Parameter estimates and volume models performance for individual total tree and sectional volume.

Section	General/site specific	Model	Parameter estimates			Error	RMSE	R^2	AIC
			a	b	c				
Total tree volume	General	6	0.00053	2.1620	—	$0.013 \times dbh^{0.5}$	0.90	0.95	92.32
		7	−0.0393	0.00102	—	$0.013 \times dbh^{0.5}$	0.85	0.95	92.06
		8	**0.000076**	**2.3488**	**0.3848**	$0.013 \times dbh^{0.5}$	0.90	0.95	91.64
		9	0.00014	0.9039	—	$0.014 \times dbh^{0.5}$	1.10	0.92	107.42
	Dindili forest	7	−0.0226	0.00090		$0.014 \times dbh^{0.5}$	0.72	0.88	45.82
		8	**0.000041**	**2.5042**	**0.4329**	$0.012 \times dbh^{0.5}$	0.74	0.89	39.36
	Rondo forest	7	−0.0760	0.0010		$0.011 \times dbh^{0.5}$	0.87	0.96	46.92
		8	0.00014	2.3176	0.1854	$0.010 \times dbh^{0.5}$	0.92	0.96	45.34
Tree branch volume	General	6	0.00024	2.1658	—	$0.013 \times dbh^{0.5}$	0.84	0.78	88.74
		7	−0.0069	0.00031	—	$0.013 \times dbh^{0.5}$	0.82	0.78	88.58
		8	**0.000034**	**2.5514**	**−0.1277**	$0.012 \times dbh^{0.5}$	0.83	0.79	89.78
		9	0.000074	0.8901	—	$0.014 \times dbh^{0.5}$	0.90	0.73	99.72
	Dindili forest	7	−0.0120	0.00046		$0.013 \times dbh^{0.5}$	0.60	0.70	39.66
		8	**0.000045**	**2.9229**	**−0.3903**	$0.012 \times dbh^{0.5}$	0.70	0.69	38.26
	Rondo forest	7	−0.015	0.00033		$0.011 \times dbh^{0.5}$	0.80	0.84	45.72
		8	**0.000045**	**2.5642**	**−0.0659**	$0.010 \times dbh^{0.5}$	0.80	0.87	43.14
Tree stem volume	General	6	0.0003	2.1452	—	$0.028 \times dbh^{0.5}$	0.68	0.89	47.90
		7	−0.0176	0.00052	—	$0.009 \times dbh^{0.5}$	0.65	0.90	47.74
		8	**0.000051**	**2.1611**	**0.5517**	$0.008 \times dbh^{0.5}$	0.65	0.90	38.70
		9	0.00007	0.9088	—	$0.008 \times dbh^{0.5}$	0.67	0.89	40.18
	Dindili forest	7	−0.0121	0.00047		$0.008 \times dbh^{0.5}$	0.42	0.79	33.9
		8	**0.0000099**	**2.0392**	**1.2855**	$0.007 \times dbh^{0.5}$	0.36	0.86	4.3
	Rondo forest	7	−0.03965	0.00059		$0.009 \times dbh^{0.5}$	0.83	0.90	33.94
		8	0.00011	2.1685	0.3038	$0.009 \times dbh^{0.5}$	0.83	0.90	35.3

3.3. Total, Branch, and Stem Tree Biomass Models. Parameter estimates of total tree AGB for general and site specific models are presented in Table 5. Model with dbh alone had the lowest R^2 and highest RMSE. Site specific models had larger R^2 and lower RMSE compared to corresponding general models. Inclusion of ht into the model (models (8) and (9)) improved the model fit marginally. On the other hand, the lowest AIC and RMSE and highest R^2 for general and site specific models were apparent for models which include WD (model (10)). Models (7), (9), and (10) were selected for further evaluation. The distribution of observed trees AGB and projected AGB by applying the selected site specific and general models against dbh is presented in Figure 4. Observed tree AGB data for Dindili forest was systematically larger than that of Rondo forest for all trees sizes.

TABLE 4: Evaluation of the selected general and site specific models for tree total, branches, and stem volume.

| Type | Tree section | Selected model | Prediction error% | | | Model efficiency | | |
			Dindili	Rondo	All	Dindili	Rondo	All
General	Total	*7*	*−2.09*	*12.78*	*8.84*	*0.87*	*0.96*	*0.94*
		8	*−0.50*	*13.27*	*6.38*	*0.90*	*0.87*	*0.87*
	Branches	7	−17.73	20.95^S	1.75	0.35	0.65	0.56
		8	−0.26	56.12^S	27.93	0.66	0.68	0.68
	Stem	7	13.72	15.61	14.66	0.88	0.89	0.88
		8	15.43	14.53	14.98	0.90	0.86	0.86
Selected site specific (Regional)	Total	7	4.08	*6.97*	—	0.87	*0.96*	—
		8	*6.68*	*3.28*	—	*0.90*	*0.96*	—
	Branches	7	15.79	17.46	—	0.58	0.66	—
		8	*14.28*	*18.90*	—	*0.73*	*0.86*	—
	Stem	*7*	*11.75*	*15.50*	—	*0.78*	*0.90*	—
		8	10.28	17.95	—	0.74	0.87	—

The best selected models are in bold and italic. ^{S}Significantly different from zero ($p < 0.05$).

TABLE 5: Parameter estimates and performance of general and site specific models for total tree aboveground biomass.

| General/sites | Model | Parameter estimates | | | Error | RMSE | R^2 | AIC |
		a	b	c				
General	6	0.6881	1.93834		$0.237 \times dbh^{2.028}$	1280.4	0.49	893.98
	7	**3.2064**	**0.6166**		$0.176 \times dbh^{2.105}$	1326.16	0.48	892.70
	8	0.3571	1.7440	0.4713	$0.274 \times dbh^{1.982}$	1214.68	0.52	893.44
	9	**0.1459**	**0.8601**		$0.323 \times dbh^{1.940}$	1169.65	0.54	892.56
	10	**0.0873**	**0.9458**		$0.559 \times dbh^{1.673}$	567.9	0.87	840.94
Dindili forest	**6**	**0.5414**	**2.0591**		$0.429 \times dbh^{1.801}$	539.98	0.83	421.38
	7	4.5076	0.6915		$0.347 \times dbh^{1.860}$	506.59	0.84	421.40
	8	0.2137	1.8004	0.6724	$0.573 \times dbh^{1.686}$	470.50	0.87	416.48
	9	**0.1568**	**0.8613**		$0.585 \times dbh^{1.683}$	478.61	0.87	415.06
	10	**0.1014**	**0.9510**		$0.638 \times dbh^{1.675}$	467.52	0.89	418.34
Rondo forest	6	0.3238	2.0673		$0.040 \times dbh^{2.430}$	1360.97	0.50	450.04
	7	**0.2816**	**1.1654**		$0.006 \times dbh^{2.916}$	1514.67	0.51	442.78
	8	**0.0542**	**1.3326**	**1.4278**	$0.165 \times dbh^{2.010}$	967.08	0.66	439.90
	9	0.0863	0.8544		$0.040 \times dbh^{2.397}$	1172.15	0.58	440.64
	10	**0.07511**	**0.9477**		$0.214 \times dbh^{1.780}$	462.47	0.92	396.58

Parameter estimates and performance of general and site specific models for biomass tree section are presented in Table 6. Similar models performance trend as that of total tree AGB models were also found for sectional biomass models. Modelling all data sets significantly reduced and increased the R^2 and RMSE, respectively, compared to site specific models. Inclusion of ht and WD reduced AIC except for Dindili where addition of WD did not improve the model fit.

The selected general biomass models were evaluated on how best they predict the tree total, branches, and stem biomass to each study site (Table 7). Overall model with dbh or a combination of dbh and ht performed poorly. The models underestimated biomass in Dindili forest and overestimated the biomass in Rondo forest. However, the magnitude of overestimation was immense in Rondo forest when compared to the magnitude of underestimation at Dindili forest. Inclusion of WD stabilized the models' prediction error

and efficiency globally. AGB models developed by Chave et al. [12] produced small mean prediction error globally (about 5%) and performed poorly when tested at site level (prediction error > 12%). Model developed by Masota [34] significantly overestimated tree biomass in all sites. The selected site specific models were found to be efficient and produced lower mean prediction error compared to best performing general model (model (10)) when tested to site level.

4. Discussion

Lowland forests in Tanzania are generally found in areas close to the coast of Indian Ocean and some areas of the inland. The locality differences as defined by the distance of the forest from the coast influence the forest structure

TABLE 6: Parameter estimates and performance of general and site specific models for biomass of tree sections.

Section	General/sites	Model	Parameter estimates			Error	RMSE	R^2	AIC
			a	b	c				
Tree branches	General	6	0.1379	2.1738		$0.027 \times dbh^{2.456}$	846.89	0.39	823.10
		7	−1.6031	0.2524		$0.042 \times dbh^{2.345}$	677.87	0.41	826.66
		8	0.1364	2.1697	0.0088	$0.028 \times dbh^{2.451}$	842.13	0.39	825.08
		9	0.0434	0.8868		$0.047 \times dbh^{2.319}$	757.86	0.40	828.52
		10	0.0270	0.9887		$0.075 \times dbh^{2.118}$	441.72	0.75	795.70
	Dindili	**6**	**0.1173**	**2.2750**		$0.023 \times dbh^{2.441}$	370.47	0.75	383.04
		7	−2.5456	0.3211		$0.033 \times dbh^{2.375}$	285.20	0.76	391.26
		8	**0.1604**	**2.35962**	**−0.2205**	$0.022 \times dbh^{2.455}$	368.64	0.76	384.92
		9	0.0483	0.9083		$0.032 \times dbh^{2.380}$	420.12	0.62	390.90
		10	0.04624	0.9549		$0.020 \times dbh^{2.510}$	296.64	0.72	388.60
	Rondo	6	0.1343	2.0777		$0.046 \times dbh^{2.277}$	619.55	0.46	422.6
		7	**0.1403**	**0.9548**		$0.041 \times dbh^{2.303}$	620.00	0.46	421.78
		8	**0.000007**	**0.5631**	**4.9456**	$0.141 \times dbh^{1.890}$	531.55	0.64	405.46
		9	0.0288	0.8912		$0.047 \times dbh^{2.249}$	594.47	0.51	416.26
		10	0.05347	0.8840		$0.205 \times dbh^{1.763}$	345.94	0.85	394.66
Tree stem	General	6	0.3859	1.7794		$0.135 \times dbh^{1.994}$	570.48	0.49	811.60
		7	−1.0523	0.2482		$0.109 \times dbh^{2.053}$	618.81	0.49	811.76
		8	0.07646	1.5073	1.0172	$0.203 \times dbh^{1.859}$	472.48	0.62	803.40
		9	**0.1683**	**0.7287**		$0.174 \times dbh^{1.903}$	494.66	0.60	801.86
		10	0.0848	0.8726		$0.333 \times dbh^{1.658}$	259.49	0.85	772.30
	Dindili	**6**	**0.2451**	**2.0119**		$0.086 \times dbh^{2.108}$	358.30	0.56	390.86
		7	−1.1567	0.2895		$0.103 \times dbh^{2.059}$	343.87	0.57	390.92
		8	**0.0395**	**1.3879**	**1.4583**	$0.181 \times dbh^{1.825}$	212.42	0.84	377.04
		9	0.0869	0.8392		$0.118 \times dbh^{1.963}$	260.6	0.77	388.60
		10	**0.0853**	**0.8798**		$0.108 \times dbh^{1.994}$	277.39	0.68	380.18
	Rondo	**6**	**0.3944**	**1.7607**		$0.200 \times dbh^{1.869}$	549.37	0.49	417.66
		7	−5.4258	0.2147		$0.104 \times dbh^{2.043}$	668.71	0.48	422.10
		8	**0.0278**	**1.2617**	**1.5116**	$0.344 \times dbh^{1.694}$	441.80	0.67	412.20
		9	0.0808	0.8026		$0.213 \times dbh^{1.830}$	493.90	0.59	412.35
		10	**0.06812**	**0.8888**		$0.779 \times dbh^{1.395}$	271.52	0.89	391.32

and conditions due to climatic and topographical differences. Therefore, in this study two sites of lowland forests were selected, that is, one near to the coast (wet lowland forest) and the second from further inland (dry lowland forest), to cater the variations associated with environmental factors. Since the tree selection for modelling was based on the tree size distribution from the information derived from the forest inventory data (Table 2), it is apparent that the modelling data in this study was representative. However, in tropical natural forests where hundreds of species exist per ha [35], it is impractical to represent every tree species for allometric model development. However, the priority was given to tree species which have high appearance frequency. Moreover, the larger trees which normally influence the trend of allometric model and also account for a very large part of the volume and AGB [4, 35] were well represented to avoid extrapolation.

4.1. Height-Diameter Relationships. Over 61% of variation on tree ht was explained by the selected site specific ht-dbh models. The coefficient of variation (R^2) in this study

corresponds to that of Mugasha et al. [18] lowland forests of Tanzania where R^2 of 0.64 was reported. However, considerable amount of variation in ht remains unexplained. This may be due to large diversity of tree species with different ht-dbh allometry. This is also evident when modelling ht-dbh of combined data from the two sites where R^2 dropped to 0.57. In addition, tree allometry was found to be different among sites as indicated by slightly taller trees in Rondo than those from Dindili forest at a given dbh (Figure 3). Due to such difference in addition to drop of R^2 and increasing of RMSE for the combined data set, it is recommended that the site specific ht-dbh models are to be applied. Furthermore, ht of trees larger than 40 cm dbh were overestimated by ht-dbh model developed by Mugasha et al. [18]. This may be due to the fact that Mugasha et al. [18] did not include trees found in areas with similar climate conditions as that of Rondo or Dindili in their model. Studies have shown that ht-dbh relationship varies significantly with climate [12, 36]. Since climate variation affects ht, this in turn affects trees AGB and volume. Therefore, as noted by Chave et al. [12], it is important

TABLE 7: Evaluation of the selected general and site specific models for tree total, branches, and stem biomass.

Type	Section	Selected model	Prediction error%			Model efficiency		
			Dindili	Rondo	All	Dindili	Rondo	All
General	Total tree biomass	7	−1.26	53[S]	50.28	0.81	−0.61	−0.10
		9	−6.59	60[S]	53.12	0.83	0.87	−0.25
		10	−14.97[S]	19.10[S]	2.06	0.81	0.91	0.87
		Chave et al., 2014 [12]	−12.59[S]	23.24[S]	−5.33	0.85	0.86	0.86
		Masota, 2015 [34]	25.07[S]	151.30[S]	88.1	0.84	−1.78	−0.84
	Branches biomass	6	−13.58	80[S]	56.66	0.60	−0.47	−0.02
		8	−13.96	87[S]	56.68	0.60	−0.47	−0.02
		10	−14.87	47.17[S]	16.16	0.59	0.84	0.74
	Stem biomass	6	−23.46[S]	38[S]	−4.40	0.30	0.48	0.44
		9	−21.95[S]	29.96[S]	4.32	0.48	0.59	0.56
		10	9.17	23.57	16.37	0.72	0.91	0.86
Site specific (regional)	Total tree biomass	*6*	*5.16*	—	—	*0.94*	—	—
		7	—	6.47	—	—	0.72	—
		8	—	−1.26	—	—	0.77	—
		9	*1.68*****	—	—	*0.94*	—	—
		10	*3.75*	*4.31*	—	*0.88*	*0.93*	—
	Branches biomass	*6*	*4.82*	—	—	*0.87*	—	—
		7	—	11.80	—	—	0.65	—
		8	*4.62*	*−4.00*	—	*0.88*	*0.76*	—
		10	5.89	7.14	—	0.69	0.69	—
	Stem biomass	*6*	*9.06*	9.39	—	*0.83*	0.72	—
		8	*7.07*	*7.17*	—	*0.93*	*0.80*	—
		10	17.52	*12.73*	—	0.71	*0.91*	—

** The best selected models are in bold and italic. [S]Significantly different from zero ($p < 0.05$).

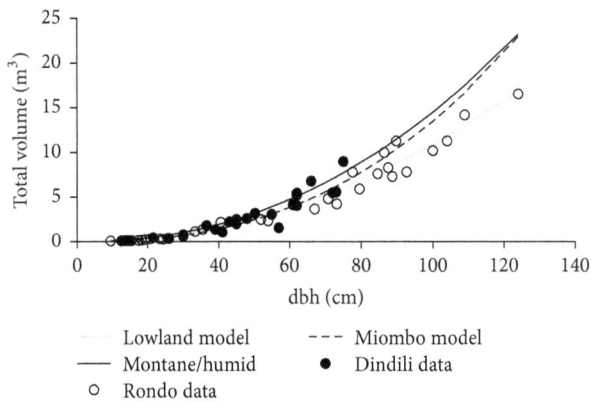

FIGURE 3: Comparison of lowland forest, miombo woodlands, and Montane/humid volume models. Lowland volume estimates computed by model (8). Total tree height was estimated using the developed ht-dbh models.

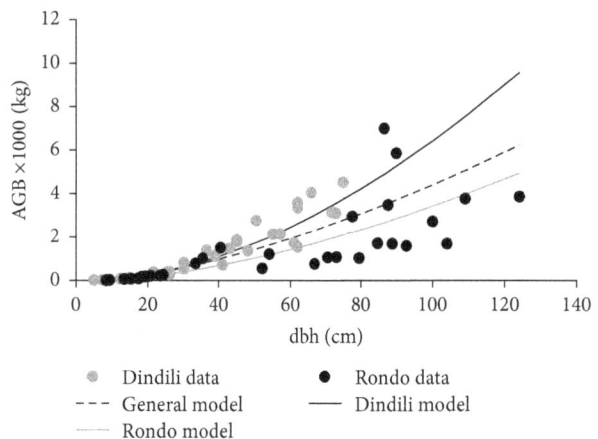

FIGURE 4: Scatter plot and biomass estimates by applying model (9). Total tree height was predicted using the developed ht-dbh models.

to include ht as an explanatory variable in AGB or volume models to accommodate variation triggered by climate and other environment factors.

4.2. Volume and Biomass Allometric Models. Over 73% of variation in tree volume was explained by dbh or by both dbh and ht. Site specific models slightly improved the model fit compared to general models. Model (7) (with only dbh)

and model (8) (with both ht and dbh) were selected for all tree section. However, model (8) with ht included outperformed model (7) with only dbh. This observation also underscores the importance of including ht in volume allometric models as also suggested by Chave et al. [12].

The comparison between the developed volume models and that of miombo woodlands and Montane/humud forest shows that the volume of trees with dbh greater than 65 cm

TABLE 8: Families and wood basic density of sample tree species for allometric model development.

Forest	Tree #	Botanical name	Local name	Family	WD ± SD (g cm^{-3})
Dindili	1	*Manilkara sulcata*	Msezi	Sapotaceae	0.70 ± 0.11
Dindili	2	*Ricinodendron heudelotii*	Mkungunolo	Euphorbiaceae	0.41 ± 0.06
Dindili	3	*Manilkara sulcata*	Msezi	Sapotaceae	0.74 ± 0.04
Dindili	4	*Combretum schumannii*	Mkatakorongo	Combretaceae	0.82 ± 0.06
Dindili	5	*Albizia gummifera*	Mkenge	Fabaceae	0.62 ± 0.06
Dindili	6	*Terminalia sambesiaca*	Mpululu	Combretaceae	0.69 ± 0.16
Dindili	7	*Tamarindus indica*	Mkwaju	Fabaceae	0.69 ± 0.04
Dindili	8	*Commiphora zimmermannii*	Mtwini	Burseraceae	0.40 ± 0.08
Dindili	9	*Pteleopsis myrtifolia*	Mngoji	Combretaceae	0.67 ± 0.04
Dindili	10	*Pteleopsis myrtifolia*	Mngoji	Combretaceae	0.70 ± 0.03
Dindili	11	*Combretum schumannii*	Mkatakorongo	Combretaceae	0.78 ± 0.04
Dindili	12	*Tamarindus indica*	Mkwaju	Fabaceae	0.70 ± 0.01
Dindili	13	*Vepris nobilis*	Mzindizi	Rutaceae	0.77 ± 0.08
Dindili	14	*Terminalia sambesiaca*	Mpululu	Combretaceae	0.58 ± 0.13
Dindili	15	*Holarrhena pubescens*	Mmelemele	Apocynaceae	0.44 ± 0.03
Dindili	16	*Sterculia appendiculata*	Mgude	Sterculiaceae	0.46 ± 0.08
Dindili	17	*Sterculia appendiculata*	Mgude	Sterculiaceae	0.49 ± 0.11
Dindili	18	*Lannea* sp.	Muumbu	Anacardiaceae	0.52 ± 0.21
Dindili	19	*Terminalia sambesiaca*	Mpululu	Combretaceae	0.70 ± 0.12
Dindili	20	*Combretum schumannii*	Mkatakorongo	Combretaceae	0.80 ± 0.06
Dindili	21	*Scorodophloeus fischeri*	Mhande	Fabaceae	0.76 ± 0.06
Dindili	22	*Terminalia sambesiaca*	Mpululu	Combretaceae	0.62 ± 0.05
Dindili	23	*Scorodophloeus fischeri*	Mhande	Fabaceae	0.70 ± 0.07
Dindili	24	*Cussonia zimmermannii*	Mkong'onolo	Araliaceae	0.40 ± 0.05
Dindili	25	*Terminalia sambesiaca*	Msezi	Sapotaceae	0.80 ± 0.04
Dindili	26	*Scorodophloeus fischeri*	Mhande	Fabaceae	0.71 ± 0.06
Dindili	27	*Celtis* sp.	Mkoma chuma	Ulmaceae	0.79 ± 0.12
Dindili	28	*Sterculia africana*	Moza	Sterculiaceae	0.81 ± 0.07
Dindili	29	*Cussonia zimmermannii*	Mkong'onolo	Araliaceae	0.63 ± 0.16
Dindili	30	*Scorodophloeus fischeri*	Mhande	Fabaceae	0.60 ± 0.17
		Mean value			**0.65 ± 0.13**
Rondo	1	*Xylopia* sp.	Nami	Annonaceae	0.51 ± 0.01
Rondo	2	*Blighia unijugata*	Mkalanga	Sapindaceae	0.44 ± 0.06
Rondo	3	*Tabernaemontana ventricosa*	Mnongoli	Apocynaceae	0.45 ± 0.02
Rondo	4	*Cussonia zimmermannii*	Mtumbitumbi	Araliaceae	0.27 ± 0.02
Rondo	5	*Parinari excelsa*	Mmula	Rosaceae	0.49 ± 0.02
Rondo	6	*Ricinodendron heudelotii*	Mtene	Euphorbiaceae	0.29 ± 0.06
Rondo	7	*Ricinodendron heudelotii*	Mtene	Euphorbiaceae	0.26 ± 0.02
Rondo	8	*Ricinodendron heudelotii*	Mtene	Euphorbiaceae	0.21 ± 0.04
Rondo	9	*Ricinodendron heudelotii*	Mtene	Euphorbiaceae	0.35 ± 0.04
Rondo	10	*Ricinodendron heudelotii*	Mtene	Euphorbiaceae	0.21 ± 0.03
Rondo	11	*Cussonia zimmermannii*	Mtumbitumbi	Araliaceae	0.32 ± 0.07
Rondo	12	*Antiaris toxicaria*	Nkalale/Nkarale	Moraceae	0.36 ± 0.13
Rondo	13	*Ricinodendron heudelotii*	Mtene	Euphorbiaceae	0.23 ± 0.03
Rondo	14	*Ricinodendron heudelotii*	Mtene	Euphorbiaceae	0.26 ± 0.03
Rondo	15	*Ricinodendron heudelotii*	Mtene	Euphorbiaceae	0.30 ± 0.03
Rondo	16	*Ricinodendron heudelotii*	Mtene	Euphorbiaceae	0.25 ± 0.07
Rondo	17	*Ricinodendron heudelotii*	Mtene	Euphorbiaceae	0.21 ± 0.04
Rondo	18	*Porlerandia penduliflora*	Nakatumbaku	Rubiaceae	0.54 ± 0.01
Rondo	19	*Trilepsium madagascariense*	Ntulumuti	Sapotaceae	0.48 ± 0.02
Rondo	20	*Euphorbia* sp.	Milembutuka Mweusi	Euphorbiaceae	0.57 ± 0.03

TABLE 8: Continued.

Forest	Tree #	Botanical name	Local name	Family	WD ± SD (g cm^{-3})
Rondo	21	*Milletia eetveldiana*	Mkunguwe	Fabaceae	0.59 ± 0.01
Rondo	22	*Manilkara discolor*	Mtondoli	Sapotaceae	0.53 ± 0.02
Rondo	23	*Dialium holtizii*	Mpepeta	Fabaceae	0.58 ± 0.02
Rondo	24	*Milletia eetveldiana*	Mkunguwe	Fabaceae	0.58 ± 0.03
Rondo	25	*Drypetes parviflora*	Mkengeda/Mnangari	Euphorbiaceae	0.50 ± 0.04
Rondo	26	*Milletia eetveldiana*	Mkunguwe	Fabaceae	0.57 ± 0.02
Rondo	27	*Dialium holtizii*	Mpepeta	Fabaceae	0.60 ± 0.03
Rondo	28	*Drypetes parviflora*	Mkengeda/Mnangari	Euphorbiaceae	0.71 ± 0.06
Rondo	29	*Milicia excelsa*	Mtunguru/Mvule	Moraceae	0.50 ± 0.08
Rondo	30	*Ricinodendron heudelotii*	Mtene	Euphorbiaceae	0.23 ± 0.02
		Mean value			**0.41 ± 0.15**

was overestimated by the miombo woodlands model while all trees sizes were overestimated by Montane/humid volume model. This pattern provides an insight into the actual volume difference between trees in miombo woodlands and Montane/humid forests and that of lowland forests at a given tree size. This variation may be attributable to the tree architectural differences since lowland forests are characterised by very tall trees as opposed to short and very wide crowned trees in miombo woodlands (Figure 2) [37]. While branching pattern for lowland forest is similar to Montane/humid forests, the biomass differences revealed in this study may be due to the fact that trees in Montane/humid forest are taller than those found in lowland forests [1, 18]. Due to large variation in branching patterns among tree sizes and species in lowland, the model fit to the tree branches was not as good as the model fit of the tree total and stem models. It can also be noted that even though the total volume models are affected by the branches, the model fits were still better than those of the stem models. The most plausible explanation for this is the fact that the demarcation point for merchantable stem relies on considerations not only on size (minimum diameter), but also on subjective stem quality assessments for timber which adds variability to the relationship between dbh and stem. Evaluation of general volume models to the sites indicates that models (7) and (8) can be reliably applied to lowland forests of Tanzania while for tree sectional tree volume the site specific volume models are recommended.

Although the selected general biomass model performed well globally, the selected site specific AGB models performed far better. The model fit improved with addition of ht and WD. In contrast to volume models, AGB varied significantly between sites. The variation is highly associated with WD (see Table 8). This explains why model (10) (with WD) performed relatively well for site specific AGB models as well as for general AGB model. Similar trend was found for biomass sectional models where inclusion of WD also improved the model fit and efficiency significantly (e.g., from R^2 values from 0.62 to 0.85 and model efficiency value from 0.80 to 0.91 from model (8) to model (10), resp., for stem general biomass model). However, the mean prediction error of general biomass model (model (10)) was large and inefficient compared to site specific models when tested at site level.

This may be due to actual differences between the two forests as a function of climate and other environmental factors which shape the forest structure and conditions [20, 38]. It is therefore recommended that, for lowland forests, the selected site specific biomass models (Table 7) be applied since their prediction error is within the acceptable range ($p > 0.05$, Table 7). For the sites which are situated inland, the AGB model developed for Dindili forest may be used and for lowland forests near the coast, the AGB model developed for Rondo forest may be used. Furthermore, for improved estimation of AGB, the model with ht and WD included is highly recommended. Model developed by Chave et al. [12] underestimated and overestimated AGB in Rondo and Dindili forest, respectively, and gave unbiased biomass estimates at global scale.

5. Conclusion

From the findings in this study, site specific ht-dbh model is recommended since the tree allometry was found to differ significantly between dry and wet lowland forests. The selected general tree total volume model may be applied in lowland forests of Tanzania since no significant difference in prediction error was found when tested to each study site. Due to biased biomass estimates of general aboveground biomass model, the application of selected site specific biomass models is recommended, that is, dry and wet lowland forests biomass models developed in Dindili and Rondo forest, respectively. Application of models with WD in addition to dbh and ht is highly recommended for improved estimates accuracy.

Conflict of Interests

The authors declare that there is no conflict of interests regarding the publication of this paper.

Acknowledgments

This work would not have been possible without the financial support from Food and Agricultural Organisation (FAO). This financial support is highly acknowledged. The authors would also like to thank Tanzania Forest Service (TFS) authorities for issuing a permission to conduct forest inventory and destructive sampling in all the study sites. The authors are also grateful to all district forest staffs and villagers of Morogoro and Lindi Region; Manager of Rondo Plantation Forest Mr. Lukas Manyoni Mabusi; and Kasimu Amani and Gelson Kyaruzi for their diligent effort during the field work. Finally, the authors thank Mr. Joshua Kalinga and Miss Diana Chacha of Sokoine University of Agriculture for assisting them in laboratory work.

References

[1] N. D. Burgess, B. Bahane, T. Clairs et al., "Getting ready for REDD+ in Tanzania: a case study of progress and challenges," *Oryx*, vol. 44, no. 3, pp. 339–351, 2010.

[2] URT, *National Forest Resources Monitoring and Assessment of Tanzania Mainland (NAFORMA): Main Results*, Ministry of Natural Resources & Tourism, Tanzania Forest Services Agency, The Government of Finland and Food and Agriculture Organization (FAO) of the United Nations, 2015.

[3] L. Schulman, L. Junikka, A. Mndolwa, I. Rajabu, and J. Lovett, *Trees of Amani Nature Reserve, NE Tanzania*, Ministry of Natural Resources and Tourism, Dar es Salaam, Tanzania, 1998.

[4] J. Chave, R. Condit, S. Aguilar, A. Hernandez, S. Lao, and R. Perez, "Error propagation and scaling for tropical forest biomass estimates," *Philosophical Transactions of the Royal Society B: Biological Sciences*, vol. 359, no. 1443, pp. 409–420, 2004.

[5] S. Milledge and B. Kaale, *Bridging the Gap: Linking Timber Trade with Infrastructure Development and Poverty Eradication Efforts in Southern Tanzania*, TRAFFIC East/Southern Africa, Dar es Salaam, Tanzania, 2003.

[6] R. E. Malimbwi, D. T. K. Shemweta, E. Zahabu, S. P. Kingazi, J. Z. Katani, and D. A. Silayo, *Summary Report of Forest Inventory for the Eleven Districts of Eastern and Southern Tanzania*, Forestry and Beekeeping Division, Dar es Salaam, Tanzania, 2005.

[7] S. Milledge, R. Elibariki, T. East, and S. Africa, "Green gold: ongoing efforts towards preventing illegal harvesting and exports of Tanzania's most valuable hardwoods," *The Arc Journal*, vol. 17, 2005.

[8] J. S. Makero, *Timber Potential Value in the Eastern-Arc Mountains, Tanzania: Nyanganje Forest Reserve*, Sokoine University of Agriculture (SUA), Morogoro, Tanzania, 2009.

[9] E. W. Mauya, W. A. Mugasha, E. Zahabu, O. M. Bollandsås, and T. Eid, "Models for estimation of tree volume in the miombo woodlands of Tanzania," *Southern Forests*, vol. 67, no. 4, pp. 209–219, 2014.

[10] M. Henry, A. Bombelli, C. Trotta et al., "GlobAllomeTree: international platform for tree allometric equations to support volume, biomass and carbon assessment," *IForest—Biogeosciences and Forestry*, vol. 6, no. 6, p. 326, 2013.

[11] Q. Molto, V. Rossi, and L. Blanc, "Error propagation in biomass estimation in tropical forests," *Methods in Ecology and Evolution*, vol. 4, no. 2, pp. 175–183, 2013.

[12] J. Chave, M. Réjou-Méchain, A. Búrquez et al., "Improved allometric models to estimate the aboveground biomass of tropical trees," *Global Change Biology*, vol. 20, no. 10, pp. 3177–3190, 2014.

[13] J. Chave, C. Andalo, S. Brown et al., "Tree allometry and improved estimation of carbon stocks and balance in tropical forests," *Oecologia*, vol. 145, no. 1, pp. 87–99, 2005.

[14] T. R. Baker, O. L. Phillips, Y. Malhi et al., "Variation in wood density determines spatial patterns in Amazonian forest biomass," *Global Change Biology*, vol. 10, no. 5, pp. 545–562, 2004.

[15] M. Henry, A. Besnard, W. A. Asante et al., "Wood density, phytomass variations within and among trees, and allometric equations in a tropical rainforest of Africa," *Forest Ecology and Management*, vol. 260, no. 8, pp. 1375–1388, 2010.

[16] R. Malimbwi and A. Mugasha, *Reconnaissance inventory of Handeni Forest Reserve in Tanga, Tanzania*, Forest and Beekeeping Division, Ministry of Natural Resources and Tourism, Dar es Salaam, Tanzania, 2001.

[17] M. M. Mpanda, E. J. Luoga, G. C. Kajembe, and T. Eid, "Impact of forestland tenure changes on forest cover, stocking and tree species diversity in amani nature reserve, Tanzania," *Forests Trees and Livelihoods*, vol. 20, no. 4, pp. 215–229, 2011.

[18] W. A. Mugasha, O. M. Bollandsås, and T. Eid, "Relationships between diameter and height of trees in natural tropical forest in Tanzania," *Southern Forests*, vol. 75, no. 4, pp. 221–237, 2013.

[19] T. R. Feldpausch, L. Banin, O. L. Phillips et al., "Height-diameter allometry of tropical forest trees," *Biogeosciences*, vol. 8, no. 5, pp. 1081–1106, 2011.

[20] T. R. Feldpausch, J. Lloyd, S. L. Lewis et al., "Tree height integrated into pantropical forest biomass estimates," *Biogeosciences*, vol. 9, no. 8, pp. 3381–3403, 2012.

[21] A. M. Masota, E. Zahabu, R. E. Malimbwi, O. M. Bollandsås, and T. H. Eid, "Volume models for single trees in tropical rainforests in Tanzania," *Journal of Energy and Natural Resources*, vol. 3, no. 5, pp. 66–76, 2014.

[22] A. R. Marshall, S. Willcock, P. J. Platts et al., "Measuring and modelling above-ground carbon and tree allometry along a tropical elevation gradient," *Biological Conservation*, vol. 154, pp. 20–33, 2012.

[23] S. Brown, *Estimating Biomass and Biomass Change of Tropical Forests: A Primer*, vol. 134, Food and Agriculture Organization, Rome, Italy, 1997.

[24] B. Husch, C. Miller, and T. Beers, *Forest Mensuration*, John Wiley & Sons, New York, NY, USA, 1982.

[25] S. Huang, S. J. Titus, and D. P. Wiens, "Comparison of nonlinear height-diameter functions for major Alberta tree species," *Canadian Journal of Forest Research*, vol. 22, no. 9, pp. 1297–1304, 1992.

[26] F. J. Richards, "A flexible growth function for empirical use," *Journal of Experimental Botany*, vol. 10, no. 2, pp. 290–301, 1959.

[27] D. A. Ratkowsky and D. E. Giles, *Handbook of Nonlinear Regression Models*, Marcel Dekker, New York, NY, USA, 1990.

[28] C. P. Winsor, "The Gompertz curve as a growth curve," *Proceedings of the National Academy of Sciences of the United States of America*, vol. 18, no. 1, pp. 1–8, 1932.

[29] M. Prodan, *Forest Biometrics*, Elsevier, Philadelphia, Pa, USA, 2013.

[30] R. C. Yang, A. Kozak, and J. H. G. Smith, "The potential of Weibull-type functions as flexible growth curves," *Canadian Journal of Forest Research*, vol. 8, no. 4, pp. 424–431, 1978.

[31] SAS, *STAT User's Guide Version 9.1*, SAS Institute, Cary, NC, USA, 2003.

[32] J. Mascaro, C. M. Litton, R. F. Hughes, A. Uowolo, and S. A. Schnitzer, "Minimizing bias in biomass allometry: model selection and log-transformation of data," *Biotropica*, vol. 43, no. 6, pp. 649–653, 2011.

[33] D. Mayer and D. Butler, "Statistical validation," *Ecological Modelling*, vol. 68, no. 1, pp. 21–32, 1993.

[34] A. M. Masota, *Management models for tropical rainforests in Tanzania [Ph.D. thesis]*, Sokoine University of Agriculture, 2015.

[35] A. Pappoe, F. Armah, E. Quaye, P. Kwakye, and G. Buxton, "Composition and stand structure of a tropical moist semi-deciduous forest in Ghana," *International Research Journal of Plant Science*, vol. 1, no. 4, pp. 95–106, 2010.

[36] L. Banin, T. R. Feldpausch, O. L. Phillips et al., "What controls tropical forest architecture? Testing environmental, structural and floristic drivers," *Global Ecology and Biogeography*, vol. 21, no. 12, pp. 1179–1190, 2012.

[37] W. A. Mugasha, T. Eid, O. M. Bollandsås et al., "Allometric models for prediction of above- and belowground biomass of trees in the miombo woodlands of Tanzania," *Forest Ecology and Management*, vol. 310, pp. 87–101, 2013.

[38] E. E. Mwakalukwa, H. Meilby, and T. Treue, "Floristic composition, structure, and species associations of dry Miombo woodland in Tanzania," *ISRN Biodiversity*, vol. 2014, Article ID 153278, 15 pages, 2014.

PERMISSIONS

LIST OF CONTRIBUTORS

Towanou Houètchégnon, Christine Ajokè Ifètayo Nougbodé Ouinsavi and Nestor Sokpon
Faculty of Agronomy, Laboratory of Forestry Studies and Research, University of Parakou, BP 123 Parakou, Benin

Dossou Seblodo Judes Charlemagne Gbèmavo
Faculty of Agronomy, Laboratory of Forestry Studies and Research, University of Parakou, BP 123 Parakou, Benin
Laboratoire de Biomathématiques et d'Estimations Forestiéres, Faculté des Sciences Agronomiques, Université d'Abomey-Calavi, 04 BP 1525 Cotonou, Benin

Neal Hockley
School of Environment,Natural Resources and Geography, Bangor University, Bangor, Gwynedd LL57 2DG, UK

Linda Chinangwa
School of Environment,Natural Resources and Geography, Bangor University, Bangor, Gwynedd LL57 2DG, UK
2Institute of Advanced Study of Sustainability, United Nations University, No. 53-70, Jingumae 5-Chome, Shibuya-ku, Tokyo 150-8925, Japan

Andrew S. Pullin
Centre for Evidence-Based Conservation, School of Environment, Natural Resources and Geography, Bangor University, Bangor, Gwynedd LL57 2DG, UK

D. K. Langat
Kenya Forestry Research Institute, P.O. Box 5199, Kisumu 40108, Kenya

E. K. Maranga and A. A. Aboud
Department of Natural Resources, Egerton University, P.O. Box 536, Njoro 20115, Kenya

J. K. Cheboiwo
Kenya Forestry Research Institute, P.O. Box 20412, Nairobi 00200, Kenya

Do Thi Ngoc Le
Tropical Silviculture & Forest Ecology, Georg-August-Universität Göttingen, Büsgenweg 1, 37077 Göttingen, Germany
Vietnam Forestry University, Xuan Mai Town, Chuong My District, Hanoi 100000, Vietnam

Nguyen Van Thinh
Tropical Silviculture & Forest Ecology, Georg-August-Universität Göttingen, Büsgenweg 1, 37077 Göttingen, Germany
Silvicultural Research Institute (SRI), Vietnamese Academy of Forest Sciences, DucThang, Bac Tu Liem District, Hanoi 100000, Vietnam

Nguyen The Dung
Vietnam Forestry University, Xuan Mai Town, Chuong My District, Hanoi 100000, Vietnam

Ralph Mitlöhner
Tropical Silviculture & Forest Ecology, Georg-August-Universität Göttingen, Büsgenweg 1, 37077 Göttingen, Germany

Dagm Fikir
University of Gondar, P.O. Box 196, Gondar, Ethiopia

Wubalem Tadesse
Ethiopian Environment and Forest Research Institute, P.O. Box 24536, 1000 Addis Ababa, Ethiopia

Abdella Gure
Hawassa UniversityWondo Genet College of Forestry and Natural Resources, P.O. Box 128, Shashemene, Ethiopia

Nickolas E. Zeibig-Kichas, Christopher W. Ardis, John-Pascal Berrill and Joseph P. King
Department of Forestry andWildland Resources, Humboldt State University, 1 Harpst St. Arcata, CA 95521, USA

Stephen F. Omondi
Kenya Forestry Research Institute, P.O. Box 20412, Nairobi 00200, Kenya
2School of Biological Sciences, University of Nairobi, P.O. Box 30197,Nairobi 00100, Kenya

DavidW. Odee
Kenya Forestry Research Institute, P.O. Box 20412, Nairobi 00200, Kenya

George O. Ongamo and James I. Kanya
School of Biological Sciences, University of Nairobi, P.O. Box 30197,Nairobi 00100, Kenya

Damase P. Khasa
Centre for Forest Research and Institute for Systems and Integrative Biology, Laval University, Sainte-Foy, QC, Canada G1V 0A6

Guofeng Wang and Jiancheng Chen
School of Economics and Management, Beijing Forestry University, Beijing 100083, China

Xiangzheng Deng
Institute of Geographic Sciences and Natural Resources Research, Chinese Academy of Sciences, Beijing 100101, China
Center for Chinese Agricultural Policy, Chinese Academy of Sciences, Beijing 100101, China

Divine O. Appiah and Sampson Yamba
Department of Geography and Rural Development, Kwame Nkrumah University of Science and Technology (KNUST), Kumasi, Ghana

John T. Bugri
Department of Land Economy, Kwame Nkrumah University of Science and Technology (KNUST), Kumasi, Ghana

Eric K. Forkuo
Department of Geomatic Engineering, Kwame Nkrumah University of Science and Technology (KNUST), Kumasi, Ghana

Aladesanmi D Agbelade
Department of Forest Resources and Wildlife Management, Faculty of Agricultural Sciences, Ekiti State University, PMB 5363, Ado Ekiti, Ekiti State, Nigeria

Jonathan C. Onyekwelu and Matthew B. Oyun
Department of Forestry and Wood Technology, School of Agriculture and Agricultural Technology, Federal University of Technology, PMB 704, Akure, Ondo State, Nigeria

Angelingis Akwilini Makatta
Ministry of Natural Resources and Tourism, Forestry Training Institute, Arusha, Tanzania

Faustin Peter Maganga and Amos Enock Majule
Institute of Resource Assessment, University of Dar es Salaam, Dar es Salaam, Tanzania

Suspense Averti Ifo
ENS, Département de Sciences et Vie de la terre, Université Marien Ngouabi, BP 69, Brazzaville, Congo

Jean-Marie Moutsambote, Félix Koubouana, Helischa Mampouya, Saint Fédriche Ndzai, Alima Brigitte Mantota, Mackline Mbemba, Dulsaint Mouanga-Sokath, Roland Odende, Lenguiya Romarick Mondzali, Yeto Emmanuel Mampouya Wenina and Brice Chérubins Ouissika
2ENSAF, Laboratoire d'Ecologie Appliquée Université Marien Ngouabi, BP 69, Brazzaville, Congo

Joseph Yoka, Leslie Nucia Orcellie Bouetou-Kadilamio, Yannick Bocko and Loumeto Jean Joel
Facult´e des Sciences, Département de Biologie et Physiologie V´egétales, Université Marien Ngouabi, Brazzaville, Congo

Yuta Uchiyama and Ryo Kohsaka
Kanazawa University Graduate School of Human and Socio-Environmental Studies, Kanazawa, Japan

Edmond Alavaisha
Institute of Resource Assessment, University of Dar es Salaam, P.O. Box 35097, Dar es Salaam, Tanzania

Mwita M. Mangora
Institute of Marine Sciences, University of Dar es Salaam, Mizingani Road, P.O. Box 668, Zanzibar, Tanzania

Lawrence H. Tanner, Megan T.Wilckens and Morgan A. Nivison
Environmental Science Systems, Le Moyne College, Syracuse, NY 13214, USA

Katherine M. Johnson
Monteverde Institute, Monteverde, Puntarenas, Costa Rica
Phinizy Center forWater Sciences, Augusta, GA 30906, USA

Luís Silva
InBIO, Rede de Investigação em Biodiversidade e Biologia Evolutiva, Laboratório Associado, CIBIO-Açores, Universidade dos Açores, Açores, 9501-801 Ponta Delgada, Portugal

Lurdes Borges Silva
InBIO, Rede de Investigação em Biodiversidade e Biologia Evolutiva, Laboratório Associado, CIBIO-Açores, Universidade dos Açores, Açores, 9501-801 Ponta Delgada, Portugal
NATURALREASON, LDA, Caminho do Meio Velho, 5-B, Açores, 9760-114 Cabo da Praia, Portugal
3CBIO, Universidade dos Açores, Faculdade de Ciências e Tecnologia, Ponta Delgada, Portugal

Mário Alves
NATURALREASON, LDA, Caminho do Meio Velho, 5-B, Açores, 9760-114 Cabo da Praia, Portugal

Rui Bento Elias
Centre for Ecology, Evolution and Environmental Changes (CE3C), Azorean Biodiversity Group and Universidade dos Açores, Faculdade de Ciências Agrárias e do Ambiente, Açores, 9700-042 Angra do Heróismo, Portugal

Chelestino Balama
Department of Wood Utilization, Sokoine University of Agriculture, P.O. Box 3014, Chuo Kikuu, Morogoro, Tanzania
Directorate of Forest Utilisation Research, Tanzania Forestry Research Institute, P.O. Box 1854, Morogoro, Tanzania

Suzana Augustino and Fortunatus B. S.Makonda
Department of Wood Utilization, Sokoine University of Agriculture, P.O. Box 3014, Chuo Kikuu, Morogoro, Tanzania

Danford Mwaiteleke and Leopord P. Lusambo
Department of Forest Economics, Sokoine University of Agriculture, P.O. Box 3011, Chuo Kikuu, Morogoro, Tanzania

Shuntaro Watanabe
Field Science Education and Research Center, Kyoto University, Oiwake-cho, Kitashirakawa, Sakyo-ku, Kyoto 606-8502, Japan
The University of Shiga Prefecture,No. 2500,Hassaka-cho,Hikone, Shiga 522-8533, Japan

Naohiko Noma
The University of Shiga Prefecture,No. 2500,Hassaka-cho,Hikone, Shiga 522-8533, Japan

Yuko Kaneko
Toyo University, 5-28-20 Hakusan, Bunkyo-ku, Tokyo 112-8606, Japan

Yuri Maesako
Osaka Sangyo University, Nakagaito, Daito, Osaka 574-8530, Japan

Peter F. Newton
Canadian Wood Fibre Centre, Canadian Forest Service, Natural Resources Canada, 1219 Queen Street East, Sault Ste. Marie, ON, Canada P6A 2E5

Wilson Ancelm Mugasha
Tanzania Forestry Research Institute (TAFORI), P.O. Box 1854, Morogoro, Tanzania

Ezekiel Edward Mwakalukwa
Department of Forest Biology, Sokoine University of Agriculture, P.O. Box 3010, Morogoro, Tanzania

Emannuel Luoga, Rogers Ernest Malimbwi and Eliakimu Zahabu
Department of Forest Mensuration and Management, Sokoine University of Agriculture, P.O. Box 3013, Morogoro, Tanzania

Dos Santos Silayo
Department of Forest Engineering, Sokoine University of Agriculture, P.O. Box 3012, Morogoro, Tanzania

Gael Sola, Philippe Crete and Matieu Henry
UN-REDD Programme, Food and Agriculture Organization of the United Nations (FAO), Viale delle Terme di Caracalla, 00153 Rome, Italy

Almas Kashindye
Forest Training Institute, Olmotonyi, P.O. Box 943, Arusha, Tanzania

Index

www.ingramcontent.com/pod-product-compliance
Lightning Source LLC
Chambersburg PA
CBHW080630200326
41458CB00013B/4569